Ergebnisse der Mathematik und ihrer Grenzgebiete

3. Folge · Band 7

A Series of Modern Surveys in Mathematics

Allan Pinkus

n-Widths in Approximation Theory

Springer-Verlag
Berlin Heidelberg New York Tokyo 1985

Allan Pinkus
Technion
Israel Institute of Technology
Department of Mathematics
Haifa 32000, Israel

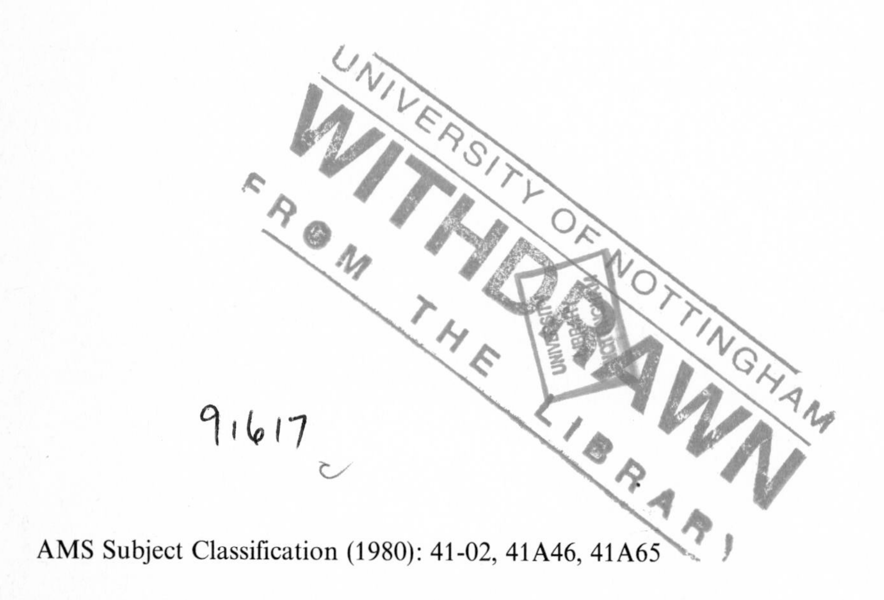

AMS Subject Classification (1980): 41-02, 41A46, 41A65

ISBN 3-540-13638-X Springer-Verlag Berlin Heidelberg New York Tokyo
ISBN 0-387-13638-X Springer-Verlag New York Heidelberg Berlin Tokyo

Library of Congress Cataloging in Publication Data
Pinkus, Allan, 1946– *N*-widths in approximation theory.
(Ergebnisse der Mathematik und ihrer Grenzgebiete; 3. Folge, Bd. 7)
Bibliography: p. Includes index.
1. Approximation theory. I. Title. II Series.
QA221.P56 1985 511′.4 84-13902
ISBN 0-387-13638-X (U.S.)

Typesetting: Daten- und Lichtsatz-Service, Würzburg
Printing and binding: Graphischer Betrieb Konrad Triltsch, Würzburg
2141/3140-543210

To Rachel

Preface

My original introduction to this subject was through conservations, and ultimately joint work with C. A. Micchelli. I am grateful to him and to Profs. C. de Boor, E. W. Cheney, S. D. Fisher and A. A. Melkman who read various portions of the manuscript and whose suggestions were most helpful. Errors in accuracy and omissions are totally my responsibility. I would like to express my appreciation to the SERC of Great Britain and to the Department of Mathematics of the University of Lancaster for the year spent there during which large portions of the manuscript were written, and also to the European Research Office of the U.S. Army for its financial support of my research endeavors. Thanks are also due to Marion Marks who typed portions of the manuscript.

Haifa, 1984 Allan Pinkus

Table of Contents

Chapter I. Introduction

In this short chapter we introduce the subject matter. We hope that this introduction whets the reader's appetite for the more systematic treatment which will follow.

Let X be a normed linear space and X_n any n-dimensional subspace of X. For each $x \in X$, $E(x; X_n)$ shall denote the distance of the n-dimensional subspace X_n from x, defined by

$$E(x; X_n) = \inf\{\|x - y\|_X : y \in X_n\}.$$

If there exists a $y^* \in X_n$ for which $E(x; X_n) = \|x - y^*\|$, then y^* is a *best approximation to x from X_n*. Problems of existence, uniqueness, and characterization of the best approximation are of central importance in approximation theory. (Since X_n is here a finite dimensional subspace of X, a best approximation always exist.) For example, consider $X = C[a, b]$, the space of real-valued continuous functions on the finite interval $[a, b]$, endowed with the uniform norm. The classical Haar Theorem delineates those n-dimensional subspaces of $C[a, b]$, called Chebyshev systems, for which there is a unique best approximation to every $f \in C[a, b]$. This unique best approximant is characterized by the fact that the error function equioscillates on at least $n + 1$ points in $[a, b]$.

Let us now suppose that instead of a single element x, we are given a subset A of X. How well does the n-dimensional subspace X_n of X approximate the subset A? A commonly used definition is to set

$$\begin{aligned} E(A; X_n) &= \sup\{E(x; X_n) : x \in A\} \\ &= \sup_{x \in A} \inf_{y \in X_n} \|x - y\|. \end{aligned}$$

$E(A; X_n)$ is the *deviation of A from X_n*. Thus $E(A; X_n)$ measures the extent to which the "worst element" of A can be approximated from X_n. Many results in approximation theory are concerned with this particular quantity for specific choices of A and X_n.

Given a subset A of X, one might also ask how well one can approximate A by n-dimensional subspaces of X. Thus, we shall consider the possibility of allowing the n-dimensional subspaces X_n to vary within X. This idea was first propounded by Kolmogorov [1936].

Definition 1. Let X be a normed linear space and A a subset of X. The *n-width, in the sense of Kolmogorov*, of A in X (or the *Kolmogorov n-width* of A in X) is given by

$$d_n(A;X) = \inf\{E(A;X_n)\colon X_n \text{ an } n\text{-dimensional subspace of } X\}$$

$$= \inf_{X_n} \sup_{x \in A} \inf_{y \in X_n} \|x - y\|_X,$$

the left-most infimum being taken over all n-dimensional subspaces X_n of X.

(Some authors prefer the expression "n-diameter" rather than "n-width". We shall always use the latter term.)

A subspace X_n of X of dimension at most n for which

$$d_n(A;X) = E(A;X_n)$$

is called an *optimal subspace* for $d_n(A;X)$.

Since $d_n(A;X)$ measures the extent to which A may be approximated by n-dimensional subspaces of X, it is, in a certain sense, a measure of the "thickness" or "massivity" of A.

It is, of course, generally impossible to obtain $d_n(A;X)$ and determine optimal subspaces X_n for $d_n(A;X)$ (if they exist) for all A and X. Nonetheless, much effort has been devoted to this task for specific choices of A and X. In this work we survey numerous choices of A and X for which both $d_n(A;X)$ and X_n are in fact obtained, or at least characterized. However in some surprisingly simple cases (see for example Chapter VI) these quantities have not as yet been calculated, though hardly through lack of effort.

It is also of considerable interest to determine the asymptotic behavior of $d_n(A;X)$ as $n \uparrow \infty$, since typically an optimal X_n cannot be explicitly obtained or if it can, too much computational effort is involved in its determination. In many cases very simple n-dimensional subspaces may approximate A in an asymptotically optimal manner. Thus the n-width, in providing a lower threshold on the degree of approximation of A by n-dimensional subspaces, tells us how well a given n-dimensional subspace (e.g., algebraic or trigonometric polynomials, splines with fixed knots, etc...) approximates A relative to the theoretical lower bound. On this basis it is then possible to judge whether the additional time, effort, and perhaps money, involved in using better but more complicated subspaces is, in fact, justified.

Kolmogorov, aside from simply defining $d_n(A;X)$, also computed this quantity in two particular instances. The first of these we give here; the second is deferred to Chapter IV.

Let $L^2 = L^2[0, 2\pi]$ denote the usual space of square integrable functions on $[0, 2\pi]$ with norm

$$\|f\| = \left(1/2\pi \int_0^{2\pi} |f(x)|^2\, dx\right)^{1/2}.$$

For any given positive integer r, let $\tilde{W}_2^{(r)}$ denote the (Sobolev) space of 2π-periodic, real-valued, $(r-1)$-times differentiable functions whose $(r-1)$st derivative is absolutely continuous and whose rth derivative is in L^2. Thus

$$\tilde{W}_2^{(r)} = \{f : f^{(r-1)} \text{ abs. cont., } f^{(r)} \in L^2,\ f^{(i)}(0) = f^{(i)}(2\pi),\ i = 0, 1, \dots, r-1\}.$$

Set

$$\tilde{B}_2^{(r)} = \{f : f \in \tilde{W}_2^{(r)},\ \|f^{(r)}\| \leqq 1\},$$

and let us consider $d_n(\tilde{B}_2^{(r)}; L^2)$.

Theorem 1 (Kolmogorov [1936]).

$$d_0(\tilde{B}_2^{(r)}; L^2) = \infty, \quad \textit{while} \quad d_{2n-1}(\tilde{B}_2^{(r)}; L^2) = d_{2n}(\tilde{B}_2^{(r)}; L^2) = n^{-r}, \qquad n = 1, 2, \dots.$$

Furthermore, an optimal subspace for $d_{2n-1}(\tilde{B}_2^{(r)}; L^2)$ *(and hence for* $d_{2n}(\tilde{B}_2^{(r)}; L^2)$*) is*

$$T_{n-1} = \text{span}\,\{1, \sin x, \cos x, \dots, \sin(n-1)x, \cos(n-1)x\}$$

i.e., trigonometric polynomials of degree less than or equal to $n-1$.

Before discussing the proof of Theorem 1, we remark that T_{n-1} is not the only optimal subspace for $d_{2n}(\tilde{B}_2^{(r)}; L^2)$. In Chapter IV, we construct an additional optimal subspace. In what follows we give a sketch of the proof of Theorem 1. The proof may easily be made rigorous and this is done in Chapter IV.

"*Proof*". Since every constant function is in $\tilde{B}_2^{(r)}$, and $d_0(\tilde{B}_2^{(r)}; L^2) = \sup\{\|f\| : f \in \tilde{B}_2^{(r)}\}$, it follows that $d_0(\tilde{B}_2^{(r)}; L^2) = \infty$. For this very same reason it may be easily seen that if $E(\tilde{B}_2^{(r)}; X_n) < \infty$, then it is necessary that the constant function be contained in X_n.

We first obtain an upper bound on $d_{2n-1}(\tilde{B}_2^{(r)}; L^2)$ by showing that $E(\tilde{B}_2^{(r)}; T_{n-1}) \leqq n^{-r}$. By definition

$$E(\tilde{B}_2^{(r)}; T_{n-1}) = \sup_{f \in \tilde{B}_2^{(r)}} \inf_{t \in T_{n-1}} \|f - t\|.$$

Assume that f is of the form $f(x) = a_0 + \sum_{k=1}^{\infty} a_k \cos kx + b_k \sin kx$. The condition $f \in \tilde{B}_2^{(r)}$ is equivalent to

$$1/2 \sum_{k=1}^{\infty} (|a_k|^2 + |b_k|^2)\, k^{2r} \leqq 1.$$

Since we wish to bound $E(\tilde{B}_2^{(r)}; T_{n-1})$ from above, it suffices to set $t(x) = a_0 + \sum_{k=1}^{n-1} a_k \cos kx + b_k \sin kx$ in the above expression (this is, in fact, the best approximation), so that

$$E(\tilde{B}_2^{(r)}; T_{n-1}) \leqq \sup \frac{\left(\sum_{k=n}^{\infty} |a_k|^2 + |b_k|^2\right)^{1/2}}{\left(\sum_{k=1}^{\infty} (|a_k|^2 + |b_k|^2)\, k^{2r}\right)^{1/2}}$$

where the supremum is taken over f for which the denominator is not zero. Since k^{2r} is an increasing sequence, this supremum is obviously attained by choosing $(|a_n|^2+|b_n|^2)\neq 0$, and $a_k=b_k=0$ for all other k. Thus $E(\tilde{B}_2^{(r)};T_{n-1})\leqq n^{-r}$.

It remains to prove the lower bound $E(\tilde{B}_2^{(r)};X_{2n})\geqq n^{-r}$ for any $2n$-dimensional subspace X_{2n} of L^2. Now,

$$E(\tilde{B}_2^{(r)};X_{2n})\geqq E(\tilde{B}_2^{(r)}\cap T_n;X_{2n})$$
$$=\sup_{t\in T_n,\ \|t^{(r)}\|\leqq 1}\ \inf_{g\in X_{2n}}\|t-g\|.$$

T_n is a subspace of dimension $2n+1$. There therefore exists a nontrivial $t^*\in T_n$ which is orthogonal to each element of X_{2n}. Furthermore, any such t^* is necessarily of the form $t^*(x)=\sum_{k=1}^{n} a_k\cos kx+b_k\sin kx$, i.e., does not contain the constant term, since as indicated above only those X_{2n} which contain the constant function are of interest.

It is a well known fact that if t^* is orthogonal to each element of X_{2n}, then

$$\inf_{g\in X_{2n}}\|t^*-g\|=\|t^*\|.$$

Thus $E(\tilde{B}_2^{(r)};X_{2n})\geqq\|t^*\|$. For each $t\in T_{2n}$ of the form $t(x)=\sum_{k=1}^{n} a_k\cos kx+b_k\sin kx$, $\|t\|=\left(1/2\sum_{k=1}^{n}(|a_k|^2+|b_k|^2)\right)^{1/2}$, while $\|t^{(r)}\|=\left(1/2\sum_{k=1}^{n}(|a_k|^2+|b_k|^2)\,k^{2r}\right)^{1/2}$. Therefore

$$E(\tilde{B}_2^{(r)};X_{2n})\geqq\inf\frac{\left(\sum_{k=1}^{n}|a_k|^2+|b_k|^2\right)^{1/2}}{\left(\sum_{k=1}^{n}(|a_k|^2+|b_k|^2)\,k^{2r}\right)^{1/2}}$$
$$=n^{-r}.$$

This "proves" the theorem. □

This proof is characteristic of many of the proofs of theorems wherein exact n-widths are obtained, in the sense that it divides into the following two parts. First an upper bound is proven by calculating $E(A;X_n)$ for a judicious choice of X_n. The second step is to show that the quantity obtained is in fact the lower bound as well. This latter problem is non-linear and is generally (but not always) the more difficult.

One property which enabled us to easily calculate $d_n(\tilde{B}_2^{(r)};L^2)$ is the fact that the best approximation operator in a Hilbert space is a linear projector. This property actually characterizes Hilbert spaces. If X_n is an n-dimensional subspace of L^2 and $\{u_i\}_{i=1}^{n}$ is an orthonormal basis for X_n, then the best approximation to $f\in L^2$ is given by

$$\sum_{i=1}^{n}(f,u_i)\,u_i.$$

The fact that linear operators of rank n do as well as best approximation operators in Hilbert spaces prompts the following definition.

Definition 2. Let X be a normed linear space and A a subset of X. The *linear n-width* of A in X is defined by

$$\delta_n(A;X) = \inf_{P_n} \sup_{x \in A} \|x - P_n(x)\|,$$

where the infimum is here taken over all continuous linear operators of rank at most n.

A linear operator is of rank n if its range is of dimension n. If $\delta_n(A;X) = \sup\{\|x - P_n(x)\| : x \in A\}$, where P_n is a continuous linear operator of rank at most n, then P_n is said to be an *optimal* linear operator for $\delta_n(A;X)$.

The linear n-width is of importance in and of itself. One is often interested in linear approximations which may be more practical and easier to calculate than best approximations, which are typically non-linear.

Of course, best approximations are no worse than linear approximations and thus $d_n(A;X) \leqq \delta_n(A;X)$ for all A and X. When δ_n is not equal to d_n, then serious problems arise in their computation. However, in many instances, these two quantities are equal even in non-Hilbert spaces. This should not be taken to mean that the error in best and linear approximation is the same for all $x \in A$. The n-widths are only measures of the "worst element" case.

Let us now consider a slightly more general problem than that discussed in Theorem 1. (To be precise, the result of Theorem 1 does not quite follow from what shall be shown.) To simplify matters, let $X = L^2[0,1]$ and let $K(x,y)$ be a fixed real-valued Hilbert-Schmidt kernel defined on $[0,1] \times [0,1]$, i.e.,

$$\int_0^1 \int_0^1 |K(x,y)|^2 \, dx \, dy < \infty.$$

Set

$$A = \left\{ f(x) : f(x) = Kh(x) = \int_0^1 K(x,y)\, h(y)\, dy,\ \|h\| \leqq 1 \right\},$$

where $\|h\|$ is the usual L^2 norm on $[0,1]$.

The choice of A as the image of a unit ball under some compact mapping will arise repeatedly in this work.

The s-numbers (or singuler values) of the operator induced by the kernel K are defined as the square roots of the eigenvalues of the self-adjoint, non-negative, compact operator induced by the kernel $K'K(x,y) \left(= \int_0^1 K(z,x)\, K(z,y)\, dz \right)$. That is, if $(K'K)\,\phi_i = \lambda_i \phi_i$, $i = 1, 2, \ldots$, where

$$\lambda_1 \geqq \lambda_2 \geqq \ldots$$

enumerate the nonzero (positive) eigenvalues associated with the kernel $K'K$, listed to their algebraic multiplicity, then the ith s-number of K is defined as $\lambda_i^{1/2}$,

$i = 1,2,\ldots$. Let the $\{\phi_i\}$ be as defined above and set $\psi_i = K\phi_i$, $i = 1,2,\ldots$. The function ψ_i is an eigenfunction, with eigenvalue λ_i, of the operator with kernel KK'.

Theorem 2. *Let A be as above. Then the Kolmogorov n-width of A in L^2 is given by*

$$d_n(A; L^2) = \lambda_{n+1}^{1/2}.$$

Furthermore, an optimal subspace for A is $X_n = \operatorname{span}\{\psi_1, \ldots, \psi_n\}$.

Proof. From the definition,

$$d_n(A; L^2) = \inf_{X_n} \sup_{\|h\| \leqq 1} \inf_{g \in X_n} \|Kh - g\|.$$

A classical duality result (reduced in this Hilbert space setting to the Cauchy-Schwarz inequality and the case of equality therein) gives

$$\inf_{g \in X_n} \|Kh - g\| = \sup_{(f, X_n) = 0} \frac{(Kh, f)}{\|f\|},$$

where by $(f, X_n) = 0$ we mean that f is orthogonal to each element of X_n and $(\cdot, \cdot)$ is the usual inner product in $L^2[0,1]$. Thus

$$\begin{aligned} d_n(A; L^2) &= \inf_{X_n} \sup_{\|h\| \leqq 1} \sup_{(f, X_n) = 0} \frac{(Kh, f)}{\|f\|} \\ &= \inf_{X_n} \sup_{(f, X_n) = 0} \sup_{\|h\| \leqq 1} \frac{(h, K'f)}{\|f\|} \\ &= \inf_{X_n} \sup_{(f, X_n) = 0} \frac{\|K'f\|}{\|f\|} \\ &= \left[\inf_{X_n} \sup_{(f, X_n) = 0} \frac{(KK'f, f)}{(f, f)} \right]^{1/2}. \end{aligned}$$

This latter expression is simply the well known Rayleigh-Ritz characterization of the $(n+1)$st eigenvalue of the symmetric operator KK' (see for example, Courant and Hilbert [1953]). Thus $d_n(A; L^2) = \lambda_{n+1}^{1/2}$. Furthermore, the infimum and supremum in the above expression are obtained by the choice $X_n = \operatorname{span}\{\psi_1, \ldots, \psi_n\}$ and $f = \psi_{n+1}$. □

In the proof of the above theorem we did not explicitly calculate the best approximation. Instead we used a duality result and then applied a well known theorem characterizing eigenvalues and eigenfunctions of self-adjoint, non-negative, compact operators. This duality result holds in a general form in any normed linear space (see Proposition 6.1 of Chapter II), and prompts the following n-width definition.

Definition 3. Let X be a normed linear space and A a subset of X. The *Gel'fand n-width* of A in X is given by

$$d^n(A;X) = \inf_{L^n} \sup_{x \in A \cap L^n} \|x\|,$$

where the infimum is taken over all subspaces L^n of X of codimension n.

(To avoid ambiguity, we shall always assume that the zero element is in A.)
A subspace L^n of X is said to be of codimension n if there exist n continuous (linearly independent) linear functionals $f_1, \ldots, f_n$ in X', the continuous dual of X, such that

$$L^n = \{x : f_i(x) = 0, \ i = 1, \ldots, n\}.$$

If L^n is a subspace of codimension at most n for which $d^n(A;X) = \sup\{\|x\| : x \in A \cap L^n\}$, then L^n is called an *optimal* subspace for the Gel'fand n-width $d^n(A;X)$.

In general, $d^n(A;X)$ and $d_n(A;X)$ are not equal (although in many cases they are). However, because of the interrelationship between these two quantities, the Gel'fand n-width is also an important quantity in the study of n-widths.

The study of the three quantities d_n, δ_n and d^n is the main topic of this monograph. (A fourth n-width, called the Bernstein n-width, is introduced in Chapter II.) While numerous other n-widths appear in the literature (see Tichomirov [1971] and Pietsch [1974] where most of them are listed) it was decided to consider only these.

Some words about the organization of this monograph. Chapter II is a more theoretical chapter in which are discussed general properties of the n-widths as well as the relationships between them. Chapter III has nothing whatsoever to do with n-widths or, for that matter, with approximation theory. It is concerned with the theory of Tchebychev systems and total positivity. These two interrelated notions are used in Chapters IV, V and VI, and it was decided to put the basic underlying tools of this theory in a separate chapter so as not to overburden the other chapters.

Chapters IV to VIII contain the main results of this monograph, and almost all the explicit examples are to be found there. In Chapter IV we consider n-widths in a Hilbert space setting. We start with a generalized form of Theorem 2 and then consider numerous variations thereof. In Sects. 5 and 6 of Chapter IV total positivity comes into play in obtaining optimal subspaces other than eigenfunction subspaces. The subject matter of Chapter V evolved from the study of the exact n-widths of Soboloev spaces. It was found that the important underlying structure was that of total positivity and by applying the tools of Chapter III, a very elegant theory has been developed, although many open problems remain.

Chapter VI is concerned with the n-widths of the set

$$A = \{Dx : \|x\|_{l_p^m} \leqq 1\}$$

in l_q^m (p, q are arbitrary numbers in $[1, \infty]$), where D is a fixed $m \times m$ matrix. Very little can be said in general unless $p = q = 2$, or $p = q = \infty$ and D is totally positive. Thus we deal, in the main, with the case where D is a diagonal matrix. Even in this situation, many problems are still unsolved. Two particular choices of p, q ($p = 1$, $q = \infty$, and $p = 2$, $q = \infty$), where explicit solutions are not known, are discussed in some detail and asymptotic estimates for these n-widths are obtained. One motivating factor behind this discussion is their importance in Chapter VII in determining asymptotic estimates of n-widths of Sobolev spaces. In Chapter VIII are discussed various problems concerned with n-widths of classes of analytic functions.

No claims are made as to the completeness of this text. We have been rather selective in what is and is not presented and there are yet too many open problems for any one to consider writing a "definitive" book on the subject. For the benefit of the researcher we have attempted to make the bibliography as complete as possible. It contains all references, known to us, on the subject of n-widths even though some of these works are not referenced in the text.

The title of this work is "n-Widths in Approximation Theory" and not simply "n-Widths". This is an important qualification as the subject of n-widths has had applications in other areas. We do not consider the topics of diametral dimension, isomorphisms of spaces, or nuclear spaces and their relationships to n-widths. Readers interested in these topics may wish to consult Dubinsky [1979], Pietsch [1972], [1980], Rolewicz [1972] and references therein.

We have also, as the reader will undoubtedly discern, a preference for exact results rather that asymptotic estimates. Thus, for example, Chapter VIII contains a discussion of exact results for n-widths of various classes of analytic functions. The associated asymptotic estimates are relegated to a perfunctory discussion at the end of the chapter in the Notes and References.

It is only in the last twenty years or so that the study of n-widths has received much attention. This renewal of interest is mainly due to work of Tichomirov, especially [1960a] and [1969]. Since then the subject has blossomed as may be seen from the bibliography which contains approximately 200 articles devoted to various aspects of the study of n-widths.

It is our hope that this monograph will bring the subject to the attention of many more mathematicians who will find in it the interest and fascination which it has brought to the author.

Chapter II. Basic Properties of n-Widths

1. Properties of d_n

Let X be a normed linear space and X_n an n-dimensional subspace of X. For each $x \in X$, $E(x; X_n)$ is the distance of the n-dimensional subspace X_n from x, defined by

$$E(x; X_n) = \inf\{\|x - y\| : y \in X_n\}.$$

Suppose now that instead of a single element x, we are given a subset A of X. How well does an n-dimensional subspace X_n of X approximate A? A commonly used definition is to set

$$\begin{aligned} E(A; X_n) &= \sup\{E(x; X_n) : x \in A\} \\ &= \sup_{x \in A} \inf_{y \in X_n} \|x - y\|. \end{aligned}$$

$E(A; X_n)$ is the *distance or deviation of A from X_n*. Thus $E(A; X_n)$ measures how well the "worst element" of A can be approximated from X_n. Many problems of approximation theory are concerned with this particular quantity for specific choices of A and X_n.

Given a subset A of X, one might also ask how well one can approximate A by n-dimensional subspaces of X. Thus, we consider the possibility of allowing the n-dimensional subspaces X_n to vary within X. This idea, introduced by Kolmogorov, is now referred to as *the n-width, in the sense of Kolmogorov*, or as *the Kolmogorov n-width* of A in X. It is defined as follows:

Definition 1.1. Let X be a real or complex normed linear space[1], and A a subset of X. *The n-width, in the sense of Kolmogorov*, of A in X is given by

$$d_n(A; X) = \inf_{X_n} \sup_{x \in A} \inf_{y \in X_n} \|x - y\|$$

where the infimum is taken over all n-dimensional subspaces X_n of X. If

$$d_n(A; X) = \sup_{x \in A} \inf_{y \in X_n} \|x - y\|$$

1 In what follows we only deal with real or complex normed linear spaces.

for some subspace X_n of dimension at most n, then X_n is said to be an *optimal* subspace for $d_n(A; X)$.

We often drop the X and write $d_n(A)$ in place of $d_n(A; X)$ if no ambiguity arises.

We list below some simple properties of d_n which are immediate consequences of the definition.

Theorem 1.1. *Let X be a normed linear space and $A \subseteq X$.*

(i) $d_n(\bar{A}) = d_n(A)$, *where $\bar{A}$ is the closure of A.*

(ii) *For every scalar α,*

$$d_n(\alpha A) = |\alpha|\, d_n(A).$$

(iii) *For each $A \subseteq X$, let $\mathrm{b}(A) = \{\alpha x\colon x \in A,\ |\alpha| \leq 1\}$ be the balanced hull of A. Then*

$$d_n(A) = d_n(\mathrm{b}(A)).$$

(iv) $d_n(\mathrm{co}\, A) = d_n(A)$, *where $\mathrm{co}\, A$ is the convex hull of A.*

(v) *For any two sets $A, B \subseteq X$, let*

$$E(A;B) = \sup_{x \in A} \inf_{y \in B} \|x - y\|.$$

Then if $B \subseteq A$, $d_n(A) - E(A;B) \leq d_n(B) \leq d_n(A)$.

(vi)
$$d_n(A) \geq d_{n+1}(A), \qquad n = 0, 1, \ldots$$

(vii) *Let X and Y be normed linear spaces, $X \subseteq Y$, and $A \subseteq X$. Then*

$$d_n(A; X) \geq d_n(A; Y).$$

On the basis of properties (i), (iii) and (iv), we can and will henceforth assume, unless otherwise stated, that A is a convex, closed, centrally symmetric subset of X. Property (vii) deserves comment. There exist examples of A, X and Y, as above, for which $d_n(A; X) > d_n(A; Y)$. Thus embedding A in some larger space may well decrease the associated n-width. This implies the important fact that an optimal subspace for $d_n(A; X)$, if it exists, need not be contained in $\overline{\mathrm{span}}(A)$, the closed subspace generated by A. One example of strict inequality in (vii) is the following:

Let X and Y be $\mathbb{R}^4$ and $\mathbb{C}^4$, respectively, endowed with the l_∞ (maximum) norm, and set

$$A = \left\{\mathbf{x} : \mathbf{x} = (x_1, x_2, x_3, x_4) \in \mathbb{R}^4,\ \sum_{i=1}^{4} |x_i| \leq 1\right\}.$$

It is known that $d_2(A; X) = 1/(1 + \sqrt{2})$, while $d_2(A; Y) = 1/(1 + \sqrt{3})$. An additional example of strict inequality holding in property (vii) is given at the end of this chapter, and another example is to be found in Tichomirov [1960a]. We now prove some further properties of d_n.

Proposition 1.2. *A is compact iff $d_n(A) \downarrow 0$ and A is bounded.*

Proof. Assume A is compact. Recall that we always assume that A is closed and we therefore do not differentiate between compact and pre-compact sets. There exists, for every $\varepsilon > 0$, a finite ε-net for A, i.e., points $\{x_1,\dots,x_N\}$ for which

$$\min\{\|x - x_i\| : i = 1,\dots,N\} \leqq \varepsilon$$

for every $x \in A$. Let $X_N = \operatorname{span}\{x_1,\dots,x_N\}$. Thus $d_N(A) \leqq \varepsilon$ and since $d_n(A)$ is a non-increasing function of n (property (vi) of Theorem 1.1), $d_n(A) \downarrow 0$.

Now assume that A is bounded and $d_n(A) \downarrow 0$. Since A is bounded $d_0(A) = \sup\{\|x\| : x \in A\} < \infty$. Given $\varepsilon > 0$, there exists an n_0 such that for all $n \geqq n_0$, $d_n(A) < \varepsilon$. By definition, this implies that there exists an n-dimensional subspace X_n and X such that $\inf\{\|x - y\| : y \in X_n\} < \varepsilon$ for all $x \in A$. Thus for each $x \in A$ there exists a $y \in X_n$ for which $\|x - y\| < \varepsilon$, and hence $\|y\| = \|x - (x - y)\| \leqq d_0(A) + \varepsilon$. The set of $y \in X_n$ for which $\|y\| \leqq d_0(A) + \varepsilon$ is a bounded subset of a finite dimensional space and is therefore compact. Hence there exist $\{y_1,\dots,y_k\}$, an ε-net for this set. It is now easily seen that $\{y_1,\dots,y_k\}$ is a 2ε-net for A. Thus A is compact. □

One of the simplest examples for which the n-widths d_n can be explicitly calculated is the following.

Proposition 1.3. *Let X be a normed linear space of dimension greater than n, and let $S(X)$ denote the closed unit ball in X. Then*

$$d_k(S(X); X) = 1, \qquad k = 0, 1, \dots, n.$$

Proof. Since $d_n(S(X); X) \leqq \dots \leqq d_0(S(X); X) = 1$, it remains to prove that $d_n(S(X); X) \geqq 1$. Let X_n be any n-dimensional subspace of X. We claim that there exists a nontrivial $x \in X$ whose best approximation from X_n is the zero element. By a suitable normalization of x, it then follows that $E(S(X); X_n) \geqq 1$. Since this inequality holds for every n-dimensional subspace X_n of X we obtain $d_n(S(X); X) \geqq 1$.

Thus it remains to prove that for every X_n, there exists an $x \in X$ whose best approximation from X_n is the zero element. Let $z \in X \backslash X_n$, and let y be any best approximation to z from X_n. Such a best approximation necessarily exists. The zero element is a best approximation to $x = z - y \in X$ and $x \neq 0$. This proves the proposition. □

In the proof of the next theorem we use the Borsuk Antipodality Theorem. Since we shall have frequent recourse to this result, we state a version of it here for easy reference.

Theorem 1.4 (Borsuk [1933]). *Let Ω be a bounded, open, symmetric neighborhood of $\mathbf{0}$ in $\mathbb{R}^m$, and T a continuous map of $\partial\Omega$ into $\mathbb{R}^{m-1}$, with T odd on $\partial\Omega$ (the boundary of Ω), i.e., $T(-\mathbf{x}) = -T(\mathbf{x})$ for all $\mathbf{x} \in \partial\Omega$. Then there exists an $\mathbf{x}^* \in \partial\Omega$ for which $T(\mathbf{x}^*) = \mathbf{0}$.*

The following theorem is a generalization of Proposition 1.3. It is an important result and is extensively used in the calculation of lower bounds for n-widths. It should be contrasted with property (vii) of Theorem 1.1.

Theorem 1.5. *Let X_{n+1} be any $(n+1)$-dimensional subspace of a normed linear space X, and let $S(X_{n+1})$ denote the unit ball of X_{n+1}. Then*

$$d_k(S(X_{n+1}); X) = 1, \quad k = 0, 1, \ldots, n.$$

Proof. We must prove that $d_n(S(X_{n+1}); X) \geqq 1$, since by definition $d_0(S(X_{n+1}); X) = 1$.

As in the proof of Proposition 1.3, it suffices to prove that for any given n-dimensional subspace X_n of X (and not necessarily of X_{n+1}), there exists a nontrivial $x \in X_{n+1}$ with zero as a best approximation from X_n.

Let $\{x_1, \ldots, x_{n+1}\}$ and $\{y_1, \ldots, y_n\}$ be bases for X_{n+1} and X_n, respectively. Thus every $x \in X_{n+1}$ may be written in the form $x = \sum_{i=1}^{n+1} a_i x_i$, and $y \in X_n$ in the form $y = \sum_{i=1}^{n} b_i y_i$ for some choice of scalars $\{a_i\}_{i=1}^{n+1}$ and $\{b_i\}_{i=1}^{n}$. In order to simplify the proof we shall assume that X is a real normed linear space, i.e., $a_i, b_i \in \mathbb{R}$. The theorem is also valid for $a_i, b_i \in \mathbb{C}$.

It obviously suffices in the proof to take $X = \operatorname{span}\{X_{n+1}, X_n\}$. We first prove the result when the norm on X is strictly convex. The strict convexity of the norm insures the uniqueness and continuity of the best approximation operator and also implies its oddness, i.e., if $y \in X_n$ is the best approximation to $x \in X$, then $-y$ is the best approximation to $-x$. Let

$$\Omega = \left\{(a_1, \ldots, a_{n+1}) \colon x = \sum_{i=1}^{n+1} a_i x_i, \ \|x\| < 1\right\}.$$

It is a simple matter to prove that Ω is a bounded, open, symmetric neighborhood of $\mathbf{0}$ in $\mathbb{R}^{n+1}$. For each $\mathbf{a} = (a_1, \ldots, a_{n+1}) \in \bar{\Omega}$, let $T(\mathbf{a}) \in \mathbb{R}^n$ denote the vector of coefficients of the best approximation to $x = \sum_{i=1}^{n+1} a_i x_i$ from X_n with respect to the basis $\{y_1, \ldots, y_n\}$. Thus if $y = \sum_{i=1}^{n} b_i y_i$ is the best approximation to x, then $T(\mathbf{a}) = (b_1, \ldots, b_n)$. T is an odd, continuous map of $\partial\Omega$ into $\mathbb{R}^n$. By the Borsuk Antipodality Theorem (Theorem 1.4), there exists an $x^* = \sum_{i=1}^{n+1} a_i^* x_i$, $\|x^*\| = 1$, for which the zero element is the best approximation from X_n.

If the norm on X is not strictly convex, we perturb it slightly to make it strictly convex, obtain the result and then perturb back. Let $\{z_1, \ldots, z_k\}$, $k \leqq 2n+1$, be a basis for $X = \operatorname{span}\{X_{n+1}, X_n\}$. For each $x = \sum_{i=1}^{k} c_i z_i$, and $\varepsilon > 0$, fixed, $\|x\|_\varepsilon = \|x\| + \varepsilon\left(\sum_{i=1}^{k} |c_i|^2\right)^{1/2}$ is a strictly convex norm on X. The finite dimensionality of X permits us to let $\varepsilon \downarrow 0$ while maintaining the validity of the theorem. □

As previously mentioned Theorem 1.5 is a powerful tool in determining lower bounds for n-widths. If a set A contains a ball of radius λ in some $(n+1)$-dimensional subspace of X, it then follows that $d_n(A; X) \geqq \lambda$. This method of obtaining a lower bound for $d_n(A; X)$ has been so frequently used that it has been codified (see e.g. Mityagin and Henkin [1963], Mityagin and Pelczynski [1968], and Tichomirov [1976]).

Definition 1.2. The *n-width, in the sense of Bernstein,* of A in X, is defined by

$$b_n(A;X) = \sup_{X_{n+1}} \sup\{\lambda\colon \lambda S(X_{n+1}) \subseteq A\}$$
$$= \sup_{X_{n+1}} \inf_{x \in \partial(A \cap X_{n+1})} \|x\|,$$

where X_{n+1} is any $(n+1)$-dimensional subspace of X, and $S(X_{n+1})$ is the unit ball of X_{n+1}.

In this definition we always assume that A is a closed, convex centrally symmetric subset of X.

Proposition 1.6. $d_n(A;\, X) \geqq b_n(A;\, X)$.

Proof. Let X_{n+1} be any $(n+1)$-dimensional subspace of X. If $\lambda S(X_{n+1})$, the ball of radius λ in X_{n+1}, is contained in A, then from property (v) of Theorem 1.1, $d_n(A;\, X) \geqq d_n(\lambda S(X_{n+1}); X)$. By Theorem 1.5 and property (ii) of Theorem 1.1, $d_n(\lambda S(X_{n+1}); X) = \lambda$. The result follows. □

Perhaps the simplest non-trivial application of Theorem 1.5 is the following result:

Let X be a normed linear space and $x_1, x_2, \dots$ a sequence of linearly independent elements of X. Let $\lambda_0 \geqq \lambda_1 \geqq \dots$ be any sequence of positive real numbers (with the understanding that some of the λ_n may be infinite). Set $X_n = \operatorname{span}\{x_1, \dots, x_n\}$ ($X_0 = \{\emptyset\}$) and define

$$A = \{x\colon E(x;X_k) \leqq \lambda_k,\ k = 0, 1, \dots\}.$$

Proposition 1.7. *$d_n(A;\, X) = \lambda_n$ and X_n is an optimal subspace.*

Proof. From the definition of A, $d_n(A;\, X) \leqq E(A;X_n) \leqq \lambda_n$. The lower bound follows from Theorem 1.5 if we can prove that $\lambda_n S(X_{n+1}) \subseteq A$. Let $x \in \lambda_n S(X_{n+1})$. Then $E(x;X_k) \leqq \|x\| \leqq \lambda_n \leqq \lambda_k$ for $k = 0, 1, \dots, n$. For $k \geqq n+1$, $E(x;X_k) = 0$. Thus $E(x;X_k) \leqq \lambda_k$ for all k and $x \in A$. □

We now present two additional lower bounds for the quantity $d_n(A;\, X)$. Each of these lower bounds is considerably more restrictive than the above result. Nonetheless we record these lower bounds here because of their elegance and simplicity.

Let B be a compact Hausdorff space and let $C(B)$ denote the set of continuous real-valued functions on B endowed with the usual uniform norm.

Proposition 1.8. *Let A be a subset of $C(B)$. Assume that there exist $n+1$ points in B, $x_1, \dots, x_{n+1}$ and a number $\varepsilon > 0$ with the following property: For every choice of signs $\delta_i = +1$ or -1, $i = 1, \dots, n+1$, there is a function $f \in A$ such that*

$$\operatorname{sgn} f(x_i) = \delta_i, \quad |f(x_i)| \geqq \varepsilon, \qquad i = 1, \dots, n+1.$$

Then,

$$d_n(A;\, C(B)) \geqq \varepsilon.$$

Proof. Let $X_n = \operatorname{span}\{g_1, \ldots, g_n\}$ be any n-dimensional subspace of $C(B)$. There exists a vector $\mathbf{c} = (c_1, \ldots, c_{n+1}) \in \mathbb{R}^{n+1}$ satisfying $\sum_{i=1}^{n+1} |c_i| = 1$, for which

$$\sum_{i=1}^{n+1} c_i g_k(x_i) = 0, \qquad k = 1, \ldots, n.$$

Let $\delta_i = +1$ or -1 be chosen such that $\delta_i c_i \geqq 0$, $i = 1, \ldots, n+1$ and let $f \in A$ satisfy the conditions of the proposition. Then for every $(a_1, \ldots, a_n) \in \mathbb{R}^n$,

$$\begin{aligned}\left\| f - \sum_{k=1}^{n} a_k g_k \right\| &\geqq \sum_{i=1}^{n+1} |c_i| \left| f(x_i) - \sum_{k=1}^{n} a_k g_k(x_i) \right| \\ &\geqq \left| \sum_{i=1}^{n+1} c_i f(x_i) - \sum_{k=1}^{n} a_k \sum_{i=1}^{n+1} c_i g_k(x_i) \right| \\ &\geqq \varepsilon \sum_{i=1}^{n+1} |c_i| = \varepsilon.\end{aligned}$$

This proves the proposition. □

This proposition is simply saying that when looking at the range of an n-dimensional subspace evaluated at the points $(x_1, \ldots, x_{n+1})$, then the range totally misses the interior of some quadrant in $\mathbb{R}^{n+1}$.

The next result is an additional application of the Borsuk Antipodality Theorem.

Let $X = X^M$ be an M-dimensional normed linear space, and let T be a linear, invertible map of X^M onto itself. Set

$$A = \{Tx \colon \|x\| \leqq 1\}.$$

Proposition 1.9. *For A and X^M as above,*

$$d_n(A; X^M) \geqq \|T^{-k}\|^{-1/k}$$

for every positive integer k satisfying $kn < M$.

Proof. We shall assume that X^M is a strictly convex normed linear space. Let X_n be any n-dimensional subspace of X^M and let $\{x_1, \ldots, x_n\}$ denote a basis for X_n. For each $x \in X^M$, $\mathbf{a}(x) = (a_1(x), \ldots, a_n(x))$ shall denote the coefficient vector of the unique best approximation to x from X_n, i.e.,

$$\min_{y \in X_n} \|x - y\| = \left\| x - \sum_{i=1}^{n} a_i(x)\, x_i \right\|.$$

The $a_i(x)$ are continuous and odd functions of x. For $nk < M$, set

$$\mathbf{f}(x) = (\mathbf{a}(Tx),\ \mathbf{a}(T^2 x), \ldots, \mathbf{a}(T^k x)).$$

$\mathbf{f}(x)$ is a vector of length $nk < M$, which is both odd and continuous. As such, it follows from the Borsuk Antipodality Theorem (Theorem 1.4), that there exists an $x_0 \in X^M$, $\|x_0\| = 1$, for which $\mathbf{f}(x_0) = \mathbf{0}$, i.e., $\mathbf{a}(T^j x_0) = \mathbf{0}$, $j = 1, \ldots, k$. Since $T^j x_0/\|T^{j-1} x_0\| \in A$, it therefore follows that

$$\begin{aligned}\sup_{\|x\| \leqq 1} \inf_{y \in X_n} \|Tx - y\| &\geqq \max_{j=1,\ldots,k} \frac{\|T^j x_0\|}{\|T^{j-1} x_0\|} \\ &\geqq \left(\prod_{j=1}^{k} \frac{\|T^j x_0\|}{\|T^{j-1} x_0\|} \right)^{1/k} \\ &= \left(\frac{\|T^k x_0\|}{\|x_0\|} \right)^{1/k} \\ &\geqq \|T^{-k}\|^{-1/k}.\end{aligned}$$

Since this lower bound is independent of X_n, we obtain $d_n(A; X^M) \geqq \|T^{-k}\|^{-1/k}$. □

Remark. If $n = M - 1$, then the result $d_{M-1}(A; X^M) \geqq \|T^{-1}\|^{-1}$ is a special case of Theorem 2.1 which follows. Equality in fact obtains.

2. Existence of Optimal Subspaces for d_n

As was previously mentioned, an optimal subspace for $d_n(A; X)$ need not be contained in $\overline{\text{span}}(A)$. In fact it is known that an optimal subspace for $d_n(A; X)$ need not exist at all in X. The interested reader may consult both Brown [1964] and Ruban [1975], wherein are constructed A and X with no optimal subspaces for $d_n(A; X)$. It is thus natural to ask for conditions on the normed linear space X and the subsets A of X which imply the existence of optimal subspaces.

For one general class of A and X, Theorem 1.5 provides an answer to this question. Namely, if A is a ball in some m-dimensional subspace of X, then any n-dimensional subspace of X is optimal for $d_n(A; X)$, $n = 0, 1, \ldots, m - 1$, while the m-dimensional subspace itself is obviously optimal for the remaining n-widths.

Another example of a situation wherein optimal subspaces always exist is the following. Let A be a proper subset of an $(n + 1)$-dimensional subspace X_{n+1} of X. As usual, we assume that A is convex, closed and centrally symmetric. From Theorem 1.5 (and Proposition 1.6), $d_n(A; X) \geqq \inf\{\|x\| : x \in \partial A\}$. We now show that equality holds and that one may choose an optimal n-dimensional subspace from X_{n+1}. Since X_{n+1} is finite dimensional, the infimum in $\inf\{\|x\| : x \in \partial A\}$ is attained, i.e., there exists an $x_0 \in \partial A$ for which $\|x_0\| = \inf\{\|x\| : x \in \partial A\}$. If $\|x_0\| = 0$, then $\dim(\text{span}(A)) \leqq n$ and $\text{span}(A)$ is an optimal subspace for $d_n(A; X)$. Assume $x_0 \neq 0$. Since A is a convex and centrally symmetric subset of an $(n + 1)$-dimensional normed linear space X_{n+1}, there exists, by the Hahn-Banach Theorem, a nontrivial linear functional f on X_{n+1} for which

$$f(x_0) = \sup\{|f(x)| : x \in A\}.$$

Set $X_n = \{x: x \in X_{n+1}, f(x) = 0\}$. Every $x \in A$ may be written in the form $x = \alpha x_0 + y$, where $y \in X_n$ and $|\alpha| \leqq 1$ (since $|f(x)| = |\alpha| f(x_0) \leqq f(x_0)$). Now,

$$\inf\{\|\alpha x_0 + y - z\|: z \in X_n\} \leqq \|\alpha x_0\| = |\alpha| \, \|x_0\| \leqq \|x_0\|.$$

Thus $d_n(A; X) \leqq E(A; X_n) \leqq \|x_0\| \leqq d_n(A; X)$. We have therefore proven the following result.

Theorem 2.1. *If A is a closed, convex, centrally symmetric proper subset of an $(n+1)$-dimensional subspace X_{n+1} of X, then*

$$d_n(A; X) = \inf\{\|x\|: x \in \partial A\}.$$

Furthermore there is an n-dimensional subspace of X_{n+1} which is optimal for $d_n(A; X)$.

It may be proven that the only optimal n-dimensional subspaces for $d_n(A; X)$ in X_{n+1} are those constructed as in the proof of Theorem 2.1. This does not preclude the existence of additional optimal subspaces in X.

Theorem 2.1 is useful but not sufficiently general. The following theorem is more generally applicable.

Theorem 2.2. *Let X be a normed linear space and let X' denote the dual of X (the space of continuous linear functionals on X). Then for every subset $A \subseteqq X'$, there exists an optimal n-dimensional subspace for $d_n(A; X')$.*

We omit the proof which is based on the compactness of the unit ball of X' in the weak*-topology. As a result of Theorem 2.2, we also obtain the following result.

Theorem 2.3. *Let X be a Banach space. Assume that there exists a projection p of X'' onto X of norm one, i.e., $p: X'' \to X$, $\|p\| = 1$. Then for every $A \subseteqq X$, there exists an optimal n-dimensional subspace for $d_n(A; X)$.*

Proof. It suffices to assume that $d_n(A; X) < \infty$. By Theorem 2.2, there exists an optimal subspace X_n for $d_n(A; X'')$. (We identify elements of X with their image under the canonical mapping J_X of X into X''.) Set $Y_n = p(X_n) \subseteqq X$. Let $x \in A$ and y be a best approximation to x from X_n. Then

$$\|x - p(y)\| = \|p(x - y)\| \leqq \|x - y\| \leqq d_n(A; X'').$$

However, from Theorem 1.1, property (vii), $d_n(A; X'') \leqq d_n(A; X)$. Thus $E(A; Y_n) \leqq d_n(A; X)$ and the theorem is proved. □

Theorem 2.3 is a genuine extension of Theorem 2.2; indeed there exist spaces which satisfy the hypotheses of Theorem 2.3, but not those of Theorem 2.2. The most notable example is $X = L^1[0, 1]$. It should be noted that there always exists a projection p of norm one of X''' onto X'.

3. Properties of d^n

We first state the definition of d^n.

Definition 3.1. Let X be a real or complex normed linear space and A a subset of X. The *n-width* of A with respect to X, *in the sense of Gel'fand*, is defined as

$$d^n(A;X) = \inf_{L^n} \sup_{x \in A \cap L^n} \|x\|$$

where the infimum is taken over all subspaces L^n of X of codimension n. If $d^n(A;X) = \sup\{\|x\|: x \in A \cap L^n\}$ where L^n is a subspace of X of codimension at most n, then L^n is an *optimal* subspace for $d^n(A;X)$.

To give meaning to the above expression we henceforth assume that the zero element is in A. We say that a subspace L^n is of codimension n if there exist n *continuous* (linearly independent) linear functionals $\{f_i\}_{i=1}^n$ on X for which

$$L^n = \{x: x \in X,\ f_i(x) = 0,\ i = 1, \ldots, n\}.$$

Certain authors, most notably Helfrich [1971], define L^n (and consequently $d^n(A;X)$) by means of n linear functionals on $\hat{X}'$, the algebraic dual of X. This definition while allowing for greater generality, has certain disadvantages. We shall use Definition 3.1.

In the same manner in which $d_n(A;X)$ depends upon $E(A;X_n)$ for n-dimensional subspaces X_n, we define

$$\Delta(A;B) = \sup\{\|x\|: x \in A \cap B\}.$$

Thus $d^n(A;X) = \inf\{\Delta(A;L^n\}: \operatorname{codim} L^n \leqq n\}$. We also drop the reference to X and write $d^n(A)$ in place of $d^n(A;X)$ if no ambiguity arises.

We list some simple properties of d^n which follow from the definition.

Theorem 3.1. *Let X be a normed linear space and $A \subseteq X$.*

(i) *For every scalar α,*

$$d^n(\alpha A) = |\alpha|\, d^n(A).$$

(ii) *For each $A \subseteq X$, let* $\mathrm{b}(A) = \{\alpha x: x \in A,\ |\alpha| \leqq 1\}$ *be the balanced hull of A. Then*

$$d^n(A) = d^n(\mathrm{b}(A)).$$

(iii) *If $B \subseteq A \subseteq X$, then*

$$d^n(B) \leqq d^n(A).$$

(iv)

$$d^n(A) \geqq d^{n+1}(A), \qquad n = 0, 1, 2, \ldots.$$

Note that various properties attributed to $d_n(A;X)$ are not included in this list. Consider the following two examples.

In this first example we show that it is possible that $d^n(A) < d^n(\bar{A})$.

Let X be $\mathbb{R}^2$ with the l_∞ (maximum norm) i.e., $l^2_\infty(\mathbb{R})$, and let

$$A = \{(x,y)\colon 0 < x < 1,\ -1 < y < 1\} \cup \{(0,0)\}.$$

Setting $L^1 = \{(x,y)\colon x = 0\}$, we have $d^1(A;X) \leqq \max\{|y|\colon (0,y) \in A\} = 0$ which implies that $d^1(A;X) = 0$. However $\bar{A} = \{(x,y)\colon 0 \leqq x \leqq 1,\ -1 \leqq y \leqq 1\}$, and it is easily seen that $d^1(\bar{A};X) = 1$, so that $d^1(A;X) < d^1(\bar{A};X)$.

In this next example we show that it is possible that $d^n(A;X) < d^n(\mathrm{co}\,A;X)$. Let X be as above and

$$A = \{(x,y)\colon |x|,\ |y| \leqq 1,\ x = y \text{ or } x = -y\}.$$

It is a simple matter to prove that $d^1(A;X) = 0$ while $d^1(\mathrm{co}\,A;X) = 1$.

Note that in the first example A is convex while in the second example A is closed, so that neither of the two conditions is sufficient in and of itself to insure that $d^n(A;X) = d^n(\bar{A};X)$ and $d^n(A;X) = d^n(\mathrm{co}\,A;X)$.

We shall henceforth assume however that A is a *closed, convex, centrally symmetric* subset of X (and if it is not, take its closed, convex, centrally symmetric hull).

Note that no analogue of property (vii) of Theorem 1.1 concerning inequalities between $d^n(A;X)$ and $d^n(A;Y)$ for $X \subseteqq Y$ has been given. This is because the following stronger result holds.

Proposition 3.2. *Assume that X and Y are normed linear spaces, and X is a subspace (and not simply a subset) of Y, then*

$$d^n(A;X) = d^n(A;Y).$$

Proof. Since the restriction to X of any $f \in Y'$ is in X', it follows from the definition of d^n that

$$d^n(A;Y) \geqq d^n(A;X).$$

Moreover, from the Hahn-Banach Theorem, there exists, for any continuous linear functional on X, a continuous linear functional on Y which is an extension. Thus $d^n(A;X) \geqq d^n(A;Y)$ and the equality holds. □

If X is only a subset and not a subspace of Y, then the example of strict inequality given after Theorem 1.1 is also valid for d^n, and thus strict inequality may in fact hold. If X and Y are normed linear spaces over the same field of scalars, then $X \subseteqq Y$ implies that X is a subspace of Y. Thus inequality may only hold in this work if X is a normed linear space over the reals and Y is a normed linear space over the complex scalars. For example, $l^n_p(\mathbb{R})$ is a subset, but not a subspace of $l^n_p(\mathbb{C})$.

More importantly, by Proposition 3.2 it suffices in the calculation of $d^n(A;X)$ to take $X = \mathrm{span}(A)$.

Analogous to Proposition 1.2 we have

Proposition 3.3. *If A is compact, then $d^n(A;X)\downarrow 0$.*

Proof. Since A is compact it has a finite ε-net. That is, given $\varepsilon > 0$ there exist $x_1,\ldots,x_N$ such that for every $x\in A$, $\min\{\|x-x_i\|: i=1,\ldots,N\}\leqq\varepsilon$. From the Hahn-Banach Theorem, there exists for each $x_i\in X$, a continuous linear functional $f_i\in X'$ for which $f_i(x_i)=\|x_i\|$, and $\|f_i\|=1$. Set $L^N=\{x: f_i(x)=0,\ i=1,\ldots,N\}$. We shall prove that $\Delta(A;L^N)\leqq 2\varepsilon$. For each $x\in A\cap L^N$ there exists an $i\in\{1,\ldots,N\}$ for which $\|x-x_i\|<\varepsilon$. Furthermore, $\|x_i\|=|f_i(x_i)|=|f_i(x_i-x)|\leqq\|f_i\|\,\|x_i-x\|\leqq\varepsilon$. Thus, $\|x\|\leqq\|x-x_i\|+\|x_i\|\leqq 2\varepsilon$. Hence $d^N(A;X)\leqq 2\varepsilon$ and the proposition is proved. □

An analogue of Theorem 1.5 holds for $d^n(A;X)$.

Theorem 3.4. *Let X_{n+1} be any $(n+1)$-dimensional subspace of a normed linear space X and let $S(X_{n+1})$ denote the unit ball in X_{n+1}. Then*

$$d^k(S(X_{n+1});X)=1,\qquad k=0,1,\ldots,n.$$

Proof. By Proposition 3.2, it suffices to prove that

$$d^n(S(X_{n+1});X)=d^n(S(X_{n+1});X_{n+1})\geqq 1.$$

In other words, we must show that given any n continuous linear functionals $f_1,\ldots,f_n$ on X_{n+1}, there exists an $x\in X_{n+1}$, $x\neq 0$, for which $f_i(x)=0$, $i=1,\ldots,n$. Since X_{n+1} is $(n+1)$-dimensional, the construction of such an x is immediate. □

On the basis of Theorem 3.4, we also have

Proposition 3.5. $d^n(A;X)\geqq b_n(A;X)$.

As in Section 2, we are also able to explicitly calculate the n-width, in the sense of Gel'fand, of any proper subset of an $(n+1)$-dimensional subspace X_{n+1} of X.

Theorem 3.6. *Let X_{n+1} be any $(n+1)$-dimensional subspace of a normed linear space X. Assume that A is a convex, closed, centrally symmetric subset of X_{n+1}. Then*

$$d^n(A;X)=\inf\{\|x\|: x\in\partial A\}.$$

The above infimum is attained by some $x_0\in\partial A$. Furthermore the subspace

$$L^n=\{x: f_i(x)=0,\ i=1,\ldots,n\}$$

is an optimal subspace for $d^n(A;X)$ if $f_1,\ldots,f_n\in X'$ are n linearly independent linear functionals on X_{n+1} for which $f_i(x_0)=0$, $i=1,\ldots,n$.

Proof. From Proposition 3.2 it suffices to consider $d^n(A;X_{n+1})$. Proposition 3.5 implies that we need only prove

$$d^n(A;X_{n+1})\leqq\inf\{\|x\|: x\in\partial A\},$$

and exhibit the optimal subspaces of codimension n.

Let $x_0 \in \partial A$ satisfy $\|x_0\| = \inf\{\|x\| : x \in \partial A\}$. If $x_0 = 0$, then A is contained in some n-dimensional subspace of X_{n+1} and the theorem follows. Assume $x_0 \neq 0$, and let $f_1, \ldots, f_n$ be n linearly independent linear functionals on X_{n+1} for which $f_i(x_0) = 0$, $i = 1, \ldots, n$. If $x \in A$ and $f_i(x) = 0$, $i = 1, \ldots, n$, then $x = \alpha x_0$ for some scalar α with $|\alpha| \leqq 1$. The result now easily follows. □

In Section 6 we consider the duality relationship between d_n and d^n. We therefore defer to Section 6 the discussion of the existence of optimal subspaces for $d^n(A;X)$.

4. Properties of δ_n

We first give the definition of δ_n, the linear n-width.

Definition 4.1. Let X be a normed linear space and let A be a subset of X. The *linear n-width* of A with respect to X is given by

$$\delta_n(A;X) = \inf_{P_n} \sup_{x \in A} \|x - P_n(x)\|$$

where the infimum is taken over all continuous linear operators P_n of X into X of rank n, i.e., for which the range of P_n is of dimension n. Any continuous linear operator of rank at most n, P_n, for which

$$\delta_n(A;X) = \sup_{x \in A} \|x - P_n(x)\|$$

is called an *optimal* linear operator for $\delta_n(A;X)$.

As previously, we drop all reference to X when no ambiguity arises. Some simple properties of δ_n are listed below.

Theorem 4.1. *For a normed linear space X and $A \subseteq X$*

(i) $\delta_n(\bar{A}) = \delta_n(A)$.

(ii) *For every scalar* α,

$$\delta_n(\alpha A) = |\alpha|\, \delta_n(A).$$

(iii) *If* $\mathrm{b}(A)$ *denotes the balanced hull of* A, *then*

$$\delta_n(\mathrm{b}(A)) = \delta_n(A).$$

(iv) $\delta_n(\mathrm{co}\, A) = \delta_n(A)$.

(v) *If $B \subseteq A \subseteq X$, then*

$$\delta_n(B) \leqq \delta_n(A).$$

(vi) $\delta_n(A) \geqq \delta_{n+1}(A)$, $n = 0, 1, 2, \ldots$.

Furthermore we also have

Proposition 4.2. *Assume that $A \subseteqq X \subseteqq Y$ and that X is a subspace of the normed linear space Y. Then*

$$\delta_n(A;X) \geqq \delta_n(A;Y).$$

Proof. Every continuous rank n linear operator P_n on X may be written in the form $P_n(x) = \sum_{i=1}^{n} f_i(x)\,x_i$ where $x_i \in X$ and $f_i \in X'$. The analogous property holds for continuous rank n linear operators on Y. Since X is a subspace of Y, it follows from the Hahn-Banach Theorem that there exists a continuous rank n linear operator $\tilde{P}_n$ on Y for which $\tilde{P}_n(x) = P_n(x)$ for all $x \in X$ and hence for all $x \in A$. The proposition follows. □

Since linear approximation is clearly no better than best approximation, we have the lower bound

$$\delta_n(A;X) \geqq d_n(A;X).$$

Thus the analogue of Theorem 1.5 holds also for $\delta_n(A;X)$. Moreover one can also prove the analogue of Theorems 2.1 and 3.6.

Theorem 4.3. *If A is a closed, convex, centrally symmetric proper subset of an $(n+1)$-dimensional subspace X_{n+1} of X, then*

$$\delta_n(A;X) = \inf\{\|x\| : x \in \partial A\}.$$

Proof. It is necessary to prove that $\delta_n(A;X) \leqq \inf\{\|x\| : x \in \partial A\}$. Let $x_0 \in \partial A$ be such that $\|x_0\| = \inf\{\|x\| : x \in \partial A\}$. Assume $x_0 \neq 0$ (if $x_0 = 0$, the result follows easily). Since A is convex and centrally symmetric, there exists a continuous linear functional $f \in X'_{n+1}$ for which

$$f(x_0) = \sup\{|f(x)| : x \in A\}.$$

Set $X_n = \{x : x \in X_{n+1}, f(x) = 0\}$. Let $x_1, \ldots, x_n$ be a basis for X_n. Thus $x_0, x_1, \ldots, x_n$ is a basis for X_{n+1}. Let $f_1, \ldots, f_n \in X'_{n+1}$ be such that $f_i(x_j) = \delta_{ij}$, $i = 1, \ldots, n$; $j = 0, 1, \ldots, n+1$. By the Hahn-Banach Theorem, we may also regard $f, f_1, \ldots, f_n$ as continuous linear functional on X. Let $P_n(x) = \sum_{i=1}^{n} f_i(x)\,x_i$. P_n is a continuous linear operator of rank n. Furthermore every $x \in A$ may be written in the form $x = \alpha x_0 + \sum_{i=1}^{n} \alpha_i x_i$, where $|\alpha| \leqq 1$. Thus

$$\|x - P_n(x)\| = \left\| \alpha x_0 + \sum_{i=1}^{n} \alpha_i x_i - \sum_{i=1}^{n} \alpha_i x_i \right\| = \|\alpha x_0\| \leqq \|x_0\|.$$

The theorem is proved. □

Note that Theorem 4.3 (together with Theorem 1.5) implies Theorem 2.1. In the next section we prove that $\delta_n \geqq d^n$, so that it also implies Theorem 3.6.

5. Inequalities Between n-Widths

From Propositions 1.6 and 3.5 and the respective definitions we have $d_n(A;X)$, $d^n(A;X) \geqq b_n(A;X)$ and $\delta_n(A;X) \geqq d_n(A;X)$. One other inequality is always valid, namely

Proposition 5.1. $\delta_n(A;X) \geqq d^n(A;X)$.

Proof. Every rank n continuous linear operator P_n of X into X may be written in the form

$$P_n(x) = \sum_{i=1}^{n} f_i(x)\, x_i.$$

where $f_i \in X'$, $i = 1, \ldots, n$, and span $\{x_1, \ldots, x_n\}$ = range P_n. Thus

$$\sup\{\|x - P_n(x)\| : x \in A\} \geqq \sup\{\|x\| : x \in A,\ f_i(x) = 0,\ i = 1, \ldots, n\}$$

whence it follows that $\delta_n(A;X) \geqq d^n(A;X)$. □

The above inequalities are the only ones generally valid in considering the four quantities δ_n, d_n, d^n and b_n. To prove this fact we consider two examples of A and X. In the first of these examples $d_n > d^n$, while in the second example $d_n < d^n$.

Example 1. Let $A = S_1^3$ be the unit ball in $l_1^3(\mathbb{R})$. Thus

$$S_1^3 = \{\mathbf{x} = (x_1, x_2, x_3) : |x_1| + |x_2| + |x_3| \leqq 1,\ x_i \in \mathbb{R}\}.$$

Set $X = l_2^3(\mathbb{R})$, i.e., $\mathbb{R}^3$ with the Euclidean norm.

To determine $d_1(S_1^3; l_2^3(\mathbb{R}))$, it is necessary to determine the best line through the origin (1-dimensional subspace) with which to approximate S_1^3 in the Euclidean norm. S_1^3 is a regular octogon and any line through the origin must cut at least two of its faces. Assume, without loss of generality, that the line cuts the face lying in the first quadrant, i.e., the triangle with vertices $(1,0,0)$, $(0,1,0)$ and $(0,0,1)$. Since at least one of these three vertices is of distance at least $\sqrt{2/3}$ from any such line, it follows that $d_1(S_1^3; l_2^3(\mathbb{R})) \geqq \sqrt{2/3}$. Equality, in fact, holds by the choice of $X_1 = \{(\alpha, \alpha, \alpha) : \alpha \in \mathbb{R}\}$.

To determine $d^1(S_1^3; l_2^3(\mathbb{R}))$ we must consider hyperplanes containing the origin, and choose the hyperplane so as to minimize the element of largest norm which is contained both in the hyperplane and in S_1^3. It is not difficult to prove that an optimal choice for a hyperplane is to choose one which stays as far away as possible from the extreme points of S_1^3, i.e., a hyperplane containing the origin parallel to any one of the faces. Set $L^1 = \{\mathbf{x} : x_1 + x_2 + x_3 = 0\}$. We may calculate $\max\{\|\mathbf{x}\|_2 : \mathbf{x} \in L^1 \cap S_1^3\} = 1/\sqrt{2}$, which provides us with the upper bound, $d^1(S_1^3; l_2^3(\mathbb{R})) \leqq 1/\sqrt{2}$. (In fact $d^1(S_1^3; l_2^3(\mathbb{R})) = 1/\sqrt{2}$.) Thus

$$d_1(S_1^3; l_2^3(\mathbb{R})) > d^1(S_1^3; l_2^3(\mathbb{R})).$$

Example 2. Let $A = S_2^3$ be the unit ball in $l_2^3(\mathbb{R})$ and set $X = l_\infty^3(\mathbb{R})$, i.e., $\mathbb{R}^3$ with the max norm

$$\|\mathbf{x}\|_\infty = \max_{i=1,2,3} |x_i|.$$

To determine the values $d_1(S_2^3; l_\infty^3(\mathbb{R}))$ and $d^1(S_2^3; l_\infty^3(\mathbb{R}))$, it is possible to proceed along the lines of Example 1. However, a more concise method of proof is to reduce this example to the previous example by the use of a duality principle.

$$\begin{aligned} d_1(S_2^3; l_\infty^3(\mathbb{R})) &= \min_{\mathbf{y}} \max_{\|\mathbf{x}\|_2 \leqq 1} \min_{\alpha} \|\mathbf{x} - \alpha \mathbf{y}\|_\infty \\ &= \min_{\mathbf{y}} \max_{\|\mathbf{x}\|_2 \leqq 1} \max_{\substack{\|\mathbf{z}\|_1 \leqq 1 \\ (\mathbf{z},\mathbf{y})=0}} (\mathbf{z}, \mathbf{x}) \\ &= \min_{\mathbf{y}} \max_{\substack{\|\mathbf{z}\|_1 \leqq 1 \\ (\mathbf{z},\mathbf{y})=0}} \max_{\|\mathbf{x}\|_2 \leqq 1} (\mathbf{z}, \mathbf{x}) \\ &= \min_{\mathbf{y}} \max_{\substack{\|\mathbf{z}\|_1 \leqq 1 \\ (\mathbf{z},\mathbf{y})=0}} \|\mathbf{z}\|_2 \\ &= d^1(S_1^3; l_2^3(\mathbb{R})) = 1/\sqrt{2}. \end{aligned}$$

On the other hand

$$\begin{aligned} d^1(S_2^3; l_\infty^3(\mathbb{R})) &= \min_{\mathbf{y}} \max_{\substack{\|\mathbf{x}\|_2 \leqq 1 \\ (\mathbf{x},\mathbf{y})=0}} \|\mathbf{x}\|_\infty \\ &= \min_{\mathbf{y}} \max_{\substack{\|\mathbf{x}\|_2 \leqq 1 \\ (\mathbf{x},\mathbf{y})=0}} \max_{\|\mathbf{z}\|_1 \leqq 1} (\mathbf{z}, \mathbf{x}) \\ &= \min_{\mathbf{y}} \max_{\|\mathbf{z}\|_1 \leqq 1} \max_{\substack{\|\mathbf{x}\|_2 \leqq 1 \\ (\mathbf{x},\mathbf{y})=0}} (\mathbf{z}, \mathbf{x}) \\ &= \min_{\mathbf{y}} \max_{\|\mathbf{z}\|_1 \leqq 1} \min_{\alpha} \|\mathbf{z} - \alpha \mathbf{y}\|_2 \\ &= d_1(S_1^3; l_2^3(\mathbb{R})) = \sqrt{2/3}. \end{aligned}$$

Thus $d^1(S_2^3; l_\infty^3(\mathbb{R})) > d_1(S_2^3; l_\infty^3(\mathbb{R}))$.

The duality between the Kolmogorov and Gel'fand n-widths as exemplified above may be cast in a more general setting. We shall return to this fact in Theorem 6.2. However, the specific instance of the duality principle used above is a fairly simple consequence of Hölder's inequality.

One particular case in which equality holds between δ_n and d_n is the following.

Proposition 5.2. *If $X = H$ is a Hilbert space, then*

$$\delta_n(A; H) = d_n(A; H) \geqq d^n(A; H).$$

Proof. The equality is a statement of the fact that the best approximation operator in a Hilbert space is a linear projection. The inequality is a re-statement of Proposition 5.1 while Example 1 demonstrates that strict inequality may in fact obtain. □

There also exist inequalities between d_n and b^n other than that given at the beginning of this section. Mityagin and Henkin [1963] proved the following two results: For A bounded, closed, convex and centrally symmetric,

$$d_n(A;X) \leqq (n+1)^2\, b_n(A;X),$$

and if $X = H$ is a Hilbert space, then

$$d_n(A;H) \leqq (n+1)\, b_n(A;H).$$

Pukhov [1979a] strengthened this latter result by proving that if H is a Hilbert space over the reals, then

$$d_n(A;H) \leqq \sqrt{e}\,\sqrt{n+1}\, b_n(A;H).$$

It is conjectured, again by Mityagin and Henkin, that the correct inequality should be

$$d_n(A;X) \leqq (n+1)\, b_n(A;X)$$

for general X, while for a Hilbert space H, the correct bound is conjectured to be

$$d_n(A;H) \leqq \sqrt{n+1}\, b_n(A;H).$$

As yet, both conjectures remain unproved. The bound in the second conjecture can certainly not be improved upon. It is known that for $X = l_2^M(\mathbb{R})$, and $A = S_1^M = \left\{\mathbf{x}\colon \mathbf{x} \in \mathbb{R}^M,\ \sum_{i=1}^{M} |x_i| \leqq 1\right\}$, $d_n(S_1^M; l_2^M(\mathbb{R})) = ((M-n)/M)^{1/2}$, and $b_n(S_1^M; l_2^M(\mathbb{R})) = (n+1)^{-1/2}$ (see Chapter VI). Since these are valid for every $M > n$, the correct bound is certainly no less than $\sqrt{n+1}$. (The bound in the first conjecture can also not be improved upon.)

We shall not prove the inequality

$$d_n(A;X) \leqq (n+1)^2\, b_n(A;X).$$

Its proof is based on Lemma 5.4, given below, and an ingenious use of the Brunn-Minkowski inequality on volumes of convex bodies. The proof of this result would lead us further afield than we wish to go and the interested reader is therefore referred to Mityagin and Henkin [1963].

We shall prove the two known bounds in the case that $X = H$ is a Hilbert space.

Proposition 5.3. *Let H be a Hilbert space and A a closed, convex, centrally symmetric bounded subset of H. Then,*

$$d_n(A;H) \leqq (n+1)\, b_n(A;H).$$

The proof is based on the following lemma which itself is independent of the Hilbert space structure.

Lemma 5.4. *Let X be a normed linear space and A a closed, convex, centrally symmetric subset of X. Let $x_1,\ldots,x_{n+1}\in A$, and set*

$$E_k = \operatorname{span}\{x_1,\ldots,x_{k-1},\ x_{k+1},\ldots,x_{n+1}\},$$

$k=1,\ldots,n+1$. *If $E(x_k;E_k)\geqq c$, $k=1,\ldots,n+1$, then*

$$b_n(A;X)\geqq c/(n+1).$$

Proof. Let $X_{n+1}=\operatorname{span}\{x_1,\ldots,x_{n+1}\}$. By the definition of b_n,

$$b_n(A;X)\geqq \sup\{\lambda\colon \lambda S(X_{n+1})\subseteqq A\}.$$

It therefore suffices to show that $(c/(n+1))\,S(X_{n+1})\subseteqq A$.

Let $x\in(c/(n+1))\,S(X_{n+1})$. Thus $x=\sum_{k=1}^{n+1}\alpha_k x_k$, and $\|x\|\leqq c/(n+1)$. We claim that $|\alpha_k|\leqq 1/(n+1)$, $k=1,\ldots,n+1$. If this is true, then since $\sum_{k=1}^{n+1}|\alpha_k|\leqq 1$, and $x_1,\ldots,x_{n+1}\in A$, where A is a convex and centrally symmetric, it follows that $x\in A$. It thus remains to prove that $|\alpha_k|\leqq 1/(n+1)$. Without loss of generality, consider $k=1$ and assume $\alpha_1\neq 0$. Then

$$c\leqq E(x_1;E_1)\leqq \left\|x_1-\sum_{k=2}^{n+1}\frac{\alpha_k}{\alpha_1}x_k\right\|\leqq c/((n+1)\,|\alpha_1|)$$

and hence $|\alpha_1|\leqq 1/(n+1)$. $\square$

Proof of Proposition 5.3. We prove Proposition 5.3 by using Lemma 5.4. The problem is therefore one of choosing X_{n+1}. For $z_1,\ldots,z_m\in H$, set

$$V_m(z_1,\ldots,z_m)=[\det((z_i,z_j))_{i,j=1}^m]^{1/2}.$$

Note that $V_m(z_1,\ldots,z_m)$ is the volume of the parallelipiped determined by the vectors $z_1,\ldots,z_m$. Thus, in the terminology of Lemma 5.4, $V_{n+1}(x_1,\ldots,x_{n+1})=E(x_1;E_1)\,V_n(x_2,\ldots,x_{n+1})$. Set

$$V=\sup\{V_{n+1}(z_1,\ldots,z_{n+1})\colon z_k\in A,\ k=1,\ldots,n+1\}.$$

Since A is bounded, $V<\infty$. *If* $V=0$, then A is contained in an n-dimensional subspace and therefore $d_n(A;H)=b_n(A;H)=0$. Assume not. For $\varepsilon>0$, sufficiently small, choose $x_1,\ldots,x_{n+1}\in A$ for which

$$V_{n+1}(x_1,\ldots,x_{n+1})>(1-\varepsilon)\,V>0.$$

(This implies that $E(x_k; E_k) > 0, k = 1, \ldots, n+1$). From the definition of $d_n(A; H)$, there exists, for each k, an $\tilde{x}_k \in A$ for which $E(\tilde{x}_k; E_k) \geqq d_n(A; H)$. Thus

$$\begin{aligned} V &\geqq V_{n+1}(x_1, \ldots, x_{k-1}, \tilde{x}_k, x_{k+1}, \ldots, x_{n+1}) \\ &= E(\tilde{x}_k; E_k)\, V_n(x_1, \ldots, x_{k-1}, x_{k+1}, \ldots, x_{n+1}) \\ &= \frac{E(\tilde{x}_k; E_k)}{E(x_k; E_k)} V_{n+1}(x_1, \ldots, x_{n+1}) \\ &\geqq \frac{d_n(A; H)}{E(x_k; E_k)} V(1 - \varepsilon). \end{aligned}$$

Therefore $E(x_k; E_k) \geqq (1 - \varepsilon)\, d_n(A; H)$, and by Lemma 5.4,

$$b_n(A; H) \geqq \frac{(1 - \varepsilon)\, d_n(A; H)}{n+1}.$$

Since this holds for all $\varepsilon > 0$, sufficiently small, it follows that

$$(n+1)\, b_n(A; H) \geqq d_n(A; H). \quad \square$$

We now turn to the better bound obtained by Pukhov [1979a] in the case that H is a Hilbert space over the reals.

Proposition 5.5. *Let $X = H$ be a real Hilbert space and A a convex, closed, centrally symmetric bounded subset of H. Then*

$$d_n(A; H) \leqq \sqrt{e}\, \sqrt{n+1}\, b_n(A; H).$$

In the proof of Proposition 5.5, we shall utilize a lemma, whose proof we do not include. Before stating this lemma we introduce the following notation.

For given $x_1, \ldots, x_m$, linearly independent elements of H, let

$$C(x_1, \ldots, x_m) = \left\{ x : x = \sum_{i=1}^{m} \alpha_i x_i,\ \sum_{i=1}^{m} \alpha_i^2 \leqq 1 \right\}.$$

$C(x_1, \ldots, x_m)$ is the m-dimensional ellipsoid subsumed by $x_1, \ldots, x_m$. By

$$V_m(C(x_1, \ldots, x_m))$$

we mean the m-volume of $C(x_1, \ldots, x_m)$.

Lemma 5.6. *Let $x_1, \ldots, x_{n+1}$ be $n+1$ linearly independent elements of H, and $x \in H$. Set $X_n = \operatorname{span}\{x_1, \ldots, x_n\}$ and assume that*

$$E(x; X_n) \geqq E(x_{n+1}; X_n).$$

Then the convex hull of

$$\{\pm (1 + 1/n)^{n/2} (n+1)^{1/2} x,\ C(x_1, \ldots, x_n)\}$$

contains an $(n+1)$*-dimensional ellipsoid with* $(n+1)$*-volume greater than or equal to* $V_{n+1}(C(x_1,\ldots,x_n,x_{n+1}))$.

Proof of Proposition 5.5. Let $V(A) = \sup\{V_{n+1}(C_{n+1}): C_{n+1} \subseteqq A\}$, where C_{n+1} is an arbitrary, centrally symmetric $(n+1)$-dimensional ellipsoid. Since A is bounded, so is $V(A)$. For fixed $\varepsilon \in (0,1)$, let $C_{n+1} \subset A$ be such that

$$V_{n+1}(C_{n+1}) > (1-\varepsilon)\,V(A).$$

Now $C_{n+1} = C(x_1,\ldots,x_{n+1})$ for some choice of $x_1,\ldots,x_{n+1} \in A$, which can certainly be chosen to be mutually orthogonal. Assume that $\|x_{n+1}\| \leqq \|x_i\|$, $i = 1,\ldots,n$. For the definition of $b_n(A;H)$, it immediately follows that

$$b_n(A;H) \geqq \|x_{n+1}\|.$$

Now, set $X_n = \operatorname{span}\{x_1,\ldots,x_n\}$. For each $x \in A$, we claim that

$$E(x;X_n) \leqq (1+1/n)^{n/2}(n+1)^{1/2}\frac{\|x_{n+1}\|}{(1-\varepsilon)}.$$

From the definition of $d_n(A;H)$ and since $\varepsilon > 0$ is arbitrarily small, and $(1+1/n)^{n/2} \leqq \sqrt{e}$, the result would then follow. Assume to the contrary that for some $x \in A$,

$$\begin{aligned} E(x;X_n) &> (1+1/n)^{n/2}(n+1)^{1/2}\frac{\|x_{n+1}\|}{(1-\varepsilon)} \\ &= (1+1/n)^{n/2}(n+1)^{1/2}\frac{E(x_{n+1};X_n)}{(1-\varepsilon)}. \end{aligned}$$

Thus by Lemma 5.6, the convex hull of $\{\pm x, C(x_1,\ldots,x_n)\}$ contains an $(n+1)$-dimensional ellipsoid with $(n+1)$-volume greater than or equal to

$$V_{n+1}\left(C\left(x_1,\ldots,x_n,\frac{x_{n+1}}{(1-\varepsilon)}\right)\right) = \frac{1}{(1-\varepsilon)}V_{n+1}(C_{n+1}) > V(A).$$

Since $C\left(x_1,\ldots,x_n,\frac{x_{n+1}}{(1-\varepsilon)}\right)$ is contained in the convex hull of $\{\pm x, C(x_1,\ldots,x_n)\}$ which is itself contained in A, this contradicts the definition of $V(A)$. □

6. Duality Between d_n and d^n

The duality between the Kolmogorov and Gel'fand n-widths rests upon the following two duality principles.

Let X be a normed linear space, and X' the (Banach) space of continuous linear functionals on X. Let L and M be linear subspaces of X and X', respectively, and set

$$L^{\perp} = \{f : f \in X', \, f(x) = 0, \text{ all } x \in L\}$$
$$M_{\perp} = \{x : x \in X, \, f(x) = 0, \text{ all } f \in M\}.$$

Proposition 6.1. 1) *For $x \in X$ and a linear subspace $L \subset X$*

$$\inf_{y \in L} \|x - y\| = \max \{|f(x)| : f \in L^{\perp}, \, \|f\| \leqq 1\}.$$

2) *For $f \in X'$ and a weak*-closed subspace M of X',*

$$\min_{g \in M} \|f - g\| = \sup \{|f(x)| : x \in M_{\perp}, \, \|x\| \leqq 1\}.$$

The duality between the Kolmogorov and Gel'fand n-widths will be considered in detail in Sections 7 and 8. In this section we shall only give one application of Proposition 6.1. For more general results, see Ioffe and Tichomirov [1968 b].

Let A be a bounded subset of the normed linear space X. We shall assume that A is closed, convex, and centrally symmetric. Proposition 3.2 implies that $d^n(A; X) = d^n(A; \operatorname{span}(A))$. Set $\hat{X} = \operatorname{span}(A)$. Assume that A contains a ball of positive radius about zero in $\hat{X}$, i.e., there exists a constant $c > 0$ for which $x \in \hat{X}$ and $\|x\| \leqq c$ implies that $x \in A$.

Let

$$p(x) = \inf \{a^{-1} : a x \in A, \, a > 0\}$$

denote the Minkowski functional on $\hat{X}$. The above assumptions imply that $p(x)$ is a norm on $\hat{X}$. Let $\hat{X}'$ denote the class of continuous linear functionals on $\hat{X}$ with respect to the original norm $\|\cdot\|_X$, and let $\hat{X}^p$ denote the continuous linear functonals on $\hat{X}$ with respect to the norm $p(x)$. Since A contains a ball of positive radius in $\hat{X}$, $f \in \hat{X}'$ if and only if $f \in \hat{X}^p$. Let $S(\hat{X}')$ denote the unit ball in $\hat{X}'$, i.e.,

$$S(\hat{X}') = \{f : f \in \hat{X}', \, \|f\|_{\hat{X}'} \leqq 1\}.$$

Theorem 6.2. *Let A be a bounded subset of a normed linear space X which contains a ball of positive radius about zero in* span(A). *Then*

$$d^n(A; X) = d_n(S(\hat{X}'); \hat{X}^p).$$

Furthermore there exists an optimal subspace of codimension n for $d^n(A; X)$.

Remark. If A is contained in some finite dimensional space, then A contains a ball of positive radius about zero in span(A).

Proof. Let $f_1,\ldots,f_n$ be elements of $\hat{X}'(\hat{X}^p)$, and set $M = \text{span}\{f_1,\ldots,f_n\}$. Then

$$\begin{aligned}\Delta(A;M_\perp) &= \sup_{x\in A\cap M_\perp} \|x\|_X = \sup_{\substack{p(x)\leqq 1\\ x\in M_\perp}} \|x\|_X \\ &= \sup_{\substack{p(x)\leqq 1\\ x\in M_\perp}} \sup_{f\in S(\hat{X}')} |f(x)| \\ &= \sup_{f\in S(\hat{X}')} \sup_{\substack{p(x)\leqq 1\\ x\in M_\perp}} |f(x)|.\end{aligned}$$

From Proposition 6.1,

$$\sup_{\substack{p(x)\leqq 1\\ x\in M_\perp}} |f(x)| = \min_{g\in M} \|f-g\|_{\hat{X}^p}.$$

Thus

$$\Delta(A;M_\perp) = \sup_{f\in S(\hat{X}')} \min_{g\in M} \|f-g\|_{\hat{X}^p}.$$

As M varies over all n-dimensional subspaces of $\hat{X}'(\hat{X}^p)$, we obtain

$$d^n(A;X) = d_n(S(\hat{X}');\hat{X}^p).$$

From Theorem 2.2, there exists an optimal subspace for $d_n(S(\hat{X}');\hat{X}^p)$. Thus there exists an optimal subspace for $d^n(A;X)$. □

7. n-Widths of Mappings of the Unit Ball

We shall only rarely deal with n-widths of an arbitrary subset A of X. It will often be the case that A is the image of the unit ball of some normed linear space X under a continuous linear mapping T from X to Y. Note here that A is a subset of Y. (We are now adopting a slightly altered notation and the reader should be aware of this fact.)

Definition 7.1. Let $L(X,Y)$ denote the class of *continuous linear operators* from X to Y, where X and Y are normed linear spaces. Let $T\in L(X,Y)$. Then

$$d_n(T(X);Y) = \inf_{X_n} \sup_{\|x\|_X\leqq 1} \inf_{y\in X_n} \|Tx-y\|_Y$$

where the infimum is taken over all n-dimensional subspaces X_n of Y.

We often write $d_n(T)$ in place of $d_n(T(X);Y)$ if no ambiguity arises.

As an example of the above, let A be the set of all functions g of the form $g(x) = \int_0^1 K(x,y)\,h(y)\,dy$, where $\|h\|_p \leqq 1$ and $K\in C([0,1]\times[0,1])$. Let T denote the linear operator given by $Th = g$ and set $X = L^p[0,1]$ and $Y = L^q[0,1]$ for

$p, q \in [1, \infty]$. A is often a variant of the above. For example, A may be of the form $A = \{Tx + u\colon \|x\|_X \leqq 1,\ x \in V,\ u \in V_p\}$ where V is some subset of X, and V_p is a fixed p-dimensional subspace of Y. In this chapter we restrict ourselves to the simpler case wherein A is the image of a unit ball.

In discussing the n-width in the sense of Gel'fand and the linear n-width for sets of the above form, a slightly altered definition is given, which could not have been previously introduced. We shall not, however, alter the notation and the reader is urged to note these differences, minor though they may be. The motivation behind these altered definitions will become apparent from the duality relationships which follow.

Definition 7.2. Let $T \in L(X, Y)$. The *n-width in the sense of Gel'fand* is defined as

$$d^n(T(X); Y) = \inf_{L^n} \sup_{\substack{\|x\|_X \leqq 1 \\ x \in L^n}} \|Tx\|_Y.$$

where the infimum is taken over subspaces L^n of X of codimension at most n.

(If we were to use the definition of d^n as given in Definition 3.1, then L^n would vary over the set of subspaces of codimension at most n of Y, rather than of X, and we would sup over $Tx \in L^n$.)

Definition 7.3. Let $T \in L(X, Y)$. Then the *linear n-width* is defined by

$$\delta_n(T(X); Y) = \inf_{P_n} \sup_{\|x\|_X \leqq 1} \|Tx - P_n x\|_Y$$

where P_n is any continuous linear operator of X into Y of rank at most n.

(This definition would agree with the former Definition 4.1 if we had restricted ourselves to a consideration of continuous rank n operators of X into Y of the form $P_n = Q_n T$, where Q_n is a continuous linear rank n operator of Y into Y.)

We often write $d^n(T)$ and $\delta_n(T)$ rather than $d^n(T(X); Y)$ and $\delta_n(T(X); Y)$, respectively. The linear n-width is often referred to, in the literature, as the nth approximation number and the symbols $a_n(T)$ and $s_n(T)$, rather than $\delta_n(T)$, are sometimes used.

Remark. $d_n(T)$, $d^n(T)$ and $\delta_n(T)$ satisfy the results of the previous sections except for Proposition 5.1 and Theorem 6.2. Proposition 5.1 will be restated for $d^n(T)$ and $\delta_n(T)$ as above. The change in the definition of $d^n(T)$ allows us to prove a result stronger than that given by Theorem 6.2.

For notational ease, we make the following definitions.

Definition 7.4. a) Let $K(X, Y)$ denote the class of *compact operators* in $L(X, Y)$, i.e., continuous linear operators $T \in L(X, Y)$ for which the closure of $\{Tx\colon \|x\|_X \leqq 1\}$ is compact in Y. (This class is sometimes referred to in the literature as the class of completely continuous operators.)

b) Let $F(X, Y)$ denote the closure of the *finite rank operators* in $L(X, Y)$, i.e., operators $T \in L(X, Y)$ for which there exist $P_n \in L(X, Y)$, P_n of rank at most n, satisfying $\|T - P_n\| \to 0$ as $n \uparrow \infty$.

Both $K(X, Y)$ and $F(X, Y)$ are closed subsets of $L(X, Y)$. $F(X, Y)$ is a subset of $K(X, Y)$ and need not equal $K(X, Y)$.

We recall that from Proposition 1.2 we have

Proposition 7.1. $d_n(T) \downarrow 0$ *iff* $T \in K(X, Y)$.

Each $T \in L(X, Y)$ has an adjoint $T' \in L(Y', X')$, defined by $\langle Tx, y' \rangle = \langle x, T' y' \rangle$ for all $x \in X$ and $y' \in Y'$, where, for ease of exposition, we are using the notation $x'(x) = \langle x, x' \rangle$ for $x \in X$, $x' \in X'$, and similarly for Y. In order to prove the analogue of Proposition 7.1 for $d^n(T)$, we make use of the following important duality result. For convenience $S(X)$ shall denote the unit ball in X.

Theorem 7.2. $d^n(T) = d_n(T')$ *for every* $T \in L(X, Y)$.

Proof. For any subspace L^n of codimension n in X,

$$\begin{aligned} \sup_{x \in L^n \cap S(X)} \| Tx \|_Y &= \sup_{x \in L^n \cap S(X)} \sup_{y' \in S(Y')} \langle Tx, y' \rangle \\ &= \sup_{y' \in S(Y')} \sup_{x \in L^n \cap S(X)} \langle x, T' y' \rangle . \end{aligned}$$

L^n may be written in the form $L^n = \{x : \langle x, f_i \rangle = 0,\ i = 1, \ldots, n,\ f_i \in X'\}$, and $(L^n)^\perp = \operatorname{span}\{f_1, \ldots, f_n\}$ is an n-dimensional subspace of X'. From Proposition 6.1

$$\sup_{y' \in S(Y')} \sup_{x \in L^n \cap S(X)} \langle x, T' y' \rangle = \sup_{y' \in S(Y')} \inf_{x' \in (L^n)^\perp} \| T' y' - x' \|_{X'} .$$

Taking the infimum over L^n we obtain $d^n(T) = d_n(T')$. □

As an immediate consequence of Theorem 2.2, we have the following result concerning the existence of optimal subspaces for $d^n(T)$.

Corollary 7.3. *For* $T \in L(X, Y)$, *there exists a subspace* L^n *of* X *of codimension at most* n *which is optimal for* $d^n(T)$.

We state (see e.g. Dunford and Schwartz [1957, p. 485]):

Schauder's Theorem. $T \in K(X, Y)$ *iff* $T' \in K(Y', X')$.

From Schauder's Theorem, Proposition 7.1 and Theorem 7.2 we obtain:

Proposition 7.4. $d^n(T) \downarrow 0$ *iff* $T \in K(X, Y)$.

From the definitions of $\delta_n(T)$ and $F(X, Y)$ we also have:

Proposition 7.5. $\delta_n(T) \downarrow 0$ *iff* $T \in F(X, Y)$.

The following are simple consequences of the definitions of $d_n(T)$, $d^n(T)$, and $\delta_n(T)$.

Proposition 7.6. *Let* s_n *denote any of the three quantities* d_n, d^n *or* δ_n.
1) *For* $T, R \in L(X, Y)$,

$$s_{m+n}(R + T) \leqq s_m(R) + s_n(T),$$

which immediately implies that

$$|s_n(T) - s_n(R)| \leqq \| T - R \| .$$

2) *For* $T \in L(X, Y)$ *and* $R \in L(Y, Z)$,

$$s_{m+n}(RT) \leqq s_m(R)\, s_n(T).$$

We also have, as a consequence of the method of proof of Proposition 5.1,

Proposition 7.7. $d^n(T) \leqq \delta_n(T)$.

8. Some Relationships Between $d_n(T)$, $d^n(T)$ and $\delta_n(T)$

In the previous section we defined $d_n(T)$, $d^n(T)$, $\delta_n(T)$, stated some simple consequences of the definitions, and reproved Proposition 5.1 and Theorem 6.2 in a strengthened form. Thus $d_n(T)$, $d^n(T) \leqq \delta_n(T)$ and $d^n(T) = d_n(T')$ for all $T \in L(X, Y)$. Because of the specific choice of $A = \{Tx: \|x\|_X \leqq 1\}$ further relationships hold between the above three quantities. We first bound δ_n from above.

Proposition 8.1. *Let* Y *be a normed linear space and assume that for every* n*-dimensional subspace* X_n *of* Y, *there exists a projection* P_n *of* Y *onto* X_n *of norm at most* α_n. *Then for* $T \in L(X, Y)$,

$$\delta_n(T) \leqq (1 + \alpha_n)\, d_n(T).$$

Proof. For each $\varepsilon > 0$, there exists an n-dimensional subspace X_n of Y for which $E(Tx; X_n) \leqq d_n(T) + \varepsilon$ for all $x \in S(X)$. Let P_n be a projection of Y onto X_n of norm at most α_n, and set $P_n^* = P_n T$. Thus for $x \in S(X)$,

$$\begin{aligned} \| Tx - P_n^* x \| &= \| Tx - P_n Tx \| \\ &= \| (Tx - y) - P_n(Tx - y) \| \end{aligned}$$

for any $y \in X_n$, since $P_n y = y$. Hence we may choose $y \in X_n$ so that

$$\begin{aligned} \| Tx - P_n^* x \| &\leqq \| Tx - y \|\, \| I - P_n \| \\ &\leqq (d_n(T) + \varepsilon)(1 + \alpha_n). \end{aligned}$$

Since $\varepsilon > 0$ is arbitrary,

$$\delta_n(T) \leqq (1 + \alpha_n)\, d_n(T). \quad \square$$

Proposition 8.2. *Let* X *be a normed linear space and assume that for any subspace* L^n *of* X *of codimension at most* n, *there exists a projection* Q_n *of* X *onto* L^n *of norm at most* β_n. *Then for* $T \in L(X, Y)$,

$$\delta_n(T) \leqq \beta_n\, d^n(T).$$

Proof. There exists a subspace L^n of X of codimension at most n for which

$$\sup_{x \in L^n \cap S(X)} \| Tx \| = d^n(T).$$

By assumption, there exists a projection Q_n of X onto L^n of norm at most β_n. Set

$$P_n^* = T(I - Q_n).$$

P_n^* is a linear operator of rank at most n of X into Y. Thus,

$$\begin{aligned}\delta_n(T) &\leqq \sup_{x \in S(X)} \| Tx - P_n^* x \| = \sup_{x \in S(X)} \| TQ_n x \| \\ &\leqq \big(\sup_{z \in L^n \cap S(X)} \| Tz \| \big) \big(\sup_{x \in S(X)} \| Q_n x \| \big) \\ &= d^n(T) \, \beta_n . \quad \square\end{aligned}$$

We now state, without proof, two results which bound α_n and β_n.

Theorem 8.3. *If Y is a normed linear space and X_n is an n-dimensional subspace of Y, then there exists a projection P_n of Y onto X_n of norm at most $\sqrt{n}$.*

Theorem 8.4. *Given $\varepsilon > 0$, a normed linear space X and a subspace L^n of X of codimension at most n, there exists a projection Q_n of X onto L^n of norm at most $\sqrt{n} + 1 + \varepsilon$.*

As a consequence, we therefore have:

Corollary 8.5. *For $T \in L(X, Y)$,*

$$\delta_n(T) \leqq (\sqrt{n} + 1) \min \{d_n(T), d^n(T)\}.$$

One consequence of Corollary 8.5 is the following.

Corollary 8.6. *The following four results are equivalent.* (i) $d_n(T) = 0$, (ii) $d^n(T) = 0$, (iii) $\delta_n(T) = 0$, (iv) *T is of rank at most n.*

Proof. If (iv) holds, then so do (i), (ii), and (iii). From Corollary 8.5, it follows that (i), (ii) and (iii) are equivalent. Thus it suffices to prove that (iii) implies (iv). This follows from the fact that the set of operators in $F(X, Y)$, of rank at most n, is closed. $\square$

An immediate application of Proposition 7.6(1) and Corollary 8.6 yields:

Corollary 8.7. *Let s_n denote any of the three quantities d_n, d^n or δ_n. For T, $T_m \in L(X, Y)$, with rank $T_m \leqq m$,*

$$s_{n+m}(T) \leqq s_n(T + T_m) \leqq s_{n-m}(T).$$

By making further assumptions on T, X or Y, it is possible to obtain additional equalities between and among the various n-widths of T and T'.

From Proposition 5.2, if Y is a Hilbert space, then $d^n(T) \leqq d_n(T) = \delta_n(T)$. The next result follows from Proposition 8.2.

Proposition 8.8. *Let X be a Hilbert space and $T \in L(X, Y)$. Then*

$$d_n(T) \leqq d^n(T) = \delta_n(T), \quad \text{and} \quad \delta_n(T) = \delta_n(T').$$

Proof. Since X is a Hilbert space, we can choose $\beta_n = 1$ in Proposition 8.2. Therefore $d^n(T) = \delta_n(T)$. From Theorem 7.2 we have $d_n(T') = d^n(T)$, and from Proposition 5.2, $d_n(T') = \delta_n(T')$. Thus

$$\delta_n(T') = d_n(T') = d^n(T) = \delta_n(T) \geqq d_n(T). \quad \square$$

The proof of the next two results use the following theorem.

Principle of Local Reflexivity. *Let Y be a normed linear space and let M be a finite dimensional subspace of Y''. Given $\varepsilon > 0$ there exists $R \in L(M, Y)$ such that $\|R\| \leqq 1 + \varepsilon$ and $RJ_Y y = y$ for all $y \in Y \cap J_Y^{-1}(M)$, where J_Y denotes the canonical mapping of Y into Y''.*

Theorem 8.9. *If $T \in K(X, Y)$ or if Y is a reflexive Banach space, then*

$$d_n(T) = d^n(T').$$

This result should be contrasted with Theorem 7.2. At the end of this section, an example is given of a non-compact T for which the above equality does not hold.

Proof. It is a simple matter to show that $d_n(T) \geqq d_n(T'')$ since X and Y may be canonically embedded in X'' and Y'', respectively. If Y is reflexive, then it follows from the definition that $d_n(T'') \geqq d_n(T)$. Since, from Theorem 7.2, $d_n(T'') = d^n(T')$, the result therefore holds in this case. It remains to prove that $d_n(T'') \geqq d_n(T)$ if T is compact.

Given $\varepsilon > 0$, there exists a subspace of dimension at most n, $X_n'' \subseteqq Y''$ for which

$$\inf_{y \in X_n''} \|T''x - y\| < d_n(T'') + \varepsilon$$

for all $x \in S(X'')$.

Furthermore, since T is compact, we also have a finite ε-net for the image of the unit ball under T, i.e., there exists $x_1, \ldots, x_k \in S(X)$, such that for each $x \in S(X)$, $\min\{\|Tx - Tx_i\| : i = 1, \ldots, k\} \leqq \varepsilon$.

Let $M \subseteqq Y''$ denote the finite dimensional subspace spanned by X_n'' and $J_Y(Tx_i)$, $i = 1, \ldots, k$. There exists, by the Principle of Local Reflexivity, an $R \in L(M, Y)$ for which $\|R\| \leqq 1 + \varepsilon$ and $RJ_Y(Tx_i) = Tx_i$, $i = 1, \ldots, k$. Set $X_n = R(X_n'')$. X_n is a subspace of Y of dimension at most n.

For each x_i, $i = 1, \ldots, k$

$$\inf_{y \in X_n''} \|T'' J_X(x_i) - y\| < d_n(T'') + \varepsilon.$$

Thus there exist $z_i' \in X_n''$, $i = 1, \ldots, k$, satisfying

$$\|T'' J_Y(x_i) - z_i'\| < d_n(T'') + \varepsilon.$$

Set $z_i = R(z_i')$, $i = 1,\ldots,k$. Hence $z_i \in X_n$, $i = 1,\ldots,k$ and

$$\begin{aligned}
\inf_{y\in X_n} \|Tx_i - y\| &= \inf_{y\in X_n} \|RJ_Y(Tx_i) - y\| \\
&= \inf_{y\in X_n} \|RT''J_X x_i - y\| \\
&\leqq \|RT''J_X x_i - z_i\| \\
&= \|RT''J_X x_i - R(z_i')\| \\
&\leqq \|R\|\,\|T''J_X x_i - z_i'\| \\
&< (1+\varepsilon)(d_n(T'')+\varepsilon).
\end{aligned}$$

Furthermore, by the choice of x_i, it is easily seen that for every $x \in S(X)$,

$$\inf_{y\in X_n} \|Tx - y\| \leqq \varepsilon + (1+\varepsilon)(d_n(T'')+\varepsilon).$$

Thus $d_n(T) \leqq d_n(T'')$, and the theorem is proved. □

The next theorem considers condition under which $\delta_n(T) = \delta_n(T')$ (see also Proposition 8.8).

Theorem 8.10. *If $T \in K(X,Y)$, or if Y is a reflexive Banach space, then*

$$\delta_n(T) = \delta_n(T').$$

It is easily deduced from the definition of $\delta_n(T)$ that $\delta_n(T') \leqq \delta_n(T)$. To prove the reverse inequality it suffices to show that $\delta_n(T) \leqq \delta_n(T'')$. This is easily shown to be true if Y is reflexive and may be deduced, much in the manner of the proof of Theorem 8.9, by an application of the Principle of Local Reflexivity, if T is compact.

If $T \in L(X,Y)$ is not compact, then equality between $\delta_n(T)$ and $\delta_n(T')$ need not obtain. The following is an example thereof.

For each $1 \leqq p \leqq \infty$, l_p shall denote the space of vectors $\mathbf{x} = (x_1, x_2, \ldots)$, $x_i \in \mathbb{R}$, with norm $\|\mathbf{x}\|_p = \left(\sum_{i=1}^{\infty} |x_i|^p\right)^{1/p} < \infty$. c_0 is the subspace of l_∞ of vectors $\mathbf{x}$ satisfying $\lim_{i\to\infty} |x_i| = 0$. Let $T = I$, the identity map, $X = l_1$ and $Y = c_0$. Thus $X' = l_\infty$ and $Y' = l_1$.

Theorem 8.11. *For all $n \geqq 1$,*

$$\delta_n(I(l_1); c_0) = 1, \quad \textit{and} \quad \delta_n(I(l_1); l_\infty) = 1/2.$$

Proof. We first prove that $\delta_n(I(l_1); c_0) = 1$ for all n. Since $\delta_0(I(l_1); c_0) = 1$, it suffices to prove that $\delta_n(I(l_1); c_0) \geqq 1$. Assume that $\delta_n(I(l_1); c_0) < 1$. There then exists a matrix P_n of rank at most n, $P_n\colon l_1 \to c_0$, for which $\|I - P_n\| < 1 - \varepsilon$ for some $\varepsilon > 0$. Since P_n is of finite rank, it is also compact. Thus there exists an $\varepsilon/4$-net $\mathbf{y}^1, \ldots, \mathbf{y}^k$ for $\{P_n\mathbf{x} : \mathbf{x} \in S(l_1)\}$. Since $\mathbf{y}^1, \ldots, \mathbf{y}^k \in c_0$ there exists a j^* for which $|y^i_{j^*}| < \varepsilon/4$ for $i = 1, \ldots, k$.

Let $P_n = (p_{ij})_{i,j=1}^{\infty}$. Since the unit vectors $\mathbf{e}^i \in S(l_1)$, $P_n\mathbf{e}^i \in c_0$ and $\|(I - P_n)\mathbf{e}^i\| < 1 - \varepsilon$. Therefore $|1 - p_{j^*j^*}| < 1 - \varepsilon$ implying that $p_{j^*j^*} > \varepsilon$. However, since $\mathbf{y}^1, \ldots, \mathbf{y}^k$ is an $\varepsilon/4$-net for $\{P_n\mathbf{x}: \mathbf{x} \in S(l_1)\}$, and $|y^i_{j^*}| < \varepsilon/4$, it easily follows that $|p_{j^*j^*}| < \varepsilon/2$. This contradiction proves that $\delta_n(I(l_1); c_0) = 1$.

We now prove that $\delta_n(I(l_1); l_\infty) = 1/2$ for all $n \geqq 1$. Set $P = (p_{ij})_{i,j=1}^{\infty}$, where $p_{ij} = 1/2$ for all $i\ j$. Then $P: l_1 \to l_\infty$, P is of rank 1, and $\delta_n(I(l_1); l_\infty) \leqq \|I - P\| = 1/2$ for all $n \geqq 1$. Assume that $\delta_n(I(l_1); l_\infty) < 1/2$. Thus there exists a matrix $Q_n: l_1 \to l_\infty$, of rank at most n, for which $\|I - Q_n\| < 1/2 - \varepsilon$ for some $\varepsilon > 0$. Since $\mathbf{e}^i \in S(l_1)$, we obtain $|1 - q_{ii}| < 1/2 - \varepsilon$ and $|q_{ij}| < 1/2 - \varepsilon$ for all $i \neq j$. Hence $\|Q_n\mathbf{e}^i - Q_n\mathbf{e}^j\| = \sup_k |q_{ki} - q_{kj}| \geqq |q_{ii} - q_{ij}| > 2\varepsilon$ for all $i \neq j$. Thus $\{Q_n\mathbf{e}^i\}_{i=1}^{\infty}$ has no convergent subsequence. This contradicts the compactness of Q_n. □

Certain situations arise wherein the linear n-width may be shown to be equal to the Kolmogorov or Gel'fand n-width. This is so, for example, in Propositions 5.2 and 8.8. Additional instances of such equalities occur as follows.

Definition 8.1. A normed linear space X is said to have the *lifting property* if for every linear operator T mapping X into a quotient space Y/N of an arbitrary normed linear space Y, and for each $\varepsilon > 0$, there is a lifting $\hat{T}$ of T from X into Y with $\|\hat{T}\| \leqq (1 + \varepsilon)\|T\|$.

$X = L_1(\mu)$ is an example of a space with the lifting property.

Definition 8.2. A normed linear space Y is said to have the *extension property* if for every $T \in L(M, Y)$, where M is a subspace of a normed linear space X, there is an extension $\hat{T}$ from X into Y with $\|T\| = \|\hat{T}\|$.

$Y = L_\infty(\mu)$ is an example of a space with the extension property.
It should be noted that the above two properties are dual in a certain sense. The following results are immediate consequences of the above properties.

Proposition 8.12. *If X has the lifting property then for every $T \in L(X, Y)$*

$$d_n(T) = \delta_n(T).$$

Proposition 8.13. *If Y has the extension property then for every $T \in L(X, Y)$*

$$d^n(T) = \delta_n(T).$$

Let us now return to the example considered in Theorem 8.11. From Propositions 8.12 and 8.13, $d_n(I(l_1); l_\infty) = d^n(I(l_1); l_\infty) = \delta_n(I(l_1); l_\infty) = 1/2$ (for $n \geqq 1$), while $d_n(I(l_1); c_0) = \delta_n(I(l_1); c_0) = 1$. From Theorem 7.2, it also follows that $d^n(I(l_1); c_0) = d_n(I(l_1); l_\infty) = 1/2$. We now have explicit counterexamples to various possible hypotheses.

a)
$$d_n(I(l_1); c_0) > d^n(I(l_1); l_\infty)$$

is an example of a non-compact operator T for which $d_n(T) \neq d^n(T')$.

b)
$$d_n(I(l_1); c_0) > d_n(I(l_1); l_\infty)$$

is another example of strict inequality in property (vii) of Theorem 1.1.

c) $$d_n(I(l_1); c_0) > d^n(I(l_1); c_0)$$

is an additional example of the fact that $d_n(A) \neq d^n(A)$ (see Example 1 of Section 5).

Notes and References

Section 1. The concept of n-widths, in the sense of Kolmogorov, was introduced by Kolmogorov in his 1936 paper Kolmogorov [1936]. The properties of d_n as listed in Theorem 1.1 may be found, for example, in Lorentz [1966a] and Singer [1970]. If the set A is not centrally symmetric, then some authors, notably Tichomirov [1960a], have defined $d_n(A; X)$ as

$$d_n(A; X) = \inf_{x_0, X_n} \sup_{x \in A} \inf_{y \in X_n} \|x - (x_0 + y)\|_X,$$

i.e., they allow for translation of the subspace. If A is centrally symmetric, then this definition agrees with the former, i.e., the optimal x_0 is the zero element. Returning to our former definition, it is easily seen that it may be rewritten in the form

$$d_n(A; X) = \inf_{X_n} \inf\{\varepsilon: \varepsilon > 0,\ A \subseteq X_n + \varepsilon S(X)\}$$

where $S(X)$ denotes the unit ball of X. This definition allows for the following generalization (see e.g. Dubinsky [1979], Johnson [1973]):

Let A, B be non-empty subsets of a normed linear space X. Assume that B absorbs A. Then, set

$$d_n(A, B; X) = \inf_{X_n} \inf\{\varepsilon: \varepsilon > 0,\ A \subseteq X_n + \varepsilon B\}.$$

Proposition 1.2 is due to Pietsch [1972]. Theorem 1.5 has an interesting history. The essence of the proof that given X_{n+1} and X_n there exists a non-zero element $x \in X_{n+1}$ for which the zero element is a best approximation from X_n, was first proved by Krein, Krasnosel'ski, Milman in [1948]. However, because this paper a difficult to obtain, the book of Gohberg, Krein [1969], where it is reproduced, is more often referenced. Its application to n-widths was first noted by Tichomirov [1960a] and Theorem 1.5 is now referred to as the Fundamental Theorem of Tichomirov. Proposition 1.7 may be found in Lorentz [1966a]. Proposition 1.8 is due to Lorentz [1960] (see also Tichomirov [1960a] where a similar result is proved). Proposition 1.9 was established by Sharygin [1972].

Section 2. Optimal subspaces are sometimes referred to as extremal subspaces. Singer [1970] calls such subspaces "best n-dimensional secants". Theorem 2.1 is a result of Brown [1964]. Theorems 2.2 and 2.3 are due to Garkavi [1962]. A proof of Theorem 2.2 may also be found in Singer [1970, p. 284].

Section 3. Proposition 3.2 is due to Helfrich [1971].

Section 4. Both Ha [1974] and Helfrich [1971], independently, proved Proposition 5.1. Examples 1 and 2 are due to Tichomirov [1965 a], although the proofs given here are somewhat different. Proposition 5.3 is, of course, to be found in Mityagin, Henkin [1963].

Ismagilov [1974] proves the following interesting result. We say that i is an extension of X if i is an isometric linear mapping of X to some normed linear space Y. The absolute linear n-width of A in X is defined by $\delta_n^a(A;X) = \inf \delta_n(i(A);Y)$, the infimum being taken over all extensions. Ismagilov proves that $\delta_n^a(A;X) = d^n(A;X)$.

Section 5. It is unclear who originally formulated Proposition 6.1. The result seems to be continually rediscovered with a frequency which is somewhat disturbing. For a history and proof of this result, see Singer [1970, p. 22]. A more general duality relationship between d_n and d^n, using polar sets, may be found in Ioffe, Tichomirov [1968 b] and Ismagilov [1974].

Section 6. Theorem 7.2 is due to a Ha [1974] and Pietsch [1974].

Section 7. Theorem 8.3 is due to Kadec, Snobar [1971], based on a result of James. Theorem 8.4 is due to Garling, Gordon [1971]. The results of Propositions 8.1 and 8.2, and Corollary 8.5 may be essentially found in Pietsch [1974], Ha [1974], and Hutton, Morrell, Retherford [1976]. The Principle of Local Reflexivity is due to Lindenstrauss, Rosenthal [1969]. Pietsch [1974] applied it to prove Theorem 8.9, and Hutton [1974] to prove Theorem 8.10. The examples of Theorem 8.11 are due to Hutton [1974]. Propositions 8.12 and 8.13 are to found in Pietsch [1974].

Chapter III. Tchebycheff Systems and Total Positivity

This chapter is not concerned with n-widths but is a discussion of Tchebycheff (T-) systems (often written Chebyshev) and total positivity which are important tools in the exact determination of n-widths and in the identification of optimal subspaces for many of the examples considered in this work. While we do assume that the reader is familiar with some of the basic results of approximation theory and related matters, we cannot assume that the reader also has a familiarity with T-systems and total positivity. It transpires that the theory of T-systems and total positivity is intimately connected with the problem of zero counting and oscillations of functions and, as such, is basic to the study of L^∞ and L^1 approximations, as shall be evident from results of succeeding chapters. Perhaps surprisingly it is also important in the L^2 theory of n-widths.

We do not give a full and complete exposition of the theory of T-systems and total positivity. The interested reader may consult the books of Gantmacher and Krein [1960], Karlin and Studden [1966], Karlin [1968], Krein and Nudelman [1977], and Zielke [1979] for a much more comprehensive treatment of these matters. We shall only present the basic theory of what is needed in the subsequent chapters.

T-systems and more particularly total positivity are subject matters heavy in notation, especially matrix and determinantal notation. It appears cumbersome and inelegant at first sight. However, after an initial investment of a little time and effort, the notational complexity recedes and one is left with some surprisingly powerful tools. In this chapter we prove, or outline the proof, of those results which are not difficult to prove and which we feel will give the reader a more basic understanding of the theory.

1. Tchebycheff Systems

In this section we always assume that $u_1, \ldots, u_n$ are n continuous, linearly independent, real-valued functions on some interval M (closed, open, or half-closed) of $\mathbb{R}$.

There exists a hierarchy of definitions of various types of Tchebycheff systems. We begin with the least restrictive of these.

Definition 1.1. We say that $u_1, \ldots, u_n$ is a *weak Tchebycheff* (*WT*-) *system* on M if

$$U\begin{pmatrix} 1, \ldots, n \\ x_1, \ldots, x_n \end{pmatrix} = \det\,(u_i(x_j))_{i,j=1}^n \tag{1}$$

is of one sign (i.e., non-negative or non-positive) for all choices of $x_i \in M$, $x_1 < \ldots < x_n$.

One of the important examples of a *WT*-system is provided by the truncated powers $\{(x-\xi_1)_+^k, \ldots, (x-\xi_n)_+^k\}$, where k is a positive integer, on any interval of $\mathbb{R}$ where they are linearly independent. Here x_+^k is defined by

$$x_+^k = \begin{cases} x^k, & x \geqq 0 \\ 0, & x < 0. \end{cases}$$

A basis for splines of degree k, with n fixed distinct knots $\xi_1, \ldots, \xi_n$, namely, $\{1, x, \ldots, x^k, (x-\xi_1)_+^k, \ldots, (x-\xi_n)_+^k\}$ is also a *WT*-system on any interval containing $\xi_1, \ldots, \xi_n$.

To each definition of the above form, there corresponds an equivalent definition based on the number of sign changes or zeros of any nontrivial "polynomial" u of $u_1, \ldots, u_n$, i.e., any function of the form $u = \sum_{i=1}^{n} a_i u_i$, with $\mathbf{a} = (a_1, \ldots, a_n) \neq \mathbf{0}$. To obtain this equivalent statement for *WT*-systems we need to count the number of sign changes of a function.

Definition 1.2. For any continuous function f on M, let $S(f)$ denote the number of sign changes of f on M. Thus, $S(f) = \sup\{N : f(x_i)\, f(x_{i+1}) < 0,\ i = 1, \ldots, N,\ x_1 < \ldots < x_{N+1}\}$.

Proposition 1.1. *$u_1, \ldots, u_n$ is a WT-system on M if and only if $S(u) \leqq n-1$ for every nontrivial "polynomial" u.*

What if we wish to construct a "polynomial" u which changes sign at given $y_1 < \ldots < y_{n-1}$? Can this always be done for a *WT*-system $u_1, \ldots, u_n$ on M? The answer is no. However one can come close.

Proposition 1.2. *Let $u_1, \ldots, u_n$ be a WT-system on M. Then given $y_1 < \ldots < y_{n-1}$ in M, there exists a nontrivial polynomial u satisfying*

$$\begin{aligned} u(x) &\geqq 0, & x &\leqq y_1 \\ u(x)(-1)^i &\geqq 0, & y_i \leqq x &\leqq y_{i+1}, \quad i = 1, \ldots, n-2 \\ u(x)(-1)^{n-1} &\geqq 0, & y_{n-1} &\leqq x. \end{aligned}$$

If we do not permit the determinants in (1) to vanish, we are then led to the notion of a Tchebycheff system.

Definition 1.3. We say that $u_1, \ldots, u_n$ is a *Tchebycheff (T-) system* on M if $U\begin{pmatrix} 1, \ldots, n \\ x_1, \ldots, x_n \end{pmatrix}$ is of one strict sign for all choices of $x_1 < \ldots < x_n$ in M. We say that $u_1, \ldots, u_n$ is a T^+*-system* if the above determinants are always positive. $u_1, \ldots, u_n$ is said to be a *complete Tchebycheff (CT-) system* (resp. CT^+*-system*) on M if $u_1, \ldots, u_k$ is a T (resp. T^+)-system for every $k = 1, \ldots, n$.

CT-systems are sometimes referred to as Markov systems.

The prototypes of *T*-systems are the monomials $\{1, x, \ldots, x^{n-1}\}$ on any interval of $\mathbb{R}$, and the trigonometric polynomials $\{1, \sin x, \cos x, \sin 2x,$

$\cos 2x, \ldots, \sin mx, \cos mx\}$ on any subinterval of $[0, 2\pi)$ (or translate thereof). These two examples may be shown to be T^+-systems by explicitly calculating the desired determinants. An equivalent definition of a T-system is often more useful.

Definition 1.4. The number of distinct zeros of a continuous function f on M is denoted by $Z(f)$.

Proposition 1.3. *$u_1, \ldots, u_n$ is a T-system on M if and only if $Z(u) \leqq n-1$ for every nontrivial polynomial u.*

The proof of this proposition is sufficiently simple to be given here.

Proof. Let $u_1, \ldots, u_n$ be a T-system on M. If there exist points $x_1 < \ldots < x_n$ in M and a polynomial $u = \sum_{i=1}^{n} a_i u_i$, $\mathbf{a} = (a_1, \ldots, a_n) \neq \mathbf{0}$, which vanishes at $x_1, \ldots, x_n$, then $U\begin{pmatrix} 1, \ldots, n \\ x_1, \ldots, x_n \end{pmatrix} = 0$, contradicting the definition of a T-system.

If $Z(u) \leqq n-1$ for every nontrivial polynomial u, then $U\begin{pmatrix} 1, \ldots, n \\ x_1, \ldots, x_n \end{pmatrix} \neq 0$ for every choice of $x_1 < \ldots < x_n$ in M. From the connectedness of M it follows that $U\begin{pmatrix} 1, \ldots, n \\ x_1, \ldots, x_n \end{pmatrix}$ cannot achieve both positive and negative values. Thus $U\begin{pmatrix} 1, \ldots, n \\ x_1, \ldots, x_n \end{pmatrix}$ is of one fixed sign in M and hence $u_1, \ldots, u_n$ is a T-system. □

Note that since $U\begin{pmatrix} 1, \ldots, n \\ x_1, \ldots, x_n \end{pmatrix} \neq 0$ for every choice of $x_1 < \ldots < x_n$ in M, we can always construct a nontrivial polynomial u which changes sign at any given $y_1 < \ldots < y_{n-1}$ in M and vanishes nowhere else in M.

This fact and Proposition 1.2 have an interesting and useful simple consequence. Let us extend the notion of the number of sign changes $S(f)$ to $f \in L^1[0,1]$. We say that $S(f) = N$ if there exist $N+1$ disjoint ordered intervals $I_1, \ldots, I_{N+1}$ (by ordered we mean that $x < y$ for all $x \in I_j$, $y \in I_{j+1}$, $j = 1, \ldots, N$) whose union is $[0,1]$, and such that $f(x)(-1)^j \varepsilon \geqq 0$ a.e. on I_j, $j = 1, \ldots, N+1$, where $\varepsilon = \pm 1$, is chosen fixed, and where $\operatorname{meas}\{x : x \in I_j,\ f(x) \neq 0\} > 0$, $j = 1, \ldots, N+1$. If no such N exists, then we set $S(f) = \infty$.

Proposition 1.4. *Let $f \in L^1[0,1]$,*

(a) *If $u_1, \ldots, u_n$ is a T-system on $[0,1]$, and if*

$$\int_0^1 f(x)\, u_i(x)\, dx = 0, \qquad i = 1, \ldots, n,$$

then $S(f) \geqq n$.

(b) *If $u_1, \ldots, u_n$ is a WT-system on $[0,1]$,* meas $\{x : f(x) = 0\} = 0$, *and*

$$\int_0^1 f(x)\, u_i(x)\, dx = 0, \qquad i = 1, \ldots, n,$$

then $S(f) \geqq n$.

The zero counting $Z(f)$ is not tight in the sense that T-systems may possess at most $n-1$ zeros even when we allow for counting of double zeros.

Definition 1.5. Let f be a continuous function on M. A zero of f at x_0 in the interior of M is said to be a *nonnodal zero* of f provided that f does not change sign at x_0, i.e., $f(x_0-\varepsilon)\,f(x_0+\varepsilon)\geqq 0$ for all ε sufficiently small. All other zeros, including zeros at the end points of M (if such points exist) are called *nodal zeros.*

At times nonnodal zeros are defined in a somewhat different manner, namely by $f(x_0-\varepsilon)\,f(x_0+\varepsilon)>0$ for all $\varepsilon\neq 0$ sufficiently small. However, since any u polynomial can have only isolated zeros if $u_1,\ldots,u_n$ is a T-system this difference is of no importance here.

Definition 1.6. Let $\tilde{Z}(f)$ count the number of zeros of a continuous function f on M, where nodal zeros are counted once and nonnodal zeros twice.

An extension of Proposition 1.3 gives us

Proposition 1.5. *$u_1,\ldots,u_n$ is a T-system on M if and only if $\tilde{Z}(u)\leqq n-1$ for every nontrivial polynomial u.*

Proof. Since $Z(u)\leqq\tilde{Z}(u)$, it follows from Proposition 1.3 that if $\tilde{Z}(u)\leqq n-1$ for every nontrivial u, then $u_1,\ldots,u_n$ is a T-system. For the converse assume that $\tilde{Z}(u)\geqq n$. Then with $x_1,\ldots,x_k$ denoting the distinct zeros of u, we have $k\leqq n-1$ by Proposition 1.3, and so at least one zero is nonnodal. For each nonnodal zero x_i, add the point $x_i+\varepsilon$, $\varepsilon>0$, small, and also $x_i-\varepsilon$ for the first nonnodal zero. This new set contains at least $n+1$ points. One may choose $n+1$ points $y_1<\ldots<y_{n+1}$ from this set with the property that $u(y_i)(-1)^i\,\delta\geqq 0$, $i=1,\ldots,n+1$ for some choice of $\delta\in\{-1,1\}$. Strict inequality must hold for at least two of these points. Consider the determinant

$$\begin{vmatrix} u(y_1)\ldots u(y_{n+1}) \\ u_1(y_1)\ldots u_1(y_{n+1}) \\ \vdots \qquad \vdots \\ u_n(y_1)\ldots u_n(y_{n+1}) \end{vmatrix}.$$

The determinant is zero since the first row is a linear combination of the remaining rows. Expanding by the first row we obtain $0=\sum_{i=1}^{n} a_i u(y_i)$, where $a_i(-1)^{i+1}\,\sigma>0$, $i=1,\ldots,n$, $\sigma\in\{-1,1\}$, from the definition of a T-system. Since $u(y_i)(-1)^i\,\delta\geqq 0$, $i=1,\ldots,n$, with at least one strict inequality, we obtain a contradiction. This proves that if $u_1,\ldots,u_n$ is a T-system, then $\tilde{Z}(u)\leqq n-1$. $\square$

Every algebraic polynomial of order n (degree $n-1$) exhibits zero counting properties stronger than those given above.

Definition 1.7. Assume that $f\in C^{n-1}(M)$. The number of zeros of f on M, counting multiplicities, is denoted by $Z^*(f)$.

We count the multiplicity of a zero of f in the usual sense, where each point of M can be a zero of multiplicity at most n.

To deal with the case of multiplicities we must have a suitable notation for the corresponding matrices. If $u_1,\ldots,u_n \in C^{n-1}(M)$, and $x_1 < \ldots < x_n$ are in M, then $U^*\begin{pmatrix}1,\ldots,n\\x_1,\ldots,x_n\end{pmatrix} \equiv U\begin{pmatrix}1,\ldots,n\\x_1,\ldots,x_n\end{pmatrix}$ is well defined. If however

$$x_1 = \ldots = x_{\alpha_1} < x_{\alpha_1+1} = \ldots = x_{\alpha_2} < \ldots < x_{\alpha_r+1} = \ldots = x_{\alpha_{r+1}},$$

$\alpha_{r+1} = n$, then by $U^*\begin{pmatrix}1,\ldots,n\\x_1,\ldots,x_n\end{pmatrix}$ we mean the determinant of the matrix

$$\begin{pmatrix} u_1(x_1)\, u_1'(x_1)\ldots u_1^{(\alpha_1-1)}(x_1)\; u_1(x_{\alpha_1+1})\ldots u_1^{(\alpha_{r+1}-\alpha_r-1)}(x_{\alpha_r+1}) \\ \vdots \quad \vdots \quad \vdots \quad \vdots \quad \vdots \\ u_n(x_1)\, u_n'(x_1)\ldots u_n^{(\alpha_1-1)}(x_1)\; u_n(x_{\alpha_1+1})\ldots u_n^{(\alpha_{r+1}-\alpha_r-1)}(x_{\alpha_r+1}) \end{pmatrix}.$$

Definition 1.8. Let $u_1,\ldots,u_n \in C^{n-1}(M)$. Then $u_1,\ldots,u_n$ is an *extended Tchebycheff (ET-) system* on M if $U^*\begin{pmatrix}1,\ldots,n\\x_1,\ldots,x_n\end{pmatrix}$ is of one strict sign for all choices of $x_1 \leqq \ldots \leqq x_n$ in M. We say that $u_1,\ldots,u_n$ is an *ET^+-system* if the above determinants are always positive. Moreover, $u_1,\ldots,u_n$ is an *extended complete Tchebycheff (ECT-) system* (resp. *ECT^+-system*) on M if $u_1,\ldots,u_k$ is an ET (resp. ET^+)-system for all $k = 1,\ldots,n$.

The following provides an equivalent definition of a ET-system.

Proposition 1.6. *$u_1,\ldots,u_n$ is an ET-system on M if and only if $Z^*(u) \leqq n-1$ for every nontrivial polynomial u.*

The proof is totally analogous to the proof of Proposition 1.3.

ECT-systems possess additional properties and characterizations. In addition to Proposition 1.6 we have the following result.

Proposition 1.7. *Let $u_1,\ldots,u_n \in C^{n-1}(M)$. Then $u_1 \ldots u_n$ is an ECT^+-system on M if and only if $W(u_1,\ldots,u_k)(x) > 0$ for all $x \in M$ and $k = 1,\ldots,n$, where $W(u_1,\ldots,u_k)$ denotes the Wronskian of the functions $u_1,\ldots,u_k$, i.e.,*

$$W(u_1,\ldots,u_k)(x) = \begin{vmatrix} u_1(x)\; u_1'(x)\ldots u_1^{(k-1)}(x) \\ u_2(x)\; u_2'(x)\ldots u_2^{(k-1)}(x) \\ \vdots \quad \vdots \quad \vdots \\ u_k(x)\; u_k'(x)\ldots u_k^{(k-1)}(x) \end{vmatrix}.$$

Remark. If $u_1,\ldots,u_n$ is an ECT-system on M, then there exist $\varepsilon_i = \pm 1$, $i = 1,\ldots,n$, such that $\varepsilon_1 u_1,\ldots,\varepsilon_n u_n$ is an ECT^+-system on M.

Let $w_1,\ldots,w_n$ be strictly positive functions on $[a,b]$, with $w_k \in C^{n+1-k}[a,b]$, $k = 1,\ldots,n$. Define

$$\begin{aligned} u_1(x) &= w_1(x) \\ u_2(x) &= w_1(x)\int_a^x w_2(\xi_1)\,d\xi_1 \\ &\vdots \\ u_n(x) &= w_1(x)\int_a^x w_2(\xi_1)\int_a^{\xi_1} w_3(\xi_2)\ldots\int_a^{\xi_{n-2}} w_n(\xi_{n-1})\,d\xi_{n-1}\ldots d\xi_1. \end{aligned} \tag{2}$$

The following characterization of ECT^+-systems is fundamental.

Theorem 1.8. *Let $u_1,\ldots,u_n \in C^n[a,b]$ satisfy the initial conditions*

(3) $$u_k^{(i)}(a) = 0, \quad i = 0, 1, \ldots, k-2; \ k = 2, \ldots, n.$$

Then $u_1,\ldots,u_n$ is an ECT^+-system on $[a,b]$ if and only if $u_1,\ldots,u_n$ has a representation of the form (2).

Remark. If $u_1,\ldots,u_n$ is an ECT-system on $[a,b]$, then there exist $v_1,\ldots,v_n$ which satisfy the initial conditions (3), which form an ECT^+-system on $[a,b]$, and such that span $\{u_1,\ldots,u_k\}$ = span $\{v_1,\ldots,v_k\}$, $k = 1,\ldots,n$.

If $D_i f = \frac{d}{dx}\left(\frac{f}{w_i}\right)$, $i = 1,\ldots,n$, for $w_1,\ldots,w_n$ as above, then $u_1,\ldots,u_n$ of the form (2) is a fundamental system of solutions for the nth order homogeneous linear differential equation $Lf = D_n D_{n-1} \ldots D_1 f = 0$. Such a differential form is said to be of Pólya type W (disconjugate) on $[a,b]$.

One the other system of functions deserves mention.

Definition 1.9. $u_1,\ldots,u_n$ is said to be a *Descartes system* on M if $u_{i_1},\ldots,u_{i_k}$ is a T^+-system on M for every $1 \leqq i_1 < \ldots < i_k \leqq n$, $k = 1,\ldots,n$.

This definition is slightly nonstandard in that we have fixed the signs of all the determinants.

Obviously every ET-system is a T-system and every T-system is a WT-system. Moreover ET-systems are dense in the class of WT-systems in a certain sense. This result will prove useful. We state it without proof.

Proposition 1.9. (a) *Let $u_1,\ldots,u_n$ be a WT-system on $[a,b]$. For each $\varepsilon > 0$, there exists a system of analytic functions $u_1^\varepsilon,\ldots,u_n^\varepsilon$, an ET-system on $[a,b]$, satisfying*

$$\lim_{\varepsilon \to 0^+} u_i^\varepsilon(x) = u_i(x), \quad i = 1,\ldots,n,$$

uniformly in x.

(b) *Let $u_1,\ldots,u_{2n+1}$ be a periodic WT-system on $[a,b)$, i.e., $u_i(a) = u_i(b)$, $i = 1,\ldots,2n+1$. For each $\varepsilon > 0$, there exists a system of analytic functions $u_1^\varepsilon,\ldots,u_{2n+1}^\varepsilon$, a periodic ET-system on $[a,b)$, satisfying*

$$\lim_{\varepsilon \to 0^+} u_i^\varepsilon(x) = u_i(x), \quad i = 1,\ldots,2n+1,$$

uniformly in x.

Remark. While there exist periodic WT-systems of n functions for n even, every periodic T-system on $[a,b)$ is composed of an odd number of functions.

2. Matrices

If $A = (a_{ij})_{i=1}^{M}{}_{j=1}^{N}$ is an $M \times N$ matrix, then we denote by

$$A\begin{pmatrix} i_1, \ldots, i_n \\ j_1, \ldots, j_n \end{pmatrix}, \qquad 1 \leqq i_1 < \ldots < i_n \leqq M; \ 1 \leqq j_1 < \ldots < j_n \leqq N.$$

the determinant of the $n \times n$ submatrix of A which has its (k, l)-entry equal to $a_{i_k j_l}$, $k, l = 1, \ldots, n$.

Definition 2.1. An $M \times N$ matrix A is said to be *sign consistent of order n* (SC_n), if there exists an $\varepsilon_n \in \{-1, 1\}$, such that

$$\varepsilon_n A\begin{pmatrix} i_1, \ldots, i_n \\ j_1, \ldots, j_n \end{pmatrix} \geqq 0$$

for all $1 \leqq i_1 < \ldots < i_n \leqq M$; $1 \leqq j_1 < \ldots < j_n \leqq N$. It is said to be *strictly sign consistent of order n* (SSC_n) if strict inequality prevails.

The usefulness of the concept of (strict) sign consistency is that if we consider A as an operator from $\mathbb{R}^N$ to $\mathbb{R}^M$, then A exhibits certain variation diminishing properties (analogous to those of WT and T-systems). To explain this statement we must consider two notions of sign changes of vectors. Namely,

Definition 2.2. Let $\mathbf{x} = (x_1, \ldots, x_n) \in \mathbb{R}^n \backslash \{\mathbf{0}\}$.

(i) $S^-(\mathbf{x})$ denotes the number of sign changes in the sequence $x_1, \ldots, x_n$ with zero terms discarded.

(ii) $S^+(\mathbf{x})$ counts the maximum number of sign changes in the sequence $x_1, \ldots, x_n$ where zero terms are arbitrarily assigned values $+1$ or -1.

For example,

$$S^-(-1, 0, 1, -1, 0, -1) = 2, \qquad \text{while } S^+(-1, 0, 1, -1, 0, -1) = 4.$$

For notational ease, we define $S^+(\mathbf{0}) = n$ for $\mathbf{0} \in \mathbb{R}^n$.

Proposition 2.1. *Let A be an $M \times N$ matrix of rank N. Then for all* $\mathbf{x} \in \mathbb{R}^N \backslash \{\mathbf{0}\}$

(a) $S^-(A\mathbf{x}) \leqq N - 1$ *if and only if A is SC_n,*

(b) $S^+(A\mathbf{x}) \leqq N - 1$ *if and only if A is SSC_n.*

Proof. We prove b). Part a) may be proven in a similar manner or via a limiting process from b).

Let us assume that A is SSC_N and that there exists an $\mathbf{x} \neq \mathbf{0}$ such that $\mathbf{y} = A\mathbf{x}$ satisfies $S^+(\mathbf{y}) \geqq N$. Set $\mathbf{y} = (y_1, \ldots, y_M)$. Thus there exist indices $1 \leqq i_1 < \ldots < i_{N+1} \leqq M$ for which $\varepsilon(-1)^k y_{i_k} \geqq 0$, $k = 1, \ldots, N+1$, for some choice of $\varepsilon \in \{-1, 1\}$. Furthermore since A is SSC_N, each set of N rows is linearly independent, and at most $N-1$ of the y_i's can be zero. Consider the determinant

$$\begin{vmatrix} a_{i_1,1} & \dots a_{i_1,N} & y_{i_1} \\ \vdots & \vdots & \vdots \\ a_{i_{N+1},1} & \dots a_{i_{N+1},N} & y_{i_{N+1}} \end{vmatrix}.$$

Since the last column is a linear combination of the preceding columns, the determinant is zero. Expanding the above determinant by the last column we obtain

$$0 = \sum_{k=1}^{N+1} (-1)^{k+N+1} y_{i_k} \quad A\begin{pmatrix} i_1,\dots,i_{k-1}, i_{k+1},\dots,i_{N+1} \\ 1,\dots,N \end{pmatrix}.$$

Since A is SSC_N, the y_{i_k} alternate in sign, and not all are zero, we obtain a contradiction.

Now assume that $S^+(A\mathbf{x}) \leqq N-1$ for all $\mathbf{x} \neq \mathbf{0}$. If $A\begin{pmatrix} i_1,\dots,i_N \\ 1,\dots,N \end{pmatrix} = 0$ for some $1 \leqq i_1 < \dots < i_N \leqq M$, then there exists an $\mathbf{x} \neq \mathbf{0}$ such that $(A\mathbf{x})_{i_k} = 0$, $k = 1,\dots,N$. However every vector $\mathbf{y} = A\mathbf{x}$ with N zeros satisfies $S^+(\mathbf{y}) \geqq N$, contradicting the hypothesis. Thus all Nth order minors of A are non-zero. To show that they all have the same sign it suffices to prove that the $N+1$ minors of Nth order generated by $N+1$ fixed rows have the same sign. Assume therefore that A is an $(N+1) \times N$ matrix and set

$$A_k = A\begin{pmatrix} 1,\dots,k-1, k+1,\dots,N+1 \\ 1,\dots,N \end{pmatrix}.$$

It suffices to prove that $A_k A_{k+1} > 0$, $k = 1,\dots,N$.

Let $y_k = (-1)^k A_k$. Then $\sum_{i=1}^{N+1} y_i a_{ij} = 0$, $j = 1,\dots,N$. Let $\mathbf{x} \neq \mathbf{0}$ satisfy $\sum_{j=1}^{N} a_{ij} x_j = 0$, $i = 1,\dots,N+1$; $i \neq k, k+1$. Thus $S^+(A\mathbf{x}) \geqq N-1$. However by hypothesis $S^+(A\mathbf{x}) \leqq N-1$ which implies that $(A\mathbf{x})_k (A\mathbf{x})_{k+1} > 0$. Now

$$0 = \sum_{j=1}^{N} \left(\sum_{i=1}^{N+1} y_i a_{ij} \right) x_j = \sum_{i=1}^{N+1} y_i \left(\sum_{j=1}^{N} a_{ij} x_j \right) = y_k (A\mathbf{x})_k + y_{k+1} (A\mathbf{x})_{k+1}.$$

Since $(A\mathbf{x})_k (A\mathbf{x})_{k+1} > 0$, and $y_l \neq 0$, for all l, $y_k y_{k+1} < 0$. □

Note that Proposition 2.1 is the finite dimensional analogue of Propositions 1.1 and 1.3.

In this same vein one may also prove the following analogue of Proposition 1.4.

Proposition 2.2. *Let A be an $M \times N$ matrix of rank N. Then*

(a) *A is SC_N if and only if $S^+(\mathbf{y}) \geqq N$ for all $\mathbf{y} \in \mathbb{R}^M$ satisfying $\mathbf{y}A = \mathbf{0}$*
(b) *A is SSC_N if and only if $S^-(\mathbf{y}) \geqq N$ for all $\mathbf{y} \in \mathbb{R}^M \backslash \{\mathbf{0}\}$ satisfying $\mathbf{y}A = \mathbf{0}$.*

The above results are too restrictive for our purposes. To obtain more general results we consider the following classes of matrices.

Definition 2.3. Let A be an $M \times N$ matrix.

(i) A is said to be *sign regular of order n* (SR_n), if A is SC_k $k = 1, \ldots, n$.
(ii) A is said to be *strictly sign regular of order n* (SSR_n), if A is SSC_k, $k = 1, \ldots, n$.
(iii) A is said to be *totally positive of order n* (TP_n), if A is SR_n and $\varepsilon_1 = \ldots = \varepsilon_n = 1$, i.e., all $k \times k$ order minors are non-negative, $k = 1, \ldots, n$.
(iv) A is said to be *strictly totally positive of order n* (STP_n) if A is SSR_n, and $\varepsilon_1 = \ldots = \varepsilon_n = 1$.

If A is SR_n, SSR_n, TP_n or STP_n and $n = \min\{M, N\}$, then we drop the subscript n.

The importance of the above notions with respect to the decrease in the number of sign changes of $A\mathbf{x}$ over $\mathbf{x}$ is the content of the next two results.

Theorem 2.3. *Assume that A is an $M \times N$ matrix of rank at least n. Then*

(i) *A is SR_n and $S^-(\mathbf{x}) \leqq n-1$ implies that $S^-(A\mathbf{x}) \leqq S^-(\mathbf{x})$.*
(ii) *A is TP_n and $S^-(\mathbf{x}) \leqq n-1$ implies that $S^-(A\mathbf{x}) \leqq S^-(\mathbf{x})$. Furthermore, if $S^-(A\mathbf{x}) = S^-(\mathbf{x})$, then the first nonzero component of $\mathbf{x}$ has the same sign as the first nonzero component of $A\mathbf{x}$.*
(iii) *If $M \geqq N$ and for all $\mathbf{x}$ satisfying $S^-(\mathbf{x}) \leqq n-1$ we have $S^-(A\mathbf{x}) \leqq S^-(\mathbf{x})$, then A is SR_n.*
(iv) *If $M \geqq N$ and for all $\mathbf{x}$ satisfying $S^-(\mathbf{x}) \leqq n-1$ we have $S^-(A\mathbf{x}) \leqq S^-(\mathbf{x})$, and furthermore if $S^-(A\mathbf{x}) = S^-(\mathbf{x})$ implies that the sign patterns agree as in* (ii), *then A is TP_n.*

Theorem 2.4. *A is an $M \times N$ matrix, and $\mathbf{x} \neq \mathbf{0}$.*

(i) *If A is SSR_n and $S^-(\mathbf{x}) \leqq n-1$, then $S^+(A\mathbf{x}) \leqq S^-(\mathbf{x})$.*
(ii) *If A is STP_n and $S^-(\mathbf{x}) \leqq n-1$, then $S^+(A\mathbf{x}) \leqq S^-(\mathbf{x})$. Furthermore if $S^+(A\mathbf{x}) = S^-(\mathbf{x})$, then the first component of $A\mathbf{x}$ (if zero, then the sign given in determining $S^+(A\mathbf{x})$) agrees in sign with the first nonzero component of $\mathbf{x}$.*
(iii) *If $M \geqq N$ then as in* (iii) *and* (iv) *of Theorem* 2.3, *the statements in* (i) *and* (ii) *are both necessary and sufficient.*

We shall only prove parts (i) and (ii) of Theorem 2.4.

Proof. To prove part (i) of Theorem 2.4, assume that A is SSR_n, $\mathbf{x} \neq \mathbf{0}$ and $S^-(\mathbf{x}) = m \leqq n-1$. The components of $\mathbf{x}$ may be divided into $m+1$ groups,

$$(x_1, \ldots, x_{\alpha_1})(x_{\alpha_1+1}, \ldots, x_{\alpha_2}) \ldots (x_{\alpha_m+1}, \ldots, x_N)$$

where we may assume (choosing $-\mathbf{x}$ if necessary) that $x_i(-1)^k \geqq 0$ for $i = \alpha_k + 1, \ldots, \alpha_{k+1}$, where $\alpha_0 = 0$, $\alpha_{m+1} = N$, $k = 0, 1, \ldots, m$, and furthermore for each $k \in \{0, 1, \ldots, m\}$ there exists at least one $i \in \{\alpha_k + 1, \ldots, \alpha_{k+1}\}$ for which $x_i \neq 0$.
Let $\mathbf{a}^j$ denote the jth column vector of A, and form the vectors

$$\mathbf{b}^k = \sum_{j=\alpha_{k-1}+1}^{\alpha_k} |x_j| \mathbf{a}^j, \qquad k = 1, \ldots, m+1.$$

Then

$$A\mathbf{x} = \sum_{j=1}^{N} x_j \mathbf{a}^j = \sum_{k=1}^{m+1} (-1)^{k+1} \left(\sum_{j=\alpha_{k-1}+1}^{\alpha_k} |x_j| \mathbf{a}^j \right)$$

$$= \sum_{k=1}^{m+1} (-1)^{k+1} \mathbf{b}^k.$$

Let B denote the $M \times (m+1)$ matrix with column vectors $\mathbf{b}^k$, $k = 1, \ldots, m+1$, and set $\mathbf{u} = (1, -1, 1, \ldots, (-1)^m)$. Hence $A\mathbf{x} = B\mathbf{u}$ and we wish to prove that $S^+(B\mathbf{u}) \leqq m$. It follows from Proposition 2.1 that it suffices to prove that B is SSC_{m+1}. Now,

$$B\begin{pmatrix} i_1, \ldots, i_{m+1} \\ 1, \ldots, m+1 \end{pmatrix} = \begin{vmatrix} \sum_{j=1}^{\alpha_1} |x_j| a_{i_1 j} & \ldots & \sum_{j=\alpha_m+1}^{N} |x_j| a_{i_1 j} \\ \vdots & & \vdots \\ \sum_{j=1}^{\alpha_1} |x_j| a_{i_{m+1} j} & \ldots & \sum_{j=\alpha_m+1}^{N} |x_j| a_{i_{m+1} j} \end{vmatrix}$$

$$= \sum_{j_1=1}^{\alpha_1} \sum_{j_2=\alpha_1+1}^{\alpha_2} \cdots \sum_{j_{m+1}=\alpha_m+1}^{N} \left[|x_{j_1}| \ldots |x_{j_{m+1}}| \, A\begin{pmatrix} i_1, \ldots, i_{m+1} \\ j_1, \ldots, j_{m+1} \end{pmatrix} \right].$$

Since $A\begin{pmatrix} i_1 \ldots i_{m+1} \\ j_1 \ldots j_{m+1} \end{pmatrix}$ is strictly of one sign $(m+1 \leqq n)$ for all $1 \leqq i_1 < \ldots < i_{m+1} \leqq M$, $1 \leqq j_1 < \ldots < j_{m+1} \leqq N$, and there exists at least one choice of $j_1, \ldots, j_{m+1}$ for which $|x_{j_1}| \ldots |x_{j_{m+1}}| \neq 0$, it follows that B is SSC_{m+1} and therefore $S^+(A\mathbf{x}) \leqq S^-(\mathbf{x})$.

To prove part (ii) of Theorem 2.4, we shall assume that $S^+(A\mathbf{x}) = S^-(\mathbf{x}) = m \leqq n-1$, and $\mathbf{x} \neq \mathbf{0}$. Reducing the problem as in the proof of part (i), we are led to the situation wherein $A\mathbf{x} = B\mathbf{u}$, B an $M \times (m+1)$ SSC_{m+1} matrix with $\varepsilon_{m+1} = 1$, and $\mathbf{u} = (1, -1, 1, \ldots, (-1)^m)$. Since $S^+(A\mathbf{x}) = m$, there exist $i_1 < \ldots < i_{m+1}$ for which $S^+((A\mathbf{x})_{i_1}, \ldots, (A\mathbf{x})_{i_{m+1}}) = m$. From the fact that $(A\mathbf{x})_{i_k} = \sum_{j=1}^{m+1} b_{i_k j}(-1)^{j+1}$, it follows by Cramer's rule that,

$$1 = \frac{\begin{vmatrix} (A\mathbf{x})_{i_1} & b_{i_1,2} & \ldots b_{i_1,m+1} \\ \vdots & \vdots & \vdots \\ (A\mathbf{x})_{i_{m+1}} & b_{i_{m+1},2} \ldots & b_{i_{m+1},m+1} \end{vmatrix}}{B\begin{pmatrix} i_1, \ldots, i_{m+1} \\ 1, \ldots, m+1 \end{pmatrix}} = \frac{\sum_{k=1}^{m+1} (-1)^{k+1} (A\mathbf{x})_{i_k} B\begin{pmatrix} i_1, \ldots, i_{k-1}, i_{k+1}, \ldots, i_{m+1} \\ 2, 3, \ldots, m+1 \end{pmatrix}}{B\begin{pmatrix} i_1, \ldots, i_{m+1} \\ 1, \ldots, m+1 \end{pmatrix}}.$$

The result follows if we can show that B is also SSC_m with $\varepsilon_m = 1$. The proof of this fact is totally analogous to the proof of the fact that B is SSC_{m+1} with $\varepsilon_{m+1} = 1$. Thus part (ii) of Theorem 2.4 is proved. □

A simple consequence of Theorem 2.4 is the following result concerning Descartes systems.

Corollary 2.5. *Let $u_1, \ldots, u_n$ be a Descartes system on $[a, b]$. Then for $u = \sum_{i=1}^{n} a_i u_i$, $\mathbf{a} = (a_1, \ldots, a_n) \neq \mathbf{0}$,*

$$\tilde{Z}(u) \leqq S^-(\mathbf{a}).$$

Furthermore, if $\tilde{Z}(u) = S^-(\mathbf{a})$ and $u(b) > 0$, then the last nonzero component of **a** *is positive.*

One of the useful properties of matrices of the above form is that SSR_n matrices are dense in the class of SR_n matrices, and similarly STP_n matrices are dense in the class of TP_n matrices. We state and prove the result for this latter class.

Proposition 2.6. *Let $A \in \mathbb{R}^{M \times N}$ be TP_n. Every neighborhood of A contains a STP_n matrix.*

Proof. The proof is based upon the following two facts, not proven here. Firstly, the $k \times k$ matrices $B_k(\varepsilon) = (\exp\{-(i-j)^2/\varepsilon\})_{i,j=1}^{k}$ are STP for all $\varepsilon > 0$, and

$$\lim_{\varepsilon \downarrow 0} \exp\{-(i-j)^2/\varepsilon\} = \delta_{ij}, \qquad i, j = 1, \ldots, k.$$

Secondly, if A is an $M \times N$ and B an $L \times M$ matrix, then for $C = BA$ we have

$$C\begin{pmatrix} i_1, \ldots, i_k \\ j_1, \ldots, j_k \end{pmatrix} = \sum_{1 \leqq \alpha_1 < \ldots < \alpha_k \leqq M} B\begin{pmatrix} i_1, \ldots, i_k \\ \alpha_1, \ldots, \alpha_k \end{pmatrix} A\begin{pmatrix} \alpha_1, \ldots, \alpha_k \\ j_1, \ldots, j_k \end{pmatrix}$$

for every choice of $1 \leqq i_1 < \ldots < i_k \leqq L$, $1 \leqq j_1 < \ldots < j_k \leqq N$. This multiplication formula is called the Cauchy-Binet formula.

Let us assume that A is TP_n and of rank at least n. Form the matrix $A(\varepsilon) = B_M(\varepsilon) \cdot A \cdot B_N(\varepsilon)$. From the STP property of the $B_k(\varepsilon)$, and since A is TP_n and of rank at least n, it follows from the Cauchy-Binet formula that $A(\varepsilon)$ is STP_n and $A(\varepsilon) \xrightarrow[\varepsilon \downarrow 0]{} A$ since $B_k(\varepsilon) \xrightarrow[\varepsilon \downarrow 0]{} I_k$, for all k.

Let us now assume that A is TP_n, but of rank $k < n$. Hence A is TP_k and of rank k and there exists a sequence of STP_k matrices $A(\varepsilon)$ of rank k which tend to A as ε decreases to zero. What we shall now do is deform $A(\varepsilon)$ so that it is TP_{k+1} and of rank $k+1$. Continuing this process we shall eventually arrive at the desired result. Let $A(\varepsilon; \delta)$ be the $M \times N$ matrix, all of whose entries agree with $A(\varepsilon)$ except that $a_{11}(\varepsilon)$ is replaced by $a_{11}(\varepsilon) + \delta$. It is easily checked that $A(\varepsilon; \delta)$ is STP_k, of rank $k+1$, and all nonzero $(k+1) \times (k+1)$ minors of $A(\varepsilon; \delta)$ are positive. The result follows. □

We noted above that the matrix $B_k(\varepsilon) = (\exp\{-(i-j)^2/\varepsilon\})_{i,j=1}^{k}$ is STP. This is equivalent to the fact that the matrix $A_k = (e^{ij})_{i,j=1}^{k}$ is STP since multiplying the ith row by $e^{i^2/\varepsilon}$ and the jth column by $e^{j^2/\varepsilon}$ has no effect on the sign of the minors, nor does the renormalization involved. We now record some additional TP and STP matrices and certain properties which they enjoy.

1) If A and B are $TP(STP)$, then AB is $TP(STP)$, if it exists.
2) If $A = (a_{ij})_{i=1}^{M}{}_{j=1}^{N}$, and $A\begin{pmatrix} i, i+1, \ldots, i+p \\ j, j+1, \ldots, j+p \end{pmatrix} > 0$, $i = 1, \ldots, M-p$; $j = 1, \ldots, N-p$; $p = 0, 1, \ldots, \min\{M, N\} - 1$, then A is STP.
3) The matrix $(\exp(x_i y_j))_{i,j=1}^{M}$ is STP if $x_1 < \ldots < x_M$ and $y_1 < \ldots < y_M$.
4) $\left(\dfrac{1}{x_i + y_j}\right)_{i,j=1}^{M}$ is an STP matrix if $0 < x_1 < \ldots < x_M$ and $0 < y_1 < \ldots < y_M$.
5) $G = (g_{ij})_{i,j=1}^{M}$ is called a Green's matrix if $g_{ij} = a_{\min\{i,j\}} b_{\max\{i,j\}}$ for some $a_1, \ldots, a_M$; $b_1, \ldots, b_M$. If $a_1, \ldots, a_M$, $b_1, \ldots, b_M$ all have the same strict sign and $\dfrac{a_1}{b_1} \leqq \dfrac{a_2}{b_2} \leqq \ldots \leqq \dfrac{a_M}{b_M}$, then G is TP.
6) $J = (a_{ij})_{i,j=1}^{M}$ is called a Jacobi matrix if $a_{ij} = 0$ for all i, j for which $|i - j| \geqq 2$. A Jacobi matrix is TP iff all its elements and principal minors are nonnegative.
7) $A = \left(\begin{pmatrix} i \\ j \end{pmatrix}\right)_{i,j=0}^{M}$ is TP where $\begin{pmatrix} i \\ j \end{pmatrix}$ is the usual binomial coefficient for $i \geqq j$, and $\begin{pmatrix} i \\ j \end{pmatrix} = 0$ for $i < j$.
8) If $u_1, \ldots, u_M$ is a Descartes system, then $U = (u_i(x_j))_{i=1}^{M}{}_{j=1}^{N}$ is STP for any $x_1 < \ldots < x_N$.

The class of STP matrices possesses, in addition to the variation diminishing properties of Theorem 2.4, some exceedingly rich eigenstructure properties. These properties are in fact enjoyed by the slightly larger class of matrices defined below.

Definition 2.4. $A \in \mathbb{R}^{M \times M}$ is said to be *oscillating* if A is TP and A^k is STP for some positive integer k.

Oscillating matrices are characterized by the following property.

Proposition 2.7. *If A is a nonsingular TP matrix, then A is oscillating iff $a_{i,i+1}$, $a_{i+1,i} > 0$ for all i.*

The eigenstructure properties of oscillatory matrices is the content of this next theorem.

Theorem 2.8. *Let A be an $M \times M$ oscillating matrix with eigenvalues $\lambda_1, \ldots, \lambda_M$, $|\lambda_1| \geqq \ldots \geqq |\lambda_M| > 0$, listed to their algebraic multiplicity.*

(i) *The eigenvalues of A are all positive and distinct, i.e.,*

$$\lambda_1 > \lambda_2 > \ldots > \lambda_M > 0.$$

(ii) *If $\mathbf{u}^k = (u_1^k, \ldots, u_M^k)$ is an eigenvector of A corresponding to λ_k, then*

$$p - 1 \leqq S^-\left(\sum_{k=p}^{q} c_k \mathbf{u}^k\right) \leqq S^+\left(\sum_{k=p}^{q} c_k \mathbf{u}^k\right) \leqq q - 1$$

for every $1 \leqq p \leqq q \leqq M$, c_i real, not all zero. In particular $S^-(\mathbf{u}^k) = S^+(\mathbf{u}^k) = k - 1$.

(iii) *Let $u^{(k)}(x)$ denote the function defined on $[1, M]$ by linear interpolation in $[i, i+1]$ to u_i^k and u_{i+1}^k at $i, i+1$, respectively. Then $u^{(k)}(x)$ has $k-1$ nodal zeros, no others, and the zeros of $u^{(k)}(x)$ and $u^{(k+1)}(x)$ strictly interlace.*

The proof is based on the simplest form of two well-known theorems. The first is the Perron-Frobenius Theorem which states that if A is a strictly positive matrix, i.e., $a_{ij} > 0$ for all i, j, then there exists a strictly positive vector $\mathbf{u}^1$ such that $A\mathbf{u}^1 = \lambda_1 \mathbf{u}^1$, ($\lambda_1 > 0$), and if λ is any other eigenvalue of A, then $|\lambda| < \lambda_1$ and the eigenvalue λ_1 is of algebraic multiplicity one. The second result, known as Kronecker's Theorem, is concerned with the eigenvalues of the compound matrix $A_{[p]}$. If A is an $M \times M$ matrix, then the pth compound matrix, $1 \leqq p \leqq M$, is the $\binom{M}{p} \times \binom{M}{p}$ matrix whose elements are of the form $A\begin{pmatrix} i_1, \ldots, i_p \\ j_1, \ldots, j_p \end{pmatrix}$, $1 \leqq i_1 < \ldots < i_p \leqq M$, $1 \leqq j_1 < \ldots < j_p \leqq M$, where all such p-tuples of indices are arranged in lexicographic order. This means that $\mathbf{i} = (i_1, \ldots, i_p)$ is less than $\mathbf{j} = (j_1, \ldots, j_p)$ if the first non-vanishing difference in the sequence $j_1 - i_1$, $j_2 - i_2, \ldots, j_p - i_p$, is positive. Kronecker's Theorem states that if $\lambda_1, \ldots, \lambda_M$ are the eigenvalues of A, listed to their algebraic multiplicity, then the eigenvalues of $A_{[p]}$ listed to their algebraic multiplicity consist of all possible products of the numbers $\lambda_1, \ldots, \lambda_M$ taken p at a time. Thus the eigenvalues of $A_{[2]}$ are $\lambda_i \lambda_j$, $i \neq j$, etc.

Proof. We prove (i) and (ii). Part (iii) is more technical and shall not be used. First note that since A^k is STP for some k, it suffices to prove the result assuming that A itself is STP. Let $\lambda_1, \ldots, \lambda_M$ be the eigenvalues of A listed to their algebraic multiplicity, and ordered so that $|\lambda_1| \geqq |\lambda_2| \geqq \ldots \geqq |\lambda_M| > 0$. Since A is STP, $A_{[p]}$ is a strictly positive matrix for each $p = 1, \ldots, M$. Thus by the Perron-Frobenius Theorem and Kronecker's Theorem, $\lambda_1 \ldots \lambda_p > 0$ for every $p = 1, \ldots, M$ implying that the λ_i are all positive, and $\lambda_1 \ldots \lambda_p > \lambda_1 \ldots \lambda_{p-1} \lambda_{p+1}$ implying that they are all distinct. Part (i) is proved.

Let $\mathbf{u}^k = (u_1^k, \ldots, u_M^k)$ be an eigenvector corresponding to λ_k. From the Perron-Frobenius Theorem $\mathbf{u}^1$ is a vector all of whose components are strictly of one sign. It is a fairly straightforward matter to prove that if $U = (u_i^k)_{i,k=1}^M$ then the eigenvector of the eigenvalue $\lambda_1 \ldots \lambda_p$ of $A_{[p]}$ is the vector with coordinates

$$U\begin{pmatrix} i_1, \ldots, i_p \\ 1, \ldots, p \end{pmatrix}, \quad 1 \leqq i_1 < \ldots < i_p \leqq M,$$

(arranged in lexicographic order). Multiplying the vectors $\mathbf{u}^1, \ldots, \mathbf{u}^M$ by ± 1, if necessary, we may assume that $U\begin{pmatrix} i_1, \ldots, i_p \\ 1, \ldots, p \end{pmatrix} > 0$ for all $1 \leqq i_1 < \ldots < i_p \leqq M$ and all $p = 1, \ldots, M$. The $M \times q$ submatrix of U composed of all rows and the first q columns is therefore SSC_q. From Proposition 2.1, part (b), we have

$$S^+\left(\sum_{i=p}^{q} c_i \mathbf{u}^i\right) \leqq q - 1$$

for every $p \leqq q$, c_i real, not all zero.

To prove the inequality $S^-\left(\sum_{i=p}^{q} c_i \mathbf{u}^i\right) \geqq p-1$, consider the left eigenvectors $\mathbf{v}^1, \ldots, \mathbf{v}^M$ corresponding to $\lambda_1, \ldots, \lambda_M$ of A, i.e., $A^T \mathbf{v}^k = \lambda_k \mathbf{v}^k$. Since A^T (the transpose of A) is STP we may assume, multiplying $\mathbf{v}^k$ by ± 1, if necessary, that

$$V\begin{pmatrix} i_1, \ldots, i_p \\ 1, \ldots, p \end{pmatrix} > 0$$

for all $1 \leqq i_1 < \ldots < i_p \leqq M$, $p = 1, \ldots, M$, where $V = (v_i^k)_{i,k=1}^M$. Furthermore for all $k \neq l$, $(\mathbf{v}^k, \mathbf{u}^l) = 0$. Thus $\left(\sum_{i=p}^{q} c_i \mathbf{u}^i, \mathbf{v}^k\right) = 0$, $k = 1, \ldots, p-1$. By Proposition 2.2, part (b), this implies that $S^-\left(\sum_{i=p}^{q} c_i \mathbf{u}^i\right) \geqq p-1$. This proves part (ii) of the theorem. □

In the succeeding chapters we shall make use of the useful and important.

Determinant Identity of Sylvester. *Let A be an $M \times M$ matrix and let $B = (b_{ij})_{i,j=p+1}^M$ be the $(M-p) \times (M-p)$ matrix given by*

$$b_{ij} = A\begin{pmatrix} 1, \ldots, p, i \\ 1, \ldots, p, j \end{pmatrix}, \qquad i, j = p+1, \ldots, M.$$

Then

$$B\begin{pmatrix} p+1, \ldots, M \\ p+1, \ldots, M \end{pmatrix} = \left[A\begin{pmatrix} 1, \ldots, p \\ 1, \ldots, p \end{pmatrix}\right]^{M-p-1} A\begin{pmatrix} 1, \ldots, M \\ 1, \ldots, M \end{pmatrix}.$$

3. Kernels

Having dealt with vectors and matrices, we now turn our attention to a consideration of kernels. In what follows we assume that $K(x, y) \in C([0,1] \times [0,1])$.

Definition 3.1. The kernel $K(x, y)$ is said to be

(i) *totally positive of order n (TP_n), if*

$$K\begin{pmatrix} x_1, \ldots, x_k \\ y_1, \ldots, y_k \end{pmatrix} = \det(K(x_i, y_j))_{i,j=1}^k \geqq 0 \tag{1}$$

for all $0 \leqq x_1 < \ldots < x_k \leqq 1$, $0 \leqq y_1 < \ldots < y_k \leqq 1$, $k = 1, \ldots, n$,

(ii) *strictly totally positive of order n (STP_n)* if strict inequality holds in (1).

We shall drop the subscript n if the kernel is TP or STP of all orders.

If $K(x, y)$ is sufficiently differentiable in x and y, then analogously to the situation for ET-systems, we define $K^*\begin{pmatrix} x_1, \ldots, x_k \\ y_1, \ldots, y_k \end{pmatrix}$ for $0 \leqq x_1 \leqq \ldots \leqq x_k \leqq 1$, $0 \leqq y_1 \leqq \ldots \leqq y_k \leqq 1$. If both the x_i and y_j are distinct, then $K^*\begin{pmatrix} x_1, \ldots, x_k \\ y_1, \ldots, y_k \end{pmatrix}$ is taken to mean $K\begin{pmatrix} x_1, \ldots, x_k \\ y_1, \ldots, y_k \end{pmatrix}$. When a block of l equal x's occurs, then the corresponding rows are determined by successive derivatives with respect to x, i.e.,

$$\left(\frac{\partial^r K(x, y_1)}{\partial x^r}, \ldots, \frac{\partial^r K(x, y_k)}{\partial x^r}\right), \; r = 0, 1, \ldots, l-1.$$

This same rule is applied to the y variable. Thus, for example

$$K^*\begin{pmatrix} x, x \\ y, y \end{pmatrix} = \begin{vmatrix} K(x,y) & \dfrac{\partial K(x,y)}{\partial y} \\ \dfrac{\partial K(x,y)}{\partial x} & \dfrac{\partial^2 K(x,y)}{\partial x \, \partial y} \end{vmatrix}.$$

With this notation we can now define

Definition 3.2. The kernel $K(x, y)$ is said to be *extended totally positive of order n (ETP_n)*, if it is sufficiently differentiable and $K^*\begin{pmatrix} x_1, \ldots, x_k \\ y_1, \ldots, y_k \end{pmatrix} > 0$ for all $0 \leqq x_1 \leqq \ldots \leqq x_k \leqq 1$, $0 \leqq y_1 \leqq \ldots \leqq y_k \leqq 1$, and $k = 1, \ldots, n$.

To decide whether a specific kernel is ETP_n, it is sufficient to consider the following determinants.

Proposition 3.1. *$K(x, y)$ is ETP_n iff*

$$K^*\begin{pmatrix} x, x, \ldots, x \\ y, y, \ldots, y \end{pmatrix} = \det\left(\frac{\partial^{i+j} K(x,y)}{\partial x^i \, \partial y^j}\right)_{i,j=0}^{k-1} > 0$$

for all $x, y \in [0, 1]$, and all $k = 1, \ldots, n$.

Corollary 3.2. *$K(x, y) = e^{xy}$ is ETP on $\mathbb{R}^2$.*

Proof.

$$\det\left(\frac{\partial^{i+j} K(x,y)}{\partial x^i \, \partial y^j}\right)_{i,j=0}^{k-1} = e^{xy}(k-1)!\,(k-2)!\ldots 2!\,1!\,0!. \quad \square$$

In Section 1 of this chapter we defined, for $f \in L^1[0,1]$, $S(f)$ the number of sign changes of f. We also defined, for $f \in C[0,1]$, $\tilde{Z}(f)$ the number of zeros of f, where zeros which are not sign changes are counted twice, and $Z^*(f)$ the number of zeros of f, counting multiplicities, for sufficiently differentiable f. Note the connection with the S^- and S^+ of the previous section. Namely, for $f \in C[0,1]$,

$$S(f) = \sup S^-(f(x_1), \ldots, f(x_n))$$
$$\tilde{Z}(f) = \sup S^+(f(x_1), \ldots, f(x_n))$$

where the suprema are taken over $0 \leqq x_1 < \ldots < x_n \leqq 1$ and n finite.

Results analogous to those of Theorems 2.3 and 2.4 hold for TP, STP, and ETP kernels. For notational ease, we write

$$Kh(x) = \int_0^1 K(x, y)\, h(y)\, dy.$$

Theorem 3.3. *Let* $K \in C([0,1] \times [0,1])$ *and* $h \in L^1[0,1]$ *satisfy* $S(h) \leqq n-1$.

(i) *If* K *is* TP_n, *then* $S(Kh) \leqq S(h)$.
(ii) *If* K *is* STP_n, *then* $\tilde{Z}(Kh) \leqq S(h)$.
(iii) *If* K *is* ETP_n, *then* $Z^*(Kh) \leqq S(h)$.

The next theorem (Theorem 3.4) is concerned with spectral and oscillatory properties of integral operators generated by totally positive kernels. It is the analogue of the corresponding theorem for oscillating matrices (Theorem 2.8). The result uses Jentzsch's Theorem and Schur's Theorem which are the continuous analogues of the Perron-Frobenius Theorem and Kronecker's Theorem, respectively. When we write that K has eigenvalue λ and corresponding eigenfunction ϕ, we mean that $K\phi = \lambda\phi$.

Jentzsch's Theorem. *Assume that the kernel* $K(x, y)$ *is non-negative and continuous on* $[0,1] \times [0,1]$, *and that* $K(x, x) > 0$ *for all* $x \in [0,1]$. *Then the kernel* $K(x, y)$ *has a positive eigenvalue* λ_0 *which is a simple root of the Fredholm determinant and which is strictly larger in modulus then all other eigenvalues of* $K(x, y)$. *Furthermore, the corresponding eigenfunction may be chosen strictly positive on* $[0,1]$.

Remark. If we only demand that $K(x, y)$ be strictly positive for *almost all* (x, y) in a neighborhood of $\{(x, x): x \in [0,1]\}$, then the results of the theorem remain valid except that the corresponding eigenfunctions may only be shown to be strictly positive at *almost all* points of $[0,1]$. In particular if $K(x, y)$ vanishes at $(0,0)$ and $(1,1)$, but otherwise satisfies the hypothesis of the theorem, then the corresponding eigenfunctions may vanish at the endpoints, but only there. This fact shall prove useful in the next chapter and also in dealing with the Green's function of a differential equation, with boundary conditions which imply that the Green's function vanishes on part of its boundary.

The theorem as stated above is a modified version of Jentzsch's [1912] original theorem and may be found in Anselone and Lee [1974].

Analogous to the compound matrices we must define compound kernels. This is done in the following manner. Set

$$\Lambda_r = \{\mathbf{x}: \mathbf{x} = (x_1, \ldots, x_r),\ 0 \leqq x_1 < \ldots < x_r \leqq 1\}.$$

The function

$$K_{[r]}(\mathbf{x}, \mathbf{y}) = K\begin{pmatrix} x_1, \ldots, x_r \\ y_1, \ldots, y_r \end{pmatrix}$$

defined on $\Lambda_r \times \Lambda_r$ is called the *compound kernel of order* r induced by K.

There is a fundamental relationship between the eigenvalue problem under consideration

$$\int_0^1 K(x, y)\, \phi(y)\, dy = \lambda \phi(x)$$

and the corresponding eigenvalue problem

$$\int_{\Lambda_r} K_{[r]}(\mathbf{x}, \mathbf{y})\, \Phi(\mathbf{y})\, d\mathbf{y} = \lambda\, \Phi(\mathbf{x})$$

for the rth compound kernel $K_{[r]}(\mathbf{x}, \mathbf{y})$.

Schur's Theorem. *Let $K \in C([0,1] \times [0,1])$. Let $\lambda_1, \lambda_2, \ldots$ be the set (possibly empty) of eigenvalues of K where each eigenvalue is listed to its multiplicity as a root of $D(\lambda)$, the Fredholm determinant of K. Then*

$$\lambda_{i_1} \lambda_{i_2} \ldots \lambda_{i_r}, \qquad 1 \leqq i_1 < \ldots < i_r$$

are the totality of eigenvalues of $K_{[r]}$ and each eigenvalue occurs to its multiplicity as a root of $D_{[r]}(\lambda)$, the Fredholm determinant of $K_{[r]}$.

With the aid of the theorems of Jentzsch and Schur it is possible to prove the following result. The proof is analogous to the proof of Theorem 2.8.

Theorem 3.4. *Assume that K is a totally positive kernel on $[0,1] \times [0,1]$ and that*

$$K\begin{pmatrix} x_1, \ldots, x_n \\ x_1, \ldots, x_n \end{pmatrix} > 0 \qquad \text{for all } 0 \leqq x_1 < \ldots < x_n \leqq 1$$

and for all $n = 1, 2, \ldots$. Consider the eigenvalue problem

$$(K\phi)(x) = \int_0^1 K(x, y)\, \phi(y)\, dy = \lambda \phi(x).$$

The eigenvalue problem has an infinite number of eigenvalues $\lambda_1, \lambda_2, \lambda_3, \ldots,$ where each eigenvalue is listed according to its multiplicity as a root of $D(\lambda)$, the Fredholm determinant of the kernel K. Furthermore,

(i) *the eigenvalues are positive and distinct*

$$\lambda_1 > \lambda_2 > \lambda_3 > \ldots > 0.$$

(ii) *If ϕ_k is an eigenfunction associated with the eigenvalue λ_k, uniquely determined up to a nonzero constant, then*

$$p - 1 \leqq S\left(\sum_{i=p}^{q} c_i \phi_i\right) \leqq \tilde{Z}\left(\sum_{i=p}^{q} c_i \phi_i\right) \leqq q - 1$$

on $[0,1]$, for any $p < q$, c_i real, not all zero. In particular $\phi_i(x)$ has exactly $(i-1)$ sign changes and no other zeros in $[0,1]$.

(iii) *The zeros of* $\phi_i(x)$ *and* $\phi_{i+1}(x)$ *strictly interlace.*

Remark. If we only assume that K is TP on $[0,1]\times[0,1]$, and $K\begin{pmatrix} x_1,\ldots,x_n \\ x_1,\ldots,x_n \end{pmatrix} > 0$ for all $0 < x_1 < \ldots < x_n < 1$ and all $n = 1,2,\ldots$ (the difference is that the endpoints are excluded), then the results of the theorem hold except that all zero counting is done on $(0,1)$ rather than on $[0,1]$.

The following is a short list of some totally positive and sign regular kernels.

1) $K(x,y) = e^{xy}$ is ETP on $(-\infty,\infty)\times(-\infty,\infty)$ (Corollary 3.2).
2) Let I, I', J, J' be connected subsets of $\mathbb{R}$ and assume that $\Phi: I' \to I$, $\Psi: J' \to J$ are continuous and strictly increasing functions. If $K(x,y)$ is TP on $I \times J$, then $K(\Phi(x), \Psi(y))$ is TP on $I' \times J'$. Thus $K(x,y) = x^y$ is TP on $(0,\infty)\times(-\infty,\infty)$ (from (1)).
3) $K(x,y) = 1/(1-xy)^\alpha$, $\alpha > 0$, is ETP on $[-1,1]\times(-1,1)$.
4) $K(x,y) = 1/(x+y)^\alpha$, $\alpha > 0$, is ETP on $[0,\infty)\times(0,\infty)$.
5) $K(x,y) = \sin xy$ is ESR on $(0,\tau)\times(0,\pi/2\tau)$ for any $\tau > 0$.
6) $K(x,y) = \cos xy$ is ESR on $(0,\tau)\times(0,\pi/2\tau)$ for any $\tau > 0$.

4. More on Kernels

One of the more important classes of totally positive kernels are the Green's functions of disconjugate nth order differential equations for a large class of separated boundary conditions. We study this class in some detail.

For given $w_i(x) > 0$, $i = 1,\ldots,n$, $x \in [0,1]$ and $w_i \in C^{n+1-i}[0,1]$, $i = 1,\ldots,n$, define

$$L = D_n \cdots D_1$$

where $D_j f = \dfrac{d}{dx}\dfrac{f}{w_j}$, $j = 1,\ldots,n$. Thus L is an nth order differential form of Pólya type W (disconjugate) on $[0,1]$. The differential equation $Lu = 0$ has a basis of solutions

$$u_1(x) = w_1(x)$$

$$u_2(x) = w_1(x)\int_0^x w_2(\xi_1)\,d\xi_1$$

$$\vdots$$

$$u_n(x) = w_1(x)\int_0^x w_2(\xi_1)\int_0^{\xi_1} w_3(\xi_2)\ldots\int_0^{\xi_{n-2}} w_n(\xi_{n-1})\,d\xi_{n-1}\ldots d\xi_1$$

which constitute an ECT-system on $[0,1]$ (see Theorem 1.8), satisfying the boundary conditions

$$D^{j-1}u_i(0) = w_i(0)\,\delta_{ij}, \qquad i,j = 1,\ldots,n$$

where $D^j = D_j \cdots D_1$, $j = 1,\ldots,n$, $D^0 = I$, the identity operator. The fundamental solution $\phi(x,y)$ of $Lu = 0$ determined by zero initial data at zero, and the characteristic jump discontinuity

$$D^{n-1}\phi(y+,y) - D^{n-1}\phi(y-,y) = w_n(y)$$

is given by

$$\phi(x,y) = \begin{cases} 0, & x < y \\ w_1(x)\int_y^x w_2(\xi_1)\int_y^{\xi_1} w_3(\xi_2)\dots\int_y^{\xi_{n-2}} w_n(\xi_{n-1})\,d\xi_{n-1}\dots d\xi_1, & x \geqq y. \end{cases}$$

$\phi(x, y)$ is a totally positive kernel and its positivity properties are explicitly given as follows:

Theorem 4.1. *The kernel $\phi(x, y)$ is totally positive of all orders, i.e.,*

$$\phi\begin{pmatrix} x_1, \dots, x_k \\ y_1, \dots, y_k \end{pmatrix} \geqq 0,$$

for increasing x_i and y_i. Assume that $0 < x_1 < \dots < x_k < 1$ and $0 < y_1 < \dots < y_k < 1$. Then strict inequality holds above if and only if

$$y_i < x_i < y_{i+n}, \quad i = 1, \dots, k.$$

(The conditions are to apply only when the subscripts are meaningful.)

Not only is the fundamental solution ϕ of $Lu = 0$, for L of the above form, totally positive of all orders, but it also possesses total positivity properties with respect to the basis of solutions $u_1, \dots, u_n$. In particular, the set of functions $\{u_1(\cdot), \dots, u_n(\cdot), \phi(\cdot, y_1), \dots, \phi(\cdot, y_k)\}$ is a *WT*-system of order $n + k$ for every $0 < y_1 < \dots < y_k < 1$ and the associated determinant at the points $0 \leqq x_1 < \dots < x_{n+k} \leqq 1$ is strictly positive iff $x_i < y_i < x_{i+n}$, $i = 1, \dots, k$. If $Lu = D^n u = u^{(n)}$, then $u_i(x) = x^{i-1}$, $i = 1, \dots, n$ and $\phi(x, y) = (x - y)_+^{n-1}$. This important example provides us with many of the basic properties of splines.

Let $A = (a_{ij})_{i=1}^{m}{}_{j=1}^{n}$ be an $m \times n$ real matrix and $B = (b_{ij})_{i=1}^{l}{}_{j=1}^{n}$ be an $l \times n$ real matrix with $m + l = n$. Consider the separated boundary conditions

$$U_i(f) = \begin{cases} \sum_{j=1}^{n} a_{ij} D^{j-1} f(0), & i = 1, \dots, m \\ \sum_{j=1}^{n} b_{i-m,j} D^{j-1} f(1), & i = m+1, \dots, n. \end{cases}$$

Let $(L, \mathscr{B})$ denote the differential operator with disconjugate differential form L and the boundary conditions

$$U_i(f) = 0, \quad i = 1, \dots, n.$$

The differential operator $(L, \mathscr{B})$ has a Green's function $G(x, y)$ iff

$$\Delta = \det(U_i(u_j))_{i,j=1}^n \neq 0$$

where the $\{u_j\}_{j=1}^n$ are as defined above. (That is, if $Lf = h$ and $U_i(f) = 0$, $i = 1, \ldots, n$, then $f(x) = \int_0^1 G(x, y)\, h(y)\, dy$.) If $\Delta \neq 0$, then

$$G(x, y) = \frac{1}{\Delta} \det \begin{pmatrix} (U_i(u_j))_{i,j=1}^n & \vdots & U_i(\phi(\cdot, y))_{i=1}^n \\ \cdots\cdots\cdots & \vdots & \cdots\cdots\cdots \\ u_1(x), \ldots, u_n(x) & \vdots & \phi(x, y) \end{pmatrix}.$$

We shall consider boundary conditions which satisfy the following requirements on A and B (recall that $m + l = n$).

Assumptions.

(a) *The $m \times n$ matrix $\tilde{A} = (a_{ij}(-1)^j)_{i=1}^m{}_{j=1}^n$ is SC_m and of rank m.*
(b) *The $l \times n$ matrix $B = (b_{ij})_{i=1}^l{}_{j=1}^n$ is SC_l and of rank l.*

It is not difficult to prove the following:

Proposition 4.2. *Let $(L, \mathscr{B})$ be the differential operator given above with A, B satisfying assumptions (a) and (b). Then $(L, \mathscr{B})$ has a Green's function if and only if there exist indices $1 \leqq i_1 < \ldots < i_m \leqq n$ and $1 \leqq j_1 < \ldots < j_l \leqq n$ for which*

$$\tilde{A}\begin{pmatrix} 1, \ldots, m \\ i_1, \ldots, i_m \end{pmatrix} \neq 0, \quad B\begin{pmatrix} 1, \ldots, l \\ j_1, \ldots, j_l \end{pmatrix} \neq 0$$

and $M_\mu \geqq \mu$, $\mu = 1, \ldots, n$, where M_μ counts the number of indices in the set $\{i_1, \ldots, i_m, j_1, \ldots, j_l\}$ which are less than or equal to μ.

The total positivity properties of $G(x, y)$ are the content of this next theorem.

Theorem 4.3. *Assume that the conditions of Propositions 4.2 hold. Then the kernel $(-1)^l\, G(x, y)$ is TP and*

$$(-1)^{kl}\, G\begin{pmatrix} x_1, \ldots, x_k \\ y_1, \ldots, y_k \end{pmatrix} > 0$$

for $0 < x_1 < \ldots < x_k < 1$, $0 < y_1 < \ldots < y_k < 1$ if and only if $y_{i-l} < x_i < y_{i+m}$, $i = 1, \ldots, k$, (whenever the subscripts are meaningful).

If $l = 0$, then $G(x, y) = \phi(x, y)$ and Theorem 4.3 reduces to Theorem 4.1.
The kernel $(-1)^l\, G(x, y)$, if it exists, does not satisfy the conditions of Theorem 3.4. However, it can be shown that Theorem 3.4 applies with minor changes if $1 \leqq l, m \leqq n - 1$ and $G(x, y)$ exists (see the remark after Theorem 3.4).

Theorem 4.4. *Assume that the Green's function $G(x, y)$ exists for the differential operator $(L, \mathscr{B})$, where the boundary conditions $\mathscr{B}$ satisfy the assumptions (a) and (b), and $1 \leqq l, m \leqq n-1$. Then the eigenvalue problem*

$$\int_0^1 G(x, y)\, \phi(y)\, dy = \lambda \phi(x)$$

has an infinite number of eigenvalues $\lambda_1, \lambda_2, \ldots$. Furthermore, if $|\lambda_1| \geqq |\lambda_2| \geqq \ldots$, where each eigenvalue is listed according to its multiplicity as a root of the Fredholm determinant of the kernel $G(x, y)$, then the following hold:

(i) *The λ_i are of sign $(-1)^l$ and distinct, i.e., $(-1)^l \lambda_1 > (-1)^l \lambda_2 > \ldots > 0$.*
(ii) *If ϕ_k is the eigenfunction associated with λ_k, uniquely determined up to a nonzero constant, then*

$$p-1 \leqq S\left(\sum_{i=p}^{q} c_i \phi_i\right) \leqq \tilde{Z}\left(\sum_{i=p}^{q} c_i \phi_i\right) \leqq q-1$$

on $(0,1)$ for any $p \leqq q$, c_i real, not all zero. In particular ϕ_i has $i-1$ sign changes and no other zeros in $(0,1)$.
(iii) *The zeros of ϕ_i and ϕ_{i+1} on $(0,1)$ strictly interlace.*

The simplest example of boundary conditions satisfying assumptions (*a*) and (*b*) are

$$D^{(i_k-1)} f(0) = 0, \qquad k = 1, \ldots, m$$
$$D^{(j_k-1)} f(1) = 0, \qquad k = 1, \ldots, l$$

where $1 \leqq i_1 < \ldots < i_m \leqq n$, $1 \leqq j_1 < \ldots < j_l \leqq n$. The Green's function exists if and only if $M_\mu \geqq \mu$ where M_μ counts the number of indices $\{i_1, \ldots, i_m, j_1, \ldots, j_l\}$ less than or equal to μ.

When dealing with periodic functions we are naturally led to a consideration of kernels of the form $K(x, y) = \phi(x-y)$, where ϕ is a 2π-periodic, integrable function. No kernel of this form may be TP_r for $r > 1$ unless it is of rank one, i.e., $\phi(x-y) = \phi_1(x)\, \phi_2(y)$ for all x, y. In fact $\phi(x-y)$ cannot be SSC_{2n} for any $n = 1, 2, \ldots$ due to its periodicity. We are thus led to an investigation of cyclic Pólya frequency (*CPF*) functions and the cyclic variation diminishing (*CVD*) property.

The consideration of these concepts necessitates the introduction of an altered form of zero counting.

Definition 4.1. Let $\mathbf{x} = (x_1, \ldots, x_n) \in \mathbb{R}^n \backslash \{\mathbf{0}\}$. The number $S_c^-(\mathbf{x})$ of cyclic variations of sign of $\mathbf{x}$ is given by

$$\begin{aligned} S_c^-(\mathbf{x}) &= \max_i S^-(x_i, x_{i+1}, \ldots, x_n,\ x_1, \ldots, x_i) \\ &= S^-(x_k, \ldots, x_n,\ x_1, \ldots, x_k) \end{aligned}$$

where k is any integer for which $x_k \neq 0$.

Obviously $S_c^-(\mathbf{x})$ is invaraint under cyclic permutations and is always an even number.

Analogous to the definition of $S(f)$, the number of sign changes of f, we define $S_c(f)$ for periodic function as follows.

Definition 4.2. Let $f(x)$ be a piecewise continuous, 2π-periodic function. We shall assume that $f(x) = \frac{1}{2}[f(x+) + f(x-)]$ for all x. Then

$$S_c(f) = \sup S_c^-(f(x_1), \ldots, f(x_m))$$

where the supremum is extended over all $x_1 < \ldots < x_m < x_1 + 2\pi$, m arbitrary.

For a real, continuous, 2π-periodic function ϕ, our concern is with the transformation

$$(\phi * h)(x) = 1/2\pi \int_0^{2\pi} \phi(x-y)\, h(y)\, dy. \tag{1}$$

Definition 4.3. The transformation (1) is said to be *cyclic variation diminishing of order* $2n$ (CVD_{2n}) if $S_c(\phi * h) \leqq S_c(h)$ for all h for which $S_c(h) \leqq 2n$.

In this case we shall also say that ϕ is a cyclic variation diminishing kernel of order $2n$, or ϕ is CVD_{2n}.

Closely related to the concept of CVD_{2n} kernels is the following class.

Definition 4.4. The continuous, 2π-periodic function ϕ is said to be a *cyclic Pólya frequency kernel of order* $2n+1$ (CPF_{2n+1}) if $K(x, y) = \phi(x-y)$ is SC_{2l+1}, with $\varepsilon_{2l+1} = 1$, $l = 0, 1, \ldots, n$.

The relationship between CVD_{2n} and CPF_{2n+1} kernels is partially contained in the following theorem.

Theorem 4.5. *Assume that ϕ is a continuous, 2π-periodic function of rank at least $2n+2$, i.e., there exist $0 \leqq y_1 < \ldots < y_{2n+2} < 2\pi$ for which $\dim(\mathrm{span}\{\phi(\cdot - y_i)\}_{i=1}^{2n+2}) = 2n+2$. Then ϕ is CVD_{2n} if and only if $\varepsilon\phi$ is CPF_{2n+1} for some $\varepsilon \in \{-1, 1\}$.*

Associated with ϕ of the above form is its Fourier series

$$\phi(x) \sim \sum_{n=-\infty}^{\infty} a_n e^{inx}, \quad a_n = \frac{1}{2\pi} \int_0^{2\pi} \phi(x) e^{-inx}\, dx.$$

Since ϕ is real, $a_{-n} = \bar{a}_n$. The CVD_{2n} property of ϕ implies certain inequalities among the Fourier coefficient $\{a_n\}_{n=-\infty}^{\infty}$.

Theorem 4.6. *Assume that ϕ is CVD_{2n} and of rank at least $2n+2$. Then*

$$|a_0| > |a_1| > \ldots > |a_n| > |a_k|, \quad k = n+1,\ n+2 \ldots.$$

Perhaps the most well-known class of *CVD* kernels is the family of de la Vallée Poussin kernels defined by

$$\phi_n(x) = \frac{1}{\binom{2n}{n}} \left(2\cos\frac{x}{2}\right)^{2n}.$$

The V means (de la Vallée Poussin means)

$$V_n(t;f) = \frac{1}{2\pi} \int_0^{2\pi} \phi_n(t-s)\, f(s)\, ds$$

are essentially the periodic analogues of the Bernstein polynomial operators. It can be shown that if $f(t)$ is 2π-periodic and continuous, then $V_n(t;f)$ converges to $f(t)$ uniformly as $n \to \infty$.

Let us consider one class of lesser known *CVD* kernels. *CVD* kernels may be constructed from Pólya frequency functions. A Pólya frequency function is any function f defined on all of $\mathbb{R}$ for which the kernel $K(x,y) = f(x-y)$ is totally positive (TP) of all orders. f is a Pólya frequency density if it is a Pólya frequency function and $\int_{-\infty}^{\infty} f(x)\,dx = 1$.

The following characterizes Pólya frequency densities.

Theorem 4.7. *A necessary and sufficient condition that f be a Pólya frequency density is that the reciprocal of its Laplace transform be of the form*

$$\psi(s) = e^{-rs^2 + \delta s} s^k \prod_{i=1}^{\infty} (1 + a_i s) e^{-a_i s} \tag{2}$$

where $r \geqq 0$, δ real, k a non-negative integer, a_i real, $0 < r + \sum_{i=1}^{\infty} a_i^2 < \infty$, and $\psi(0) = 1$.

On the basis of Theorem 4.7, the following result may be obtained.

Theorem 4.8. *Let f be a Pólya frequency density. Then the function*

$$\phi(t) = 2\pi \sum_{n=-\infty}^{\infty} f(t - 2\pi n)$$

is a regular CVD kernel on $[0, 2\pi)$.

As an example, consider

$$f(x) = \frac{1}{\sqrt{2\pi}\,\sigma} e^{-\frac{x^2}{2\sigma^2}}, \qquad \sigma > 0.$$

f is a Pólya frequency density. The function

$$\begin{aligned}\phi(t;\sigma) &= \frac{\sqrt{2\pi}}{\sigma}\sum_{n=-\infty}^{\infty}\exp\left\{-\frac{1}{2\sigma^2}(t-2\pi n)^2\right\}\\ &= \sum_{n=-\infty}^{\infty} e^{-\frac{n^2\sigma^2}{2}} e^{int}\\ &= 1+2\sum_{n=1}^{\infty} e^{-\frac{n^2\sigma^2}{2}}\cos nt\end{aligned}$$

is a *CVD* kernel. This function may be obtained from the heat equation and is used in the smoothing of periodic *WT*-systems.

As a second example consider the function

$$\begin{aligned}\phi_\beta(x) &= 1+2\sum_{k=1}^{\infty}\frac{\cos kz}{\cosh k\beta}\\ &= \sum_{k=-\infty}^{\infty}\frac{2}{e^{k\beta}+e^{-k\beta}}e^{ikz}.\end{aligned}$$

From Theorems 4.7 and 4.8, $\phi_\beta(x)$ is *CVD* if $\psi(in)=\dfrac{e^{n\beta}+e^{-n\beta}}{2}$ is of the form (2), i.e., $\psi(s)=\cos\beta s$ is of the form (2). Setting $r=\delta=k=0$ and $a_l=\dfrac{-2\beta}{(2l+1)\pi}$, $l=0,\pm 1,\pm 2,\ldots$ in (2), we see that

$$\cos\beta s=\prod_{l=-\infty}^{\infty}\left(1-\frac{2\beta s}{(2l+1)\pi}\right)\exp(2\beta s/(2l+1)\pi)$$

which implies the result. The function $\phi_\beta(x)=1+2\sum_{k=1}^{\infty}\dfrac{\cos kx}{\cosh k\beta}$ arises in the consideration of real-valued 2π-periodic functions on $\mathbb{R}$ which are analytic in the strip

$$S_\beta=\{z\in\mathbb{C}\colon |\mathrm{Im}\, z|\leqq\beta\}.$$

Chapter IV. n-Widths in Hilbert Spaces

1. Introduction

The organization of this chapter is as follows. The most general theorem concerning n-widths in Hilbert spaces is the main content of Section 2. Let T be a compact operator mapping H_1 to H_2, where both H_1 and H_2 are Hilbert spaces. Then the Kolmogorov, linear, Gel'fand, and Bernstein n-width of the set

$$A = \{T\phi : \|\phi\|_{H_1} \leqq 1\}$$

as a subset of H_2, is simply the $(n+1)$st singular value of T and optimal subspaces are easily constructed in terms of eigenvector subspaces. In Section 3, we consider variations on this problem and many examples. The n-widths and optimal subspaces are easily identified if T is given by convolution against a fixed periodic function. This is discussed in Section 4. Sections 5 and 6 are different in nature. In those sections we consider integral operators whose kernels are either totally positive or cyclic variation diminishing (and variants thereof). In these cases we are able to determine additional optimal subspaces of an elementary form.

2. n-Widths of Compact Linear Operators

Let H_1, H_2 be Hilbert spaces with inner products $(\cdot,\cdot)_1$ and $(\cdot,\cdot)_2$, respectively, and associated norms. Let $T \in L(H_1, H_2)$, i.e., T is a linear operator mapping H_1 into H_2. Recall that in Chapter II we defined $K(H_1, H_2)$, the class of compact operators in $L(H_1, H_2)$. A compact operator, often referred to as a completely continuous operator, is simply one which maps bounded sets to sets whose closures are compact, i.e., precompact sets.

A noteworthy class of compact operators mapping $L^2(a,b)$ into $L^2(a,b)$ (where (a,b) is a finite or infinite real interval) are the integral operators given by

$$(Tf)(x) = \int_a^b K(x,y)\, f(y)\, dy,$$

where $K(x,y)$ is a *Hilbert-Schmidt* kernel. (A complex-valued function $K(x,y)$,

$a < x, y < b$, is called a Hilbert-Schmidt kernel if

$$\int_a^b \int_a^b |K(x, y)|^2 \, dx \, dy < \infty.)$$

To each $T \in L(H_1, H_2)$, there exists a $T' \in L(H_2, H_1)$, called the *adjoint* of T, which is defined by

$$(Tx, y)_2 = (x, T'y)_1, \qquad \text{for all } x \in H_1, \; y \in H_2.$$

If $T \in L(H, H)$ and $T = T'$, then we say that T is *self-adjoint*. It is a well-known fact that if $T \in K(H, H)$ is self-adjoint, then all its eigenvalues are real, and any solution ϕ of $(T - \lambda I)^n \phi = 0$ for some eigenvalue λ and positive integer n, is an eigenvector of T. A self-adjoint compact operator T is said to be *non-negative* if all its eigenvalues are non-negative. In what follows $v = v(T)$ shall denote the sum of the algebraic multiplicities of the non-zero eigenvalues of T and $\{\lambda_i(T)\}_{i=1}^{v(T)}$ $(\{\lambda_i\}_{i=1}^{v})$ shall denote the sequence of non-zero eigenvalues of T, listed to their algebraic multiplicity and ordered so that

$$|\lambda_1(T)| \geqq |\lambda_2(T)| \geqq \ldots > 0.$$

The essential facts concerning $T \in K(H, H)$ which we shall use are the following:

Theorem 2.1. *Let H be a Hilbert space, and $T \in K(H, H)$ be self-adjoint and non-negative. Let $\{\lambda_i\}_{i=1}^{v}$ denote the non-zero eigenvalues of T listed to their algebraic multiplicities and given in non-increasing order of magnitude, i.e.,*

$$\lambda_1 \geqq \lambda_2 \geqq \ldots > 0.$$

Let $\{\phi_i\}_{i=1}^{v}$ be an orthonormal system of eigenvectors of T satisfying

$$T\phi_i = \lambda_i \phi_i, \qquad i = 1, \ldots, v.$$

Then $\{\phi_i\}_{i=1}^{v}$ is complete in the range of T, and the eigenvalues and eigenvectors satisfy the following min-max and max-min properties for $n \leqq v$:

(a) $$\lambda_n = \min_{\zeta_1, \ldots, \zeta_{n-1}} \max \left\{ \frac{(T\phi, \phi)}{(\phi, \phi)} : \phi \in H, \; (\phi, \zeta_i) = 0, \; i = 1, \ldots, n-1, \; \phi \neq 0 \right\}$$

where the minimum is taken over all sets of $n-1$ vectors $\{\zeta_i\}_{i=1}^{n-1}$ in H. Furthermore, the above min-max is attained for $\zeta_i = \phi_i$, $i = 1, \ldots, n-1$ and $\phi = \phi_n$.

(b) $$\lambda_n = \max_{\zeta_1, \ldots, \zeta_n} \min \left\{ \frac{(T\phi, \phi)}{(\phi, \phi)} : \phi \in \operatorname{span}\{\zeta_1, \ldots, \zeta_n\}, \; \phi \neq 0 \right\}$$

where the maximum is taken over all sets of n linearly independent vectors $\{\zeta_i\}_{i=1}^{n}$ in H. Furthermore, the above max-min is attained for $\zeta_i = \phi_i$, $i = 1, \ldots, n$, and $\phi = \phi_n$.

Let $T \in K(H_1, H_2)$. Then $T'T$ and TT' are self-adjoint, non-negative compact operators. The s-numbers (singular values) of T are defined by

$$s_n(T) = [\lambda_n(T'T)]^{1/2}, \qquad n = 1, 2, \ldots, \nu(T'T).$$

It is easily proven that $s_n(T) = s_n(T')$ for all T.

s-Numbers were first introduced by E. Schmidt [1907]. Schmidt was interested in the following problem for a Hilbert-Schmidt kernel $K(x, y)$. Determine

$$\min\left\{\int_a^b\int_a^b \left|K(x,y) - \sum_{i=1}^n u_i(x)\, v_i(y)\right|^2 dx\, dy \colon u_i, v_i \in L^2(a,b),\ i = 1,\ldots,n\right\}.$$

His solution to this problem is the following. The operator $K'K$ (i.e., the operator with kernel $K'K(x,y) = \int_a^b K(z,x)\, \overline{K(z,y)}\, dz$) is self-adjoint, non-negative, compact, and has eigenvalues $\lambda_1 \geqq \lambda_2 \geqq \ldots > 0$, and associated orthonormal eigenvectors ϕ_i, $i = 1, 2, \ldots, \nu$. Set $\psi_i(x) = (K\phi_i)(x) = \int_a^b K(x,y)\, \phi_i(y)\, dy$. The functions $\{\psi_i\}_{i=1}^{\nu}$ are the eigenvectors of KK'. The "Hilbert-Schmidt decomposition" of $K(x, y)$ is given by

$$K(x,y) = \sum_{i=1}^{\nu} \psi_i(x)\, \phi_i(y), \qquad \text{a.e.}$$

Schmidt proved that

$$\min\left\{\int_a^b\int_a^b \left|K(x,y) - \sum_{i=1}^n u_i(x)\, v_i(y)\right|^2 dx\, dy \colon u_i, v_i \in L^2(a,b),\ i = 1,\ldots,n\right\}$$
$$= \int_a^b\int_a^b \left|K(x,y) - \sum_{i=1}^n \psi_i(x)\, \phi_i(y)\right|^2 dx\, dy$$
$$= \sum_{i=n+1}^{\nu} \lambda_i.$$

Let $T \in K(H_1, H_2)$. Thus $T'T \in K(H_1, H_1)$ is self-adjoint and non-negative. Let $\{s_n(T)\}$ denote the totality of s-numbers (given in non-increasing order of magnitude) and let $\{\phi_n\}$ denote the associated orthonormal eigenvectors. Set $\psi_n = T\phi_n$. The $\{\psi_n\}$ are eigenvectors of $TT' \in K(H_2, H_2)$ and $(\psi_i, \psi_j)_2 = \delta_{ij}[s_j(T)]^2$.

The relevance of s-numbers in the calculation of n-widths of compact operators on Hilbert spaces is the content of this next result.

Theorem 2.2. *Let $T \in K(H_1, H_2)$ and let $\{s_n(T)\}$, $\{\phi_n\}$ and $\{\psi_n\}$ be as defined above. Then*

$$d_n(T(H_1); H_2) = d^n(T(H_1); H_2) = \delta_n(T(H_1); H_2) = b_n(T(H_1); H_2) = s_{n+1}(T),$$
$$n = 0, 1, 2, \ldots.$$

Furthermore,

(a) *the subspace* $X_n = \text{span}\{\psi_1,\ldots,\psi_n\}$ *is optimal for* $d_n(T(H_1);H_2)$,
(b) *the subspace*

$$L^n = \{\phi\colon \phi \in H_1,\ (\phi,\phi_i)_1 = 0,\ i=1,\ldots,n\}$$

is optimal for $d^n(T(H_1);H_2)$,
(c) *the linear operator*

$$P_n\phi = \sum_{i=1}^{n} (\phi,\phi_i)_1\,\psi_i$$

is optimal for $\delta_n(T(H_1);H_2)$,
(d) *the subspace* $X_{n+1} = \text{span}\{\psi_1,\ldots,\psi_{n+1}\}$ *is optimal for* $b_n(T(H_1);H_2)$.

Remark. We call the reader's attention to the fact that we are here using the definitions of $d^n(T(H_1);H_2)$ and $\delta_n(T(H_1);H_2)$ as given in Section 7 of Chapter II. If we were to use the earlier definitions of the Gel'fand and linear n-widths as given in Sections 3 and 4 of Chapter II, then the values of the n-widths would be the same, but the optimal subspace of codimension n and the optimal rank n linear operator would, of necessity, be altered.

Remark. Note that this theorem provides us with optimal subspaces which enjoy the property of inclusion, i.e., $X_n \subseteqq X_{n+1}$, $L^{n+1} \subseteqq L^n$. It should be mentioned that these subspaces are in no way unique, i.e., there are other subspaces which are optimal. Other optimal subspaces are considered in Sections 5 and 6.

Proof. From Propositions 3.5 and 5.2 of Chapter II, $d_n(T(H_1);H_2) = \delta_n(T(H_1);H_2) \geqq d^n(T(H_1);H_2) \geqq b_n(T(H_1);H_2)$.

The proof of the fact that $b_n(T(H_1);H_2) = s_{n+1}(T)$ and $X_{n+1} = \text{span}\{\psi_1,\ldots,\psi_{n+1}\}$ is optimal for b_n is a consequence of (b) of Theorem 2.1. The upper bound $\delta_n(T(H_1);H_2) \leqq s_{n+1}(T)$ easily follows from the explicit form of P_n as given in (c). Thus all the n-widths equal $s_{n+1}(T)$, and (a), (c) and (d) hold. The statement (b) is both a direct consequence of Theorem 2.1, (a) and of the fact that $L^n = \{\phi\colon P_n\phi = 0\}$. □

Theorem 2.2 is the basic result of this chapter. In one sense it tells us all about the n-widths by calculating them and giving us explicit optimal subspaces and operators. In another sense it tells us very little since we must now determine the eigenvalues and eigenvectors of a certain operator. In the remaining sections of this chapter we consider generalizations and variations of this result, and exhibit numerous examples where more explicit answers are obtained. As a first simple example consider the following:

Example 2.1. Set $\Delta_1 = \{z\colon |z| < 1\}$, the unit disk in $\mathbb{C}$. Let h^2 denote the class of real-valued harmonic functions $u(z)$ in Δ_1 whose boundary values $u(e^{i\theta})$ lie in $L^2[0,2\pi]$ (see Duren [1970]). Such functions may be uniquely recovered from their boundary values via the Poisson integral

$$u(z) = u(re^{i\theta}) = \frac{1}{2\pi}\int_0^{2\pi} P(r;\theta - t)\,u(e^{it})\,dt,$$

$0 \leqq r < 1$, where

$$P(r;\theta) = \frac{1-r^2}{1-2r\cos\theta + r^2}.$$

For fixed $R, 0 < R \leqq 1$, we wish to approximate $u(z)$ of the above form on Δ_R (the disk of radius R) in the area norm. To be more explicit, we are interested in $d_n(A_2; H(\Delta_R))$, where

$$A_2 = \left\{u\colon u \in h^2, \frac{1}{2\pi}\int_0^{2\pi} |u(e^{i\theta})|^2 \, d\theta \leqq 1\right\}$$

and

$$\|u\|_{H(\Delta_R)} = \left[\frac{1}{2\pi}\int_0^{2\pi}\int_0^R |u(re^{i\theta})|^2 \, r\,dr\,d\theta\right]^{1/2}.$$

In order to calculate the n-widths, we shall determine T, the compact linear operator from $L^2(\partial\Delta_1)$ to $H(\Delta_R)$, the space of real-valued harmonic functions in h^2, with norm $\|\cdot\|_{H(\Delta_R)}$. T is easily seen to be given by

$$(Tu)(re^{i\theta}) = \frac{1}{2\pi}\int_0^{2\pi} P(r;\theta - t)\, u(e^{it})\, dt.$$

We claim that $T'\colon H(\Delta_R) \to L^2(\partial\Delta_1)$ is given by

$$(T'v)(e^{it}) = \frac{1}{2\pi}\int_0^{2\pi}\int_0^R v(re^{i\theta})\, P(r;\theta - t)\, r\,dr\,d\theta.$$

To prove this claim we must show that $(Tu, v)_{H(\Delta_R)} = (u, T'v)_{L^2(\partial\Delta_1)}$ for all $u \in L^2(\partial\Delta_1)$ and $v \in H(\Delta_R)$. This fact is readily proved and its verification is left to the reader.

To obtain n-widths and optimal n-dimensional subspaces it suffices, from Theorem 2.2, to calculate the eigenvalues and eigenfunctions of the operator

$$TT'\colon H(\Delta_R) \to H(\Delta_R).$$

Lemma 2.3. *The functions* $v_k(re^{i\theta}) = r^{|k|} e^{ik\theta}$, $k = 0, \pm 1, \pm 2, \ldots$, *are the totality of eigenfunctions, with associated non-zero eigenvalues* $\frac{R^{2|k|+2}}{2|k|+2}$, *of the operator* TT'.

Proof. Let $v \in H(\Delta_R)$. By the above formulae

$$\begin{aligned}
(TT'v)(re^{i\theta}) &= \frac{1}{2\pi}\int_0^{2\pi} P(r;\theta - t)\,(T'v)(e^{it})\,dt \\
&= \frac{1}{2\pi}\int_0^{2\pi} P(r;\theta - t)\left[\frac{1}{2\pi}\int_0^{2\pi}\int_0^R v(se^{i\phi})\, P(s;\phi - t)\, s\,ds\,d\phi\right] dt \\
&= \frac{1}{2\pi}\int_0^{2\pi}\int_0^R \left[\frac{1}{2\pi}\int_0^{2\pi} P(r;\theta - t)\, P(s;\phi - t)\,dt\right] v(se^{i\phi})\, s\,ds\,d\phi.
\end{aligned}$$

Since $P(r;\theta) = \sum_{-\infty}^{\infty} r^{|k|} e^{ik\theta}$, it follows that

$$\frac{1}{2\pi} \int_0^{2\pi} P(r;\theta - t)\, P(s;\phi - t)\, dt = P(rs;\theta - \phi)$$

and thus

$$(TT'v)(re^{i\theta}) = \frac{1}{2\pi} \int_0^{2\pi} \int_0^R P(rs;\theta - \phi)\, v(se^{i\phi})\, s\, ds\, d\phi.$$

Using the above formula and the Fourier series representation of $P(rs;\theta - \phi)$, it is a simple matter to prove that

$$(TT'v_k)(re^{i\theta}) = \frac{R^{2|k|+2}}{2|k|+2} v_k(re^{i\theta}).$$

It remains to prove that these are all the eigenfunctions of TT', with non-zero eigenvalues. Assume that $TT'v = \lambda v \neq 0$, and $(v, v_k)_{H(\Delta_R)} = 0$ for all $k = 0, \pm 1, \pm 2, \ldots$. Thus

$$0 = (TT'v, v_k)_{H(\Delta_R)} = (T'v, T'v_k)_{L^2(\partial\Delta_1)}.$$

Now $(T'v_k)(e^{i\theta}) = \frac{R^{2|k|+2}}{2|k|+2} e^{ik\theta}$, which implies that $(T'v, e^{ik\theta})_{L^2(\partial\Delta_1)} = 0$, $k = 0, \pm 1, \pm 2, \ldots$. Thus $T'v \equiv 0$ a.e., and $TT'v \equiv 0$, a contradiction. □

We have therefore obtained the following consequence of Theorem 2.2.

Proposition 2.4. *For A_2 and $H(\Delta_R)$ as above, $d_0(A_2; H(\Delta_R)) = \frac{R}{\sqrt{2}}$, while $d_{2n-1}(A_2; H(\Delta_R)) = d_{2n}(A_2; H(\Delta_R)) = \frac{R^{n+1}}{\sqrt{2n+2}}$. Furthermore,*

$$\operatorname{span}\{1, re^{i\theta}, re^{-i\theta}, \ldots, r^{n-1}e^{i(n-1)\theta}, r^{n-1}e^{-i(n-1)\theta}\}$$

is an optimal $(2n-1)$-dimensional subspace for $d_{2n-1}(A_2; H(\Delta_R))$, and thus for $d_{2n}(A_2; H(\Delta_R))$.

The above may well be considered within the following framework. To avoid too many details, we do not consider the most general possible situation (see Shapiro [1979]).

Let D_1, D_2 be bounded domains in $\mathbb{R}^N$ or $\mathbb{C}^N$, with $\bar{D}_2 \subset D_1$. Let H_1 and H_2 be Hilbert spaces of functions on D_1 and D_2, respectively. Assume that H_2 is realizable as a subspace of $L^2(D_2, d\mu)$ for some positive measure μ on D_2. Assume that H_1 has a reproducing kernel function. That is, for each $w \in D_1$ there is a "reproducing element" $K_w \in H_1$ such that

$$f(w) = (f, K_w)_{H_1}, \quad f \in H_1.$$

For $z, w \in D_1$, write $K(z, w) = K_w(z)$. $K(\cdot, \cdot)$ is a hermitian function on $D_1 \times D_1$ (and since H_1 has a reproducing kernel function, the point evaluations are bounded linear functionals). Finally, assume that the restriction of $f \in H_1$ to $w \in D_2$ is a function in H_2, and let T denote this restriction.

To determine the n-widths of the set

$$A = \{Tf \colon \|f\|_{H_1} \leqq 1\}$$

as a subset of H_2, it is necessary to consider the eigenvalue-eigenfunction problem $TT'\psi = \lambda\psi$. The definition of T implies that this may be rewritten in the form

$$(T'\psi)(w) = \lambda\psi(w), \qquad w \in D_2.$$

Now, for $w \in D_2$,

$$\begin{aligned}(T'\psi)(w) &= (T'\psi, K_w)_{H_1} = (\psi, TK_w)_{H_2}\\ &= \int_{D_2} \psi(z)\, \overline{K(z,w)}\, d\mu(z)\\ &= \int_{D_2} K(w,z)\, \psi(z)\, d\mu(z).\end{aligned}$$

Thus, we have reduced the relevant eigenvalue-eigenfunction problem to

$$\int_{D_2} K(w,z)\, \psi(z)\, d\mu(z) = \lambda\psi(w), \qquad w \in D_2$$

where $K(z, w)$ is the reproducing kernel of H_1.

The result of Theorem 2.2 did not so much depend on the compactness of T as on the fact, implied by the compactness, that the eigenvectors $\{\psi_i\}_{i=1}^{\nu}$ of TT' form a complete orthogonal basis for the range of TT'. The next theorem, which we shall not prove, is useful.

Theorem 2.5. *Let $T \in L(H_1; H_2)$. Assume that $TT'\psi_i = \lambda_i\psi_i$, $i = 1, 2, \ldots$, where $\lambda_1 \geqq \lambda_2 \geqq \ldots \geqq \lambda_{n+1} \geqq \lambda_k > 0$, $k \geqq n+1$, and that the $\{\psi_i\}_{i=1}^{\nu}$ form an orthogonal basis for the range of TT'. Then*

$$d_n(T(H_1); H_2) = \sqrt{\lambda_{n+1}}$$

and span $\{\psi_1, \ldots, \psi_n\}$ *is an optimal subspace.*

Of course it is not in general necessary that a self-adjoint, non-negative, continuous linear operator possess eigenvectors.

It is often the case, see e.g., Lorentz [1966a], Tichomirov [1976], that a very simple form of Theorem 2.5 is given and used. We record this simple form for easy reference.

Corollary 2.6. *Let $\{\psi_i\}_{i=1}^{\infty}$ be an orthonormal basis for a Hilbert space H. Let $\{\mu_i\}_{i=1}^{\infty}$ be a sequence of non-increasing positive numbers, i.e., $\mu_1 \geqq \mu_2 \geqq \ldots > 0$. Set*

$$A = \left\{ \sum_{i=1}^{\infty} a_i \psi_i : \left(\sum_{i=1}^{\infty} \mu_i^{-2} |a_i|^2 \right)^{1/2} \leqq 1 \right\}.$$

Then $d_n(A; H) = \mu_{n+1}$, and $X_n = \operatorname{span}\{\psi_1, \ldots, \psi_n\}$ is an optimal subspace.

As a simple consequence of Corollary 2.6, we have the following:

Example 2.2. Let $H^2(R)$ denote the class of power series $\sum_{k=0}^{\infty} a_k z^k$, with $\left\| \sum_{k=0}^{\infty} a_k z^k \right\|_R = \left(\sum_{k=0}^{\infty} |a_k|^2 R^{2k} \right)^{1/2} < \infty$. Thus $H^2(R)$ is the class of analytic functions in the disk Δ_R of radius R, whose boundary values, correctly interpreted, lie in L^2.

Let $0 < r \leqq R$, and set

$$A_R = \{f : f \in H^2(R), \ \|f\|_R \leqq 1\}.$$

Then $d_n(A_R; H^2(r)) = (r/R)^n$, $n = 0, 1, 2, \ldots$, and $X_n = \operatorname{span}\{1, z, \ldots, z^{n-1}\}$ is an optimal subspace.

It may be that the n-widths and optimal subspaces are more readily found by using the method of proof of Theorem 2.2, than by applying Theorem 2.2 directly. This is often the case if the operator T, associated with the specific problem, is not explicitly given. In the next two examples we find the n-widths, in the sense of Kolmogorov, and optimal subspaces, in a simple and straightforward manner without bothering to exhibit the associated operator T. (In the first example, T can be easily obtained.)

Example 2.3. Let $L^2(\mathbb{R})$ denote the space of complex-valued square integrable functions on $\mathbb{R} = (-\infty, \infty)$, with the usual L^2-norm. A function $f(t) \in L^2(\mathbb{R})$ is said to be strictly "band-limited" to $(-\sigma, \sigma)$ if its Fourier transform $\hat{f}(x)$ vanishes off $(-\sigma, \sigma)$. Let $\mathscr{B}_\sigma$ be those functions in $L^2(\mathbb{R})$ strictly band-limited to $(-\sigma, \sigma)$, i.e., $\mathscr{B}_\sigma = \{f : f \in L^2(\mathbb{R}), \hat{f}(x) = 0 \text{ for } |x| \geqq \sigma\}$. (This class is considered again in Examples 2.4 and 3.5.)

Every $f \in \mathscr{B}_\sigma$ has the following absolutely and uniformly convergent cardinal series representation

$$f(t) = \sum_{j=-\infty}^{\infty} f(jh) \frac{\sin \sigma(t - jh)}{\sigma(t - jh)}$$

where $\sigma h = \pi$. Let

$$\phi_j(t) = \frac{\sin \sigma(t - jh)}{\sigma(t - jh)}, \qquad \text{for every integer } j.$$

Then $\phi_j(kh) = \delta_{jk}$. This expansion is also an orthogonal representation for $f \in \mathscr{B}_\sigma$, i.e.,

$$\int_{-\infty}^{\infty} \phi_j(t)\, \phi_k(t)\, dt = h \delta_{jk}.$$

Thus

$$\| f \|^2 = \int_{-\infty}^{\infty} |f(t)|^2 \, dt = h \sum_{j=-\infty}^{\infty} |f(jh)|^2.$$

Let k be a positive integer, and set $E_{\sigma,k} = \{f\colon \|t^k f(t)\| \leqq 1,\ f \in \mathscr{B}_\sigma\}$. Note that since $f \in \mathscr{B}_\sigma$, then $t^k f(t) \in \mathscr{B}_\sigma$.

Proposition 2.7. *For $\sigma h = \pi$, and k a positive integer*

$$d_{2n-1}(E_{\sigma,k}; L^2) = d_{2n}(E_{\sigma,k}; L^2) = (nh)^{-k}, \quad n = 1, 2, \ldots.$$

Furthermore, $X_{2n-1} = \operatorname{span}\{\phi_i\}_{i=-(n-1)}^{n-1}$ is an optimal subspace for $d_{2n-1}(E_{\sigma,k}; L^2)$ (and thus for $d_{2n}(E_{\sigma,k}; L^2)$).

Proof. For $f \in E_{\sigma,k}$,

$$t^k f(t) = \sum_{j=-\infty}^{\infty} (jh)^k f(jh)\, \phi_j(t)$$

while

$$f(t) = \sum_{j=-\infty}^{\infty} f(jh)\, \phi_j(t).$$

Thus

$$\| f \|^2 = h \sum_{j=-\infty}^{\infty} |f(jh)|^2$$

and

$$\| t^k f(t) \|^2 = h \sum_{j=-\infty}^{\infty} (jh)^{2k} |f(jh)|^2.$$

Now,

$$\begin{aligned} E(E_{\sigma,k}; X_{2n-1}) &= \sup_{f \in E_{\sigma,k}} \inf_{g \in X_{2n-1}} \| f - g \| \\ &= \sup_{f \in E_{\sigma,k}} \Big\| \sum_{|j| \geqq n} f(jh)\, \phi_j \Big\| \\ &= \sup_{f \in E_{\sigma,k}} \Big(h \sum_{|j| \geqq n} |f(jh)|^2\Big)^{1/2} \\ &= \sup \left[\frac{\sum_{|j| \geqq n} |f(jh)|^2}{\sum_{j=-\infty}^{\infty} (jh)^{2k} |f(jh)|^2} \right]^{1/2}. \end{aligned}$$

Since $(jh)^{2k}$ is an increasing sequence in $|j|$, we obtain

$$E(E_{\sigma,k}; X_{2n-1}) = (nh)^{-k},$$

which implies that $d_{2n-1}(E_{\sigma,k}; L^2) \leqq (nh)^{-k}$.

The lower bound argument is also standard. Set

$$M_{2n+1} = \left\{ \sum_{j=-n}^{n} a_j \phi_j(t) \colon \left\| \sum_{j=-n}^{n} a_j \phi_j \right\| \leqq (nh)^{-k} \right\}.$$

To prove that $d_{2n}(E_{\sigma,k}; L^2) \geqq (nh)^{-k}$, it suffices, by Proposition 1.6 of Chapter II to prove that $M_{2n+1} \subseteqq E_{\sigma,k}$. The proof of this fact is immediate.

Example 2.4. *Let* $\mathscr{B}_\sigma$ be as defined in Example 2.3. For $\mu_1, \ldots, \mu_s$, strictly positive real numbers, consider the "Sobolev" inner product for $f, g \in \mathscr{B}_\sigma$ defined by

$$(f,g)_s = \int_{-T}^{T} (f(t)\,\bar{g}(t) + \mu_1 f'(t)\,\bar{g}'(t) + \ldots + \mu_s f^{(s)}(t)\,\bar{g}^{(s)}(t))\,dt$$

for some fixed positive T. Let $H_s^T = \{f\colon f, f', \ldots, f^{(s)} \in L^2(-T, T)\}$ normed by $\| f \|_s = [(f,f)_s]^{1/2}$, and set

$$A_\sigma = \{ f \in \mathscr{B}_\sigma \colon \| f \| \leqq 1 \},$$

where $\| f \| = \left(\int_{-\infty}^{\infty} |f(t)|^2\,dt \right)^{1/2}$. We wish to determine the values $d_n(A_\sigma; H_s^T)$ and construct optimal subspaces. ($\mathscr{B}_\sigma$ is, by the Paley-Wiener Theorem, a subset of H_s^T.)

It is known that for $f, g \in A_\sigma$

$$(f,g)_s = \int_{-\sigma}^{\sigma} \int_{-\sigma}^{\sigma} \frac{\sin T(x-y)}{\pi(x-y)} (1 + \mu_1 xy + \ldots + \mu_s x^s y^s)\,\hat{f}(x)\,\overline{\hat{g}(y)}\,dx\,dy.$$

Let

$$(K\hat{f})(y) = \int_{-\sigma}^{\sigma} \frac{\sin T(x-y)}{\pi(x-y)} (1 + \mu_1 xy + \ldots + \mu_s x^s y^s)\,\hat{f}(x)\,dx,$$

for $|x| \leqq \sigma$. Thus,

$$\| f \|_s^2 = \int_{-\sigma}^{\sigma} (K\hat{f})(y)\,\overline{\hat{f}(y)}\,dy.$$

K is a self-adjoint non-negative Hilbert-Schmidt kernel and thus the eigenvalue problem

$$K\phi = \lambda\phi$$

has solutions $\lambda_1 \geqq \lambda_2 \geqq \ldots > 0$, listed to their algebraic multiplicity, with associated eigenfunctions $\phi_1, \phi_2, \ldots$, which are orthonormal and complete in $L^2(-\sigma, \sigma)$. Set

$$\psi_k(t) = \frac{1}{\sqrt{2\pi}} \int_{-\sigma}^{\sigma} e^{ixt}\,\phi_k(x)\,dx.$$

Then the $\{\psi_k\}_1^\infty$ are orthonormal in $L^2(-\infty, \infty)$, and $(\psi_j, \psi_k)_s = \lambda_j \delta_{jk}$. The $\{\psi_k\}_1^\infty$ are orthogonal in H_s^T, and complete over the space of which A_σ is the unit ball.

We may now prove, in a manner totally analogous to the method of proof of Theorem 2.2, the following result.

Proposition 2.8. *For all* $n = 0, 1, 2, \ldots,$

$$d_n(A_\sigma; H_s^T) = \sqrt{\lambda_{n+1}},$$

and span $\{\psi_1, \ldots, \psi_n\}$ *is an optimal subspace.*

An application of the methods of this section also proves:

Proposition 2.9. *Let* $T \in K(H_1, H_2)$ *and let* $\{s_n(T)\}_{n=1}^{\infty}$ *denote the s-numbers of* T, *given in non-increasing order of magnitude. For* $q \in [1, \infty]$, *set*

$$B = \{Th \colon \| Th \|_{H_2}^q + \| h \|_{H_1}^q \leqq 1\}.$$

Then

$$d_n(B; H_2) = s_{n+1}(T)/(s_{n+1}^q(T) + 1)^{1/q}.$$

3. n-Widths, with Constraints

We analyze, in this section, a number of variations of the problem considered in the previous section. To fix our notation let $T \in K(H_1, H_2)$ and set

$$A = \{T\phi \colon \| \phi \|_{H_1} \leqq 1\}.$$

We will generally suppress the subscripts H_1 or H_2 on the norms and inner products. For convenience, we will often write $d_n(A)$ in place of $d_n(T(H_1); H_2)$. In what follows we only consider n-widths in the sense of Kolmogorov. This is done for ease of exposition. Statements concerning the linear, Gel'fand and Bernstein n-widths are generally (but not always) of the same form.

3.1 Restricted Approximating Subspaces

In the previous section we determined the n-widths and optimal subspaces of A as a subset of H_2. Let us now consider $B_r = \{T\phi + u \colon \| \phi \| \leqq 1, u \in U_r\}$, where U_r is a fixed r-dimensional subspace of H_2. We will determine the n-widths of B_r in H_2. We prove that finding the n-widths of B_r is equivalent to determining the n-widths of A, for $n \geqq r$, where we restrict ourselves to approximating with n-dimensional subspaces containing U_r. To deal with this latter problem we define a different compact operator T_r, related to T in the following manner.

Let $U_r = \text{span}\{u_1, \ldots, u_r\}$ where, for convenience, we assume that the $\{u_i\}_{i=1}^r$ are orthonormal, i.e., $(u_i, u_j) = \delta_{ij}$, $i, j = 1, \ldots, r$. Let Q_r denote the orthogonal projection of H_2 onto U_r. Thus

$$Q_r \psi = \sum_{i=1}^{r} (\psi, u_i) u_i.$$

Set $T_r = T - Q_r T$ and

$$A_r = \{T_r \phi \colon \|\phi\| \leqq 1\}.$$

Note that $(T - Q_r T)' u_i = 0$, $i = 1, \ldots, r$. The n-widths of A_r are calculable (to the extent to which those of A were calculable).

Theorem 3.1. *For B_r and A_r as above*

$$d_n(B_r) = \begin{cases} \infty, & n < r \\ d_{n-r}(A_r), & n \geqq r. \end{cases}$$

Furthermore, for $n \geqq r$, span $\{u_1, \ldots, u_r, \psi_1, \ldots, \psi_{n-r}\}$ *is an optimal subspace for $d_n(B_r)$ whenever* span $\{\psi_1, \ldots, \psi_{n-r}\}$ *is an optimal subspace for $d_{n-r}(A_r)$.*

Proof. For $n < r$ and any n-dimensional subspace X_n of H_2 there exists a $u \in U_r$ for which $E(u; X_n) \neq 0$. Thus

$$\sup \{E(u; X_n) \colon u \in U_r\} = \infty$$

and $d_n(B_r) = \infty$. Furthermore, by totally analogous reasoning

$$\sup \{E(T\phi + u; X_n) \colon \|\phi\| \leqq 1, u \in U_r\} < \infty$$

if and only if X_n contains the subspace U_r. Thus if $n \geqq r$, then the determination of the n-widths of B_r is equivalent to the determination of the n-widths of A where we restrict ourselves to approximating subspaces containing U_r. In other words

$$d_n(B_r) = \inf_{X_{n-r}} \sup \{E(T\phi; X_{n-r} \cup U_r) \colon \|\phi\| \leqq 1\}.$$

Since we are approximating in a Hilbert space and the $\{u_i\}_{i=1}^r$ are orthonormal,

$$\begin{aligned} E(T\phi; X_{n-r} \cup U_r) &= \inf_{g \in X_{n-r}} \left\| T\phi - \sum_{i=1}^{r} (T\phi, u_i)\, u_i - g \right\| \\ &= E(T_r \phi; X_{n-r}). \end{aligned}$$

Thus $d_n(B_r) = d_{n-r}(A_r)$ for $n \geqq r$, and the theorem follows. $\square$

The following explicit formula for $(T_r' T_r)\phi$ is often useful. Let us here assume that $\{u_1, \ldots, u_r\}$ is a basis for U_r, which is not necessarily orthonormal. Let $G(u_1, \ldots, u_r) = \det((u_i, u_j))_{i,j=1}^r$ denote the Grammian of $u_1, \ldots, u_r$. Then

$$(T_r' T_r)\phi = \frac{\begin{vmatrix} T'T\phi & T'u_1 \ldots T'u_r \\ (T\phi, u_1) & (u_1, u_1) \ldots (u_r, u_1) \\ \vdots & \vdots \qquad \vdots \\ (T\phi, u_r) & (u_1, u_r) \ldots (u_r, u_r) \end{vmatrix}}{G(u_1, \ldots, u_r)}.$$

There exist simple inequalities relating $d_n(B_r)$ and $d_n(A)$. These inequalities are sometimes useful in estimating one of the above quantities on the assumption that the other is known. It should be noted that these inequalities do not depend on the Hilbert space structure nor do they depend on the fact that T is a compact operator.

Proposition 3.2. *For A and B_r as above,*

$$d_{n+r}(B_r) \leqq d_n(A) \leqq d_n(B_r)$$

for all n.

Compare this with Corollary 8.7 of Chapter II.

Simple criteria which guarantee equalities in the above inequalities are easily obtained.

Corollary 3.3. *Let A and B_r be as above. Assume that $n \geqq r$, and X_n is an optimal subspace for $d_n(A)$. If $U_r \subseteq X_n$, then $d_n(B_r) = d_n(A)$.*

Corollary 3.4. *Let A and B_r be as above. Let $\{\psi_i\}_{i=1}^{n+1}$ denote the first $n+1$ eigenvectors of the operator TT'. If $(u, \psi_i) = 0$, $i = 1, \ldots, n+1$ for all $u \in U_r$, then*

$$d_{n+r}(B_r) = d_n(A).$$

Example 3.1. The Sobolev space of functions $W_2^{(r)}[0,1]$ is defined by

$$W_2^{(r)} = W_2^{(r)}[0,1] = \{f\colon f^{(r-1)} \text{ abs. cont., } f^{(r)} \in L^2[0,1]\}.$$

Setting $B_2^{(r)} = \{f\colon f \in W_2^{(r)},\ \|f^{(r)}\|_2 \leqq 1\}$, we wish to calculate $d_n(B_2^{(r)}; L^2)$. This is an example due to Kolmogorov, which was considered is his seminal work [1936], and alluded to in Chapter I.

Every function $f \in W_2^{(r)}$ may be written in the form

$$f(x) = \sum_{i=0}^{r-1} \frac{f^{(i)}(0)}{i!} x^i + \frac{1}{(r-1)!} \int_0^1 (x-y)_+^{r-1} f^{(r)}(y)\, dy,$$

where

$$x_+^k = \begin{cases} x^k, & x \geqq 0 \\ 0, & x < 0. \end{cases}$$

This is simply Taylor's formula with remainder in integral form. Note that $B_2^{(r)}$ is explicitly of the form

$$B_2^{(r)} = \{U_r + Th\colon \|h\|_2 \leqq 1\},$$

where $U_r = \operatorname{span}\{1, \ldots, x^{r-1}\}$ and $Th = \dfrac{1}{(r-1)!} \displaystyle\int_0^1 (x-y)_+^{r-1} h(y)\, dy$. Thus we may apply Theorem 3.1 to obtain the desired n-widths. In this example the eigenvalue problem $T_r' T_r h = \lambda h$ is equivalent to the eigenvalue problem

$$(-1)^r y^{(2r)}(x) = \mu y(x)$$

with boundary conditions

$$y^{(i)}(0) = y^{(i)}(1) = 0, \qquad i = 0, 1, \ldots, r-1,$$

and with the identification $y = T_r' T_r h$, and $\lambda = 1/\mu$. (See Theorem 4.4 of Chapter III.) The following result therefore holds.

Corollary 3.5. *Let $B_2^{(r)}$ be as defined above.*

$$d_n(B_2^{(r)}; L^2) = \begin{cases} \infty, & n < r \\ \lambda_{r,n-r+1}^{1/2}, & n \geqq r, \end{cases}$$

where $\lambda_{r,n-r+1}$ is the $(n-r+1)$st eigenvalue (arranged in decreasing order of magnitude) of the eigenvalue problem

$$\begin{aligned} (-1)^r \lambda y^{(2r)}(x) &= y(x) \\ y^{(i)}(0) = y^{(i)}(1) &= 0, \qquad i = 0, 1, \ldots, r-1. \end{aligned}$$

Furthermore, $X_n = \operatorname{span}\{1, x, \ldots, x^{r-1}, y_{r,1}(x), \ldots, y_{r,n-r}(x)\}$ is an optimal subspace for $d_n(B_2^{(r)}; L^2)$ where $y_{r,k}(x)$ is the eigenfunction associated with $\lambda_{r,k}$ in the above eigenvalue problem.

Example 3.2. Let $H^2 (= H^2(1)$ of Example 2.2) denote the class of power series $\sum_{k=0}^{\infty} a_k z^k$ with $\left\| \sum_{k=0}^{\infty} a_k z^k \right\| = \left(\sum_{k=0}^{\infty} |a_k|^2 \right)^{1/2} < \infty$. Set

$$B_m = \{f \colon f \in H^2, \ \|f^{(m)}\| \leqq 1\}.$$

If $f(z) = \sum_{k=0}^{\infty} a_k z^k$, then

$$f^{(m)}(z) = \sum_{k=m}^{\infty} \frac{k!\, a_k}{(k-m)!} z^{k-m}, \quad \text{and} \quad \|f^{(m)}\| = \left(\sum_{k=m}^{\infty} \left| \frac{k!\, a_k}{(k-m)!} \right|^2 \right)^{1/2}.$$

Proposition 3.6. *For H^2 and B_m as above*

$$d_n(B_m; H^2) = \begin{cases} \infty, & n < m \\ \dfrac{(n-m)!}{n!}, & n \geqq m \end{cases}$$

and $X_n = \operatorname{span}\{1, z, \ldots, z^{n-1}\}$ is an optimal subspace for $n \geqq m$.

(This proposition should also be considered as an example of a generalization of Corollary 2.6 to the case where some of the μ_k (i.e., $\mu_1, \ldots, \mu_m$) are infinite.)

3.2 Restricting the Unit Ball and Optimal Recovery

Let $v_1,\ldots,v_r$ denote r linearly independent functions in H_1, and set

$$C^r = \{T\phi \colon \|\phi\| \leqq 1,\ (\phi, v_i) = 0,\ i = 1,\ldots,r\}.$$

We now determine the n-widths $d_n(C^r)$. Without loss of generality assume that $\{v_i\}_{i=1}^r$ is an orthonormal set of functions in H_1. Let P_r denote the orthogonal projection of H_1 onto $V_r = \operatorname{span}\{v_1,\ldots,v_r\}$. Then

$$P_r\phi = \sum_{i=1}^{r} (\phi, v_i)\, v_i.$$

Set $T^r = T - TP_r$ and $A^r = \{T^r\phi \colon \|\phi\| \leqq 1\}$. Note that $T^r v_i = 0$, $i = 1,\ldots,r$.

Theorem 3.7. *Let C^r be defined as above. Then*

$$d_n(C^r) = d_n(A^r), \qquad n = 0, 1, 2 \ldots.$$

Furthermore X_n is an optimal subspace for $d_n(C^r)$ if and only if it is an optimal subspace for $d_n(A^r)$.

Proof. The proof is an immediate consequence of the fact that the sets $\{\phi \colon \|\phi\| \leqq 1, (\phi, v_i) = 0,\ i = 1,\ldots,r\}$, and $\{\phi - P_r\phi \colon \|\phi\| \leqq 1\}$ are identical since $\|\phi - P_r\phi\| \leqq \|\phi\|$. □

Analogous to Proposition 3.2 and Corollaries 3.3 and 3.4 we have

Proposition 3.8. *For A and C^r as above*

$$d_{n+r}(A) \leqq d_n(C^r) \leqq d_n(A).$$

Furthermore,

(a) *if X_{n+r} is an optimal subspace for $d_{n+r}(A)$ and $T(V_r) \subseteqq X_{n+r}$, then $d_{n+r}(A) = d_n(C^r)$,*

(b) *if $\{\phi_i\}_{i=1}^{n+1}$ are the first $n+1$ eigenvectors of the operator $T'T$, and if $(v, \phi_i) = 0$, $i = 1,\ldots,n+1$ for all $v \in V_r$, then $d_n(A) = d_n(C^r)$.*

There exists a simple duality between the results of Theorems 3.1 and 3.7. To explain, let $U_r = \operatorname{span}\{u_1,\ldots,u_r\}$ be an r-dimensional subspace of H_2. From Theorem 3.1, we have $d_{n+r}(B_r) = d_n(A_r)$, where

$$A_r = \{T_r\phi \colon \|\phi\| \leqq 1\}$$

and $T_r = T - Q_r T$, Q_r being the orthogonal projection onto U_r. Thus the n-width $d_n(A_r)$ is simply the square root of the $(n+1)$st eigenvalue of $T_r' T_r = T'(I - Q_r)T$ or equivalently of $T_r T_r' = (I - Q_r)\, TT'(I - Q_r)$. Moreover the associated eigenvectors $\psi_1,\ldots,\psi_n$ of $T_r T_r'$ span an optimal subspace for $d_n(A_r)$. Let

$\bar{C}^r = \{T'\psi: \|\psi\| \leqq 1,\ (\psi, u_i) = 0,\ i = 1,\ldots,r\}$. From Theorem 3.7, $d_n(\bar{C}^r)$ is obtained by considering the $(n+1)$st eigenvalue of $(S^r)' S^r$ or $S^r(S^r)'$, where $S^r = T' - T'Q_r$. Thus $(S^r)' S^r = (I - Q_r)\, TT'(I - Q_r)\,(= T_r T_r')$, and $S^r(S^r)' = T'(I - Q_r)\, T\,(= T_r' T_r)$. The n-widths $d_n(A_r)$ and $d_n(\bar{C}^r)$ are the same, however the optimal subspaces differ.

The n-widths $d_n(C^r)$ may be applied to obtain an attainable lower bound in a problem of *optimal recovery (estimation)*.

n-Widths, in general, play an important role in the theory of optimal recovery. This work shall not, however, concern itself with the general problem of optimal recovery. The interested reader is referred to the paper of Micchelli and Rivlin [1977] and the book of Traub and Wozniakowski [1980], where the use of n-widths as a theoretical tool in the evaluation of lower bounds for optimal recovery problems is discussed in great detail.

Example 3.3. Let $u_1,\ldots,u_r$ be fixed elements of H_2, and assume, for convenience, that dim span $\{T'u_i\}_{i=1}^r = r$. As above, set

$$A = \{T\phi: \|\phi\| \leqq 1\}.$$

Assume that we are given the values $(T\phi, u_i) = c_i(\phi)$, $i = 1,\ldots,r$, i.e., the values of n linear functionals in H_2. On the basis of these r values we wish to determine an optimal method of recovering the function $T\phi$.

To be more precise, consider any map $S\colon \mathbb{C}^r \to H_2$, whether it be linear or non-linear, continuous or non-continuous, and define

$$E(S) = \sup\{\|T\phi - S(\mathbf{c}(\phi))\|: \|\phi\| \leqq 1,\ \mathbf{c}(\phi) = (c_1(\phi),\ldots,c_r(\phi))\}.$$

The problem of optimal recovery, or optimal estimation, in this setting, is one of determining

$$\inf_S E(S)$$

and finding an S^* which attains the above infimum, if such exists.

The solution to this problem is not at all difficult to ascertain. Let

$$E^* = \sup\{\|T\phi\|: \|\phi\| \leqq 1,\ \mathbf{c}(\phi) = \mathbf{0}\}.$$

We claim that $E^* = \inf_S E(S)$.

Assume that $\|\phi\| \leqq 1$ and $\mathbf{c}(\phi) = \mathbf{0}$. Then

$$\begin{aligned}\|T\phi\| &\leqq \tfrac{1}{2}\|T\phi - S(\mathbf{0})\| + \tfrac{1}{2}\|T\phi + S(\mathbf{0})\| \\ &\leqq \sup\{\|T\phi - S(\mathbf{c}(\phi))\|: \|\phi\| \leqq 1\} \\ &= E(S).\end{aligned}$$

Thus $E^* \leqq \inf_S E(S)$. To prove that E^* is in fact attained, let P_r denote the orthogonal projection of H_1 onto span $\{T'u_i\}_{i=1}^r$. Since

$$TP_r\phi = -\frac{\begin{vmatrix} 0 & TT'u_1 & \dots & TT'u_r \\ (T'u_1,\phi) & (T'u_1,T'u_1) & \dots & (T'u_1,T'u_r) \\ \vdots & \vdots & & \vdots \\ (T'u_r,\phi) & (T'u_r,T'u_1) & \dots & (T'u_r,T'u_r) \end{vmatrix}}{G(T'u_1,\dots,T'u_r)}$$

we see that $TP_r\phi$ is of the form $\tilde{S}(\mathbf{c}(\phi))$ for some admissible $\tilde{S}$. Hence

$$\inf_S E(S) \leqq E(\tilde{S}) = \sup\{\|T\phi - TP_r\phi\| : \|\phi\| \leqq 1\}.$$

Now, $((T - TP_r)\phi, u_i) = 0, i = 1,\dots,r$, since $(I - P_r)T'u_i = 0, i = 1,\dots,r$. In addition, $\|\phi - P_r\phi\| \leqq \|\phi\|$. Thus,

$$\sup\{\|T\phi - TP_r\phi\| : \|\phi\| \leqq 1\} = \sup\{\|T\phi\| : \|\phi\| \leqq 1,\ \mathbf{c}(\phi) = \mathbf{0}\} = E^*,$$

and

$$E^* = \inf_S E(S).$$

The quantity E^* is simply $d_0(C_1^r)$, where

$$C_1^r = \{T\phi : \|\phi\| \leqq 1,\ (\phi, T'u_i) = 0,\ i = 1,\dots,r\}.$$

From Proposition 3.8, $d_r(A)$ is a lower bound on the above quantity independent of the choice of $\{u_i\}_{i=1}^r$, and this lower bound is attained if $TT'u_i \in X_r, i = 1,\dots,r$, where X_r is an optimal subspace for $d_r(A)$. We have therefore proven

Theorem 3.9. *Let $T \in K(H_1,H_2)$, and $c_i(\phi) = (T\phi,u_i)$, $i = 1,\dots,r$, for some fixed $u_1,\dots,u_r \in H_2$. Then*

$$d_r(A) \leqq \inf_S \sup\{\|T\phi - S(\mathbf{c}(\phi))\| : \|\phi\| \leqq 1\},$$

and the infimum is attained if $TT'u_i \in X_r$, $i = 1,\dots,r$, where X_r is an optimal subspace for $d_r(A)$.

Let us now complicate our problem somewhat. Assume that we are given r fixed data of the form $c_i(\phi) = (T\phi,u_i)$, $i = 1,\dots,r$, and that we have the option of choosing n additional arbitrary data, i.e., we choose $v_1,\dots,v_n$ to obtain $d_i(\phi) = (T\phi,v_i)$, $i = 1,\dots,n$, and ask, a priori, for the best choice of $v_1,\dots,v_n$ so as to "get the most information", i.e., minimize the error in the associated optimal recovery scheme. The above analysis leads us to the problem

$$\inf_{v_1,\dots,v_n} \sup\{\|T\phi\| : \|\phi\| \leqq 1,\ \mathbf{c}(\phi) = \mathbf{0},\ \mathbf{d}(\phi) = \mathbf{0}\}.$$

Let P_r, as above, be the orthogonal projection of H_1 onto span$\{T'u_i\}_{i=1}^r$, and $T^r = T - TP_r$. The above quantity is easily seen to equal $d_n(T^r(H_1); H_2)$. This value is attained, for example, by the choice of $v_i = \psi_i^r$, $i = 1, \ldots, n$, where the $\{\psi_i^r\}_{i=1}^n$ are the n eigenvectors of the n largest eigenvalues of $(T^r)(T^r)'$.

3.3 n-Widths Under a Pair of Constraints

As above, let H_2 denote a Hilbert space with norm $\|\cdot\|_2$ and inner product $(\cdot,\cdot)_2$. In place of the H_1 considered above, we assume that H_0 and H_1 are two Hilbert semi-normed spaces, over the same linear space M. Let $\|\cdot\|_0$, $\|\cdot\|_1$ denote the semi-norms, and $(\cdot,\cdot)_0$, $(\cdot,\cdot)_1$ the inner products on H_0, H_1, respectively. $\|\cdot\|_* = \max\{\|\cdot\|_0, \|\cdot\|_1\}$ defines a semi-norm on M and we denote by E this semi-normed linear space.

We shall derive a method of obtaining the n-widths and optimal subspaces for the set

$$D = \{T\phi\colon \|\phi\|_i \leqq 1,\ i = 0, 1\}$$

where T is a continuous linear operator from E into H_2, i.e., $T \in L(E; H_2)$.

For each

$$0 \leqq \lambda \leqq 1,$$

set

$$\|\cdot\|_\lambda = (\lambda \|\cdot\|_0^2 + (1-\lambda)\|\cdot\|_1^2)^{1/2}.$$

$\|\cdot\|_\lambda$ defines a semi-norm on M induced by the inner product $(\cdot,\cdot)_\lambda = \lambda(\cdot,\cdot)_0 + (1-\lambda)(\cdot,\cdot)_1$. Let H_λ denote the associated Hilbert semi-normed space, and set

$$D_\lambda = \{T\phi\colon \|\phi\|_\lambda \leqq 1\}.$$

Theorem 3.10.

$$d_n(D) = \inf_{0 \leqq \lambda \leqq 1} d_n(D_\lambda).$$

Moreover, if $d_n(D) = d_n(D_\mu)$ and X_n is an optimal subspace for D_μ, then it is also an optimal subspace for D.

Before proving Theorem 3.10, we need some ancillary results.

Let $\mathbf{y} = (y_1, y_2) \in \mathbb{R}^2$ and as usual set $\|\mathbf{y}\|_\infty = \max\{|y_1|, |y_2|\}$, $\|\mathbf{y}\|_1 = |y_1| + |y_2|$, and $\mathbf{x}\cdot\mathbf{y} = x_1 y_1 + x_2 y_2$, for $\mathbf{x}, \mathbf{y} \in \mathbb{R}^2$. Also for any set $A \subseteq \mathbb{R}^2$, co(A) denotes the convex hull of A and $\partial(\text{co}(A))$ the boundary thereof. Then,

Lemma 3.11. *If A is a subset of $\mathbb{R}^2$ for which $\partial(\text{co}(A)) \subseteq A$, and $\mathbf{0} \notin$ interior $(\text{co}(A))$, then*

$$\inf_{\mathbf{a}\in A} \|\mathbf{a}\|_\infty = \max_{\|\boldsymbol{\lambda}\|_1 \leqq 1} \inf_{\mathbf{a}\in A} \boldsymbol{\lambda}\cdot\mathbf{a}.$$

Proof. Since $\partial(\mathrm{co}(A)) \subseteq A$ and $\mathbf{0} \notin$ interior $(\mathrm{co}(A))$, it follows that

$$\inf_{\mathbf{a}\in A} \|\mathbf{a}\|_\infty \geqq \inf_{\mathbf{a}\in \mathrm{co}(A)} \|\mathbf{a}\|_\infty = \inf_{\mathbf{a}\in \partial(\mathrm{co}(A))} \|\mathbf{a}\|_\infty \geqq \inf_{\mathbf{a}\in A} \|\mathbf{a}\|_\infty$$

while $\inf_{\mathbf{a}\in A} \boldsymbol{\lambda}\cdot\mathbf{a} = \inf_{\mathbf{a}\in \mathrm{co}(A)} \boldsymbol{\lambda}\cdot\mathbf{a}$. It therefore suffices to prove that

$$\inf_{\mathbf{a}\in \mathrm{co}(A)} \|\mathbf{a}\|_\infty = \max_{\|\boldsymbol{\lambda}\|_1\leqq 1} \inf_{\mathbf{a}\in \mathrm{co}(A)} \boldsymbol{\lambda}\cdot\mathbf{a}.$$

The proof of this fact is a simple consequence of the existence of separating hyperplanes. Let $\delta = \inf_{\mathbf{a}\in \mathrm{co}(A)} \|\mathbf{a}\|_\infty$. Obviously, $\max_{\|\boldsymbol{\lambda}\|_1\leqq 1} \inf_{\mathbf{a}\in \mathrm{co}(A)} \boldsymbol{\lambda}\cdot\mathbf{a} \leqq \delta$. We wish to prove the existence of a $\boldsymbol{\lambda}_0$ for which $\|\boldsymbol{\lambda}_0\|_1 = 1$ and $\boldsymbol{\lambda}_0\cdot\mathbf{a} \geqq \delta$ for all $\mathbf{a}\in \mathrm{co}(A)$. If $\delta = 0$, there is nothing to prove. Assume $\delta > 0$. For $0 < \varepsilon < \delta$, the set $B_\varepsilon = \{\mathbf{a}\colon \|\mathbf{a}\|_\infty \leqq \delta - \varepsilon\}$ is convex and disjoint from $\mathrm{co}(A)$. Hence there exists a $\boldsymbol{\lambda}_\varepsilon$, $\|\boldsymbol{\lambda}_\varepsilon\|_1 = 1$ and a constant b_ε such that $\boldsymbol{\lambda}_\varepsilon\cdot\mathbf{a} \leqq b_\varepsilon$ for all $\mathbf{a}\in B_\varepsilon$ and $\boldsymbol{\lambda}_\varepsilon\cdot\mathbf{a} \geqq b_\varepsilon$ for all $\mathbf{a}\in \mathrm{co}(A)$. Furthermore it is easily seen that $b_\varepsilon \geqq \delta - \varepsilon$. Letting $\varepsilon\to 0^+$, through a subsequence if necessary, we obtain the desired result. □

How do we determine when $\partial(\mathrm{co}(A)) \subseteq A$ for a given set $A \subseteq \mathbb{R}^2$? An equivalent formulation of the above property will prove helpful.

Lemma 3.12. *Let A be a closed subset of $\mathbb{R}^2$. Then $\partial(\mathrm{co}(A)) \subseteq A$ if and only if whenever a supporting hyperplane to A touches A at two points then the line segment joining these two points lies in A.*

Lemmas 3.11 and 3.12 are used to prove the following result.

Proposition 3.13. *Let T, H_0, H_1, and H_2, be as defined earlier. Then*

$$\sup_{\substack{\|\phi\|_i\leqq 1\\ i=0,1}} \|T\phi\|_2 = \min_{0\leqq\tau\leqq 1} \sup_{\|\phi\|_\tau\leqq 1} \|T\phi\|_2.$$

Proof. Set

$$C = \{(\|\phi\|_0^2, \|\phi\|_1^2)\colon \|T\phi\|_2 = 1\}.$$

Assume that $\partial(\mathrm{co}(C)) \subseteq C$. Because $\mathbf{0} \notin \mathrm{int}(\mathrm{co}(C))$, we obtain from Lemma 3.11

$$\inf_{\mathbf{a}\in C} \|\mathbf{a}\|_\infty = \max_{\|\boldsymbol{\lambda}\|_1\leqq 1} \inf_{\mathbf{a}\in C} \boldsymbol{\lambda}\cdot\mathbf{a}.$$

Since C lies in the first quadrant, the maximum is taken by some $\|\bar{\boldsymbol{\lambda}}\|_1 = 1$ for which $\bar{\lambda}_1, \bar{\lambda}_2 \geqq 0$. We now reciprocate both sides of the above equality to obtain the desired result. It thus suffices to prove that $\partial(\mathrm{co}(C)) \subseteq C$. We shall apply Lemma 3.12 to prove this result.

Let $\mathbf{a}(\phi) = (\|\phi\|_0^2, \|\phi\|_1^2)$, and suppose $\mathbf{a}(\phi), \mathbf{a}(\psi) \in C$. For each $\mu \in \mathbb{R}$ define $\mathbf{a}(\mu) = \mathbf{a}(\rho(\mu))$, where

$$\rho(\mu) = \frac{\mu\phi + (1-\mu)\psi}{\|T(\mu\phi + (1-\mu)\psi)\|_2}.$$

Since $\| T\phi \|_2 = \| T\psi \|_2 = 1$, we have $|(T\phi, T\psi)_2| \leqq 1$, and

$$\| T(\mu\phi + (1-\mu)\psi) \|_2 = 0$$

if and only if $\mu = 1/2$ and $(T\phi, T\psi)_2 = -1$. We first assume that $(T\phi, T\psi)_2 \neq -1$. In this case $\mathbf{a}(\mu)$ is a continuous curve lying in C joining $\mathbf{a}(\phi)$ to $\mathbf{a}(\psi)$. Now, let $\boldsymbol{\lambda} \cdot \mathbf{a} \geqq b$ be a half-space supporting C and touching it at $\mathbf{a}(\phi)$ and $\mathbf{a}(\psi)$. Thus $\| T(\mu\phi + (1-\mu)\psi) \|_2^2 (\boldsymbol{\lambda} \cdot \mathbf{a}(\mu) - b)$ is a quadratic polynomial in μ which is non-negative for all $\mu \in \mathbb{R}$ and which vanishes at $\mu = 0, 1$. Hence it is identically zero and $\boldsymbol{\lambda} \cdot \mathbf{a}(\mu) = b$ for all $\mu \in \mathbb{R}$. $\mathbf{a}(\mu)$, $0 \leqq \mu \leqq 1$, is the required line segment joining $\mathbf{a}(\phi)$ to $\mathbf{a}(\psi)$.

If $(T\phi, T\psi)_2 = -1$, then we replace ϕ by $-\phi$. Since $\mathbf{a}(-\phi) = \mathbf{a}(\phi)$ and $(T(-\phi), T\psi)_2 = 1$, the proof follows as above. □

We can now prove Theorem 3.10.

Proof of Theorem 3.10. Let X_n be an n-dimensional subspace of H_2 and let P_n be the orthonormal projection of H_2 onto X_n. Set $T^n = T - P_n T$ (T^n depending of course on X_n). Thus

$$\| T^n \phi \|_2 = \inf_{y \in X_n} \| T\phi - y \|_2 .$$

From Proposition 3.13 we have

$$\sup_{\substack{\|\phi\|_i \leqq 1 \\ i=0,1}} \| T^n \phi \|_2 = \min_{0 \leqq \lambda \leqq 1} \sup_{\|\phi\|_\lambda \leqq 1} \| T^n \phi \|_2 .$$

Therefore

$$\begin{aligned} d_n(D) &= \inf_{X_n} \sup_{\substack{\|\phi\|_i \leqq 1 \\ i=0,1}} \| T^n \phi \|_2 \\ &= \inf_{X_n} \min_{0 \leqq \lambda \leqq 1} \sup_{\|\phi\|_\lambda \leqq 1} \| T^n \phi \|_2 \\ &= \inf_{0 \leqq \lambda \leqq 1} \inf_{X_n} \sup_{\|\phi\|_\lambda \leqq 1} \| T^n \phi \|_2 \\ &= \inf_{0 \leqq \lambda \leqq 1} d_n(D_\lambda) . \end{aligned}$$

Now, assume that

$$d_n(D_\mu) = \sup_{\|\phi\|_\mu \leqq 1} \| T^n \phi \| = d_n(D)$$

for some particular n-dimensional space X_n. Hence

$$\sup_{\|\phi\|_\mu \leqq 1} \| T^n \phi \|_2 \leqq \sup_{\|\phi\|_\lambda \leqq 1} \| T^n \phi \|_2$$

for all $\lambda \in [0,1]$. Thus

$$d_n(D) = \sup_{\|\phi\|_\mu \leqq 1} \|T^n\phi\|_2$$

$$= \min_{0 \leqq \lambda \leqq 1} \sup_{\|\phi\|_\lambda \leqq 1} \|T^n\phi\|_2$$

$$= \sup_{\substack{\|\phi\|_i \leqq 1 \\ i=0,1}} \|T^n\phi\|_2,$$

whence X_n is an optimal subspace for D. □

Example 3.4. *Approximation of smooth functions of bounded extensions.*

Set

$$H_2^1 = H_2^1(\mathbb{R}) = L^2(\mathbb{R}) \cap W_2^{(1)}(\mathbb{R}),$$

where as previously

$$W_2^{(1)}(\mathbb{R}) = \left\{ f\colon f \text{ abs. cont. on } \mathbb{R},\ \int_{-\infty}^{\infty} |f'(t)|^2\,dt < \infty \right\}.$$

Let T, ε be fixed positive numbers and set

$$\mathscr{S}_\varepsilon^T = \left\{ f \in H_2^1\colon \int_{|t|>T} |f(t)|^2\,dt \leqq \varepsilon^2,\ \int_{-\infty}^{\infty} |f'(t)|^2\,dt \leqq 1 \right\}.$$

We shall compute the n-widths of $\mathscr{S}_\varepsilon^T$ as a subset of $L^2(-T,T)$.

Note that

$$d_0(\mathscr{S}_\varepsilon^T; L^2(-T,T)) = \sup_{f \in \mathscr{S}_\varepsilon^T} \left(\int_{-T}^{T} |f(t)|^2\,dt \right)^{1/2}$$

is the "concentration problem" for the class $\mathscr{S}_\varepsilon^T$. That is, it tells us the degree to which a smooth function, small outside $(-T,T)$, can be concentrated in $(-T,T)$. Fairly exact upper and lower bounds for $d_n(\mathscr{S}_\varepsilon^T; L^2(-T,T))$ may be easily obtained. However, these bounds are independent of ε.

Let

$$B_2^{(1)}(-T,T) = \left\{ f \in W_2^{(1)}(-T,T)\colon \int_{-T}^{T} |f'(t)|^2\,dt \leqq 1 \right\}.$$

Since $\mathscr{S}_\varepsilon^T \subseteqq B_2^{(1)}(-T,T)$, it follows that

$$d_n(\mathscr{S}_\varepsilon^T; L^2(-T,T)) \leqq d_n(B_2^{(1)}(-T,T); L^2(-T,T)).$$

As may be deduced from Corollary 3.5 (the eigenvalue problem is easily solved in this case),

$$d_n(B_2^{(1)}(-T,T); L^2(-T,T)) = \frac{2T}{n\pi}, \qquad n = 1,2\ldots.$$

Thus an upper bound is obtained for $d_n(\mathscr{S}_\varepsilon^T; L^2(-T, T))$.

Now, let

$$f_j(t) = \begin{cases} \sin j \dfrac{\pi(t+T)}{2T}, & |t| < T \\ 0, & |t| \geqq T \end{cases}$$

and set

$$M_{n+1} = \left\{ \sum_{j=1}^{n+1} a_j f_j(t) \colon \left(\int_{-T}^{T} \left| \sum_{j=1}^{n+1} a_j f_j(t) \right|^2 dt \right)^{1/2} \leqq \frac{2T}{(n+1)\pi} \right\}.$$

It follows from Proposition 1.6 of Chapter II that if $M_{n+1} \subseteqq \mathscr{S}_\varepsilon^T$, then

$$d_n(\mathscr{S}_\varepsilon^T; L^2(-T, T)) \geqq \frac{2T}{(n+1)\pi}.$$

To prove that $M_{n+1} \subseteqq \mathscr{S}_\varepsilon^T$, it suffices to show that

$$\int_{-\infty}^{\infty} \left| \sum_{j=1}^{n+1} a_j f_j'(t) \right|^2 dt \leqq 1$$

for every $\sum_{j=1}^{n+1} a_j f_j(t) \in M_{n+1}$.

Since

$$\int_{-\infty}^{\infty} \left| \sum_{j=1}^{n+1} a_j f_j(t) \right|^2 dt = \sum_{j=1}^{n+1} |a_j|^2 T$$

and

$$\int_{-\infty}^{\infty} \left| \sum_{j=1}^{n+1} a_j f_j'(t) \right|^2 dt = \sum_{j=1}^{n+1} |a_j|^2 \frac{|j\pi|^2}{4T} \leqq \frac{(n+1)^2}{4T} \pi^2 \sum_{j=1}^{n+1} |a_j|^2,$$

it follows that $M_{n+1} \subseteqq \mathscr{S}_\varepsilon^T$. Thus

$$\frac{2T}{(n+1)\pi} \leqq d_n(\mathscr{S}_\varepsilon^T; L^2(-T, T)) \leqq \frac{2T}{n\pi}, \qquad n = 0, 1, 2 \ldots.$$

Let us now determine the exact value of $d_n(\mathscr{S}_\varepsilon^T; L^2(-T, T))$. From Theorems 3.10 and 2.2 we are led to a consideration of the eigenvalue problem

$$\chi_T f = \mu((\lambda/\varepsilon^2)(1 - \chi_T) f - (1 - \lambda) f'')$$

where χ_T is the characteristic function of the interval $(-T, T)$.

The eigenfunctions of this problem are given by

$$f(t) = \begin{cases} e^{-\sqrt{\delta}(t-T)}, & t \geqq T, \\ \dfrac{\cos \sqrt{\rho}\, t}{\cos \sqrt{\rho}\, T}, & 0 \leqq t \leqq T, \end{cases} \qquad f(t) = f(-t),$$

and by

$$g(t) = \begin{cases} e^{-\sqrt{\delta}(t-T)}, & t \geqq T, \\ \dfrac{\sin \sqrt{\rho t}}{\sin \sqrt{\rho T}}, & 0 \leqq t \leqq T, \end{cases} \qquad g(t) = -g(-t),$$

where $\delta = \lambda/\varepsilon^2(1-\lambda)$ and $\rho = 1/\mu(1-\lambda)$. The eigenvalues for the eigenfunctions of the form f are determined by the equation

$$(\mu\lambda/\varepsilon^2)^{1/2} = \tan(T/(\mu(1-\lambda))^{1/2}) \tag{1}$$

while those corresponding to the eigenfunctions of the form g are determined by the equation

$$(\mu\lambda/\varepsilon^2)^{1/2} = -\cot(T/(\mu(1-\lambda))^{1/2}). \tag{2}$$

Equations (1) and (2) insure that f' and g' are continuous at T.

Let $x_1(\lambda) > \ldots > x_n(\lambda) > \ldots$ denote the positive roots of equation (1), and let $y_1(\lambda) > \ldots > y_n(\lambda) > \ldots$ denote the positive roots of equation (2). Let $z_1(t) < \ldots < z_n(t) < \ldots$ denote the positive roots of

$$z \tan z = t, \ (n-1)\pi < z_n(t) < (n-(1/2))\pi,$$

and let $w_1(t) < \ldots < w_n(t) < \ldots$ denote the positive roots of

$$-w \cot w = t, \ (n-(1/2))\pi < w_n(t) < n\pi.$$

Then

$$x_n^{1/2}(\lambda) = (T^2 + \varepsilon^2 t^2)^{1/2}/z_n(t), \quad \text{and} \quad y_n^{1/2}(\lambda) = (T^2 + \varepsilon^2 t^2)^{1/2}/w_n(t),$$

where $t = T\lambda^{1/2}/\varepsilon(1-\lambda)^{1/2}$. It follows that

$$d_{2n}(\mathscr{S}_\varepsilon^T; L^2(-T, T)) = \min_{0 \leqq \lambda \leqq 1} x_{n+1}^{1/2}(\lambda), \qquad n = 0, 1, \ldots,$$

and

$$d_{2n+1}(\mathscr{S}_\varepsilon^T; L^2(-T, T)) = \min_{0 \leqq \lambda \leqq 1} y_{n+1}^{1/2}(\lambda), \qquad n = 0, 1, \ldots.$$

Minimizing $x_n(\lambda)$ and $y_n(\lambda)$ over $\lambda \in [0, 1]$, we obtain

$$d_{2n}(\mathscr{S}_\varepsilon^T; L^2(-T, T)) = (T^2 + \varepsilon^2 \sigma_{n+1}^2)^{1/2}/z_{n+1},$$

and

$$d_{2n+1}(\mathscr{S}_\varepsilon^T; L^2(-T, T)) = (T^2 + \varepsilon^2 \tau_{n+1}^2)^{1/2}/w_{n+1},$$

where (σ_n, z_n) and (τ_n, w_n) are uniquely determined by the equations

$$z_n \tan z_n = \sigma_n, \ (n-1)\pi < z_n < (n-(1/2))\pi$$
$$\sigma_n^3 + \sigma_n z_n^2 = (T/\varepsilon)^2,$$

and

$$-w_n \cot w_n = \tau_n, \quad (n-(1/2))\pi < w_n < n\pi$$
$$\tau_n^3 + \tau_n w_n^2 = (T/\varepsilon)^2.$$

The following example may be deduced as a consequence of Theorems 2.5 and 3.10. However, we prefer to compute its n-widths directly.

Example 3.5. *n-Widths of the Space of Essentially Time-Limited and Band-Limited Signals.*

As in Example 2.3, let $L^2(\mathbb{R})$ denote the space of complex-valued square integrable functions on $\mathbb{R}$ with the usual L^2-norm. In problems of information theory $f(t)$ is often known as the "signal" and $\|f\|$ as its "total power". A "signal" $f(t)$ is said to be strictly "time-limited" to $(-T, T)$ if $f(t)$ vanishes outside $(-T, T)$. It is said to be strictly "band-limited" to $(-\sigma, \sigma)$ if its Fourier transform $\hat{f}(x)$ vanishes outside $(-\sigma, \sigma)$. Signals are for all practical physical purposes strictly time-limited and band-limited. However, mathematically it may be proven that only the zero function in $L^2(\mathbb{R})$ is both time-limited and band-limited. One is then, from a mathematical point of view, naturally led to the question of determining the "size" of the set of functions which are "essentially" time- and band-limited.

Let

$$\mathscr{D}_T = \mathscr{D} = \{f\colon f \in L^2(\mathbb{R}),\ f(t) = 0,\ |t| \geqq T\}$$

and

$$\mathscr{B}_\sigma = \mathscr{B} = \{f\colon f \in L^2(\mathbb{R}),\ \hat{f}(x) = 0,\ |x| \geqq \sigma\}.$$

We define D and B as the projections of $L^2(\mathbb{R})$ onto $\mathscr{D}$ and $\mathscr{B}$, respectively, given by

$$Df(t) = \begin{cases} f(t), & |t| < T \\ 0, & |t| \geqq T, \end{cases}$$

and

$$Bf(t) = \int_{-\infty}^{\infty} f(s) \frac{\sin \sigma(t-s)}{\pi(t-s)}\, ds,$$

i.e.,

$$(\widehat{Bf})(x) = \begin{cases} \hat{f}(x), & |x| < \sigma \\ 0, & |x| \geqq \sigma. \end{cases}$$

Now, set

$$G = \{f\colon f \in L^2(\mathbb{R}),\ \|(I-D)f\| \leqq \varepsilon,\ \|(I-B)f\| \leqq \eta\}.$$

G may be regarded as the class of functions time-limited to $\mathbb{R}\backslash(-T, T)$ at level ε, and band-limited to $\mathbb{R}\backslash(-\sigma, \sigma)$ at level η. We shall compute the n-widths of G as a subset of $L^2(\mathbb{R})$.

To compute these quantities we use certain known facts (see e.g. Slepian and Pollack [1961]).

Lemma 3.14. *Then space $\mathscr{D} + \mathscr{B}$ is closed. Any $f \in L^2(\mathbb{R})$ may be written in the form $f = d + b + g$, where $d \in \mathscr{D}$, $b \in \mathscr{B}$, and $Dg = Bg = 0$.*

Lemma 3.15. *BD is a self-adjoint, non-negative, compact operator given by*

$$BDf(t) = \int_{-T}^{T} f(s) \frac{\sin \sigma(t-s)}{\pi(t-s)} ds.$$

Let $\{\lambda_i\}_{i=1}^{\infty}$ *and* $\{\psi_i\}_{i=1}^{\infty}$ *denote the eigenvalues listed in decreasing order, to their multiplicity, and their associated orthonormal eigenfunctions. Then*

(i) $1 > \lambda_1 > \lambda_2 > \ldots > \lambda_n \to 0$ *as* $n \uparrow \infty$
(ii) *the* $\{\psi_i\}_1^{\infty}$ *is a basis for* $\mathscr{B}$
(iii) $\left\{\frac{1}{\sqrt{\lambda_i}} D\psi_i\right\}_1^{\infty}$ *is an orthonormal basis for* $\mathscr{D}$.

In what follows set $\phi_i = \frac{1}{\sqrt{\lambda_i}} D\psi_i$.

Lemma 3.16. *Every* $f \in L^2(\mathbb{R})$ *has the representation*

$$f(t) = \sum_{i=1}^{\infty} [d_i \phi_i(t) + b_i \psi_i(t)] + g(t)$$

where $Dg = Bg = 0$. *Furthermore, if* $f \in G$, *i.e.,* $\|(I-D)f\| \leqq \varepsilon$, $\|(I-B)f\| \leqq \eta$, *then*

$$\sum_{i=1}^{\infty} |b_i|^2 (1-\lambda_i) + \|g\|^2 \leqq \varepsilon^2, \quad \sum_{i=1}^{\infty} |d_i|^2 (1-\lambda_i) + \|g\|^2 \leqq \eta^2.$$

Proof. The first part of the lemma is clear. Let us prove that if $f \in G$, then $\sum_{i=1}^{\infty} |b_i|^2 (1-\lambda_i) + \|g\|^2 \leqq \varepsilon^2$. The second inequality follows in an analogous fashion.

$$(f - Df)(t) = \sum_{i=1}^{\infty} d_i \phi_i(t) + \sum_{i=1}^{\infty} b_i \psi_i(t) + g(t) - \sum_{i=1}^{\infty} d_i D\phi_i(t)$$
$$- \sum_{i=1}^{\infty} b_i D\psi_i(t) - Dg(t).$$

Now $D\phi_i(t) = \phi_i(t)$, $D\psi_i(t) = \sqrt{\lambda_i}\, \phi_i(t)$, and $Dg(t) = 0$. Thus

$$(f - Df)(t) = \sum_{i=1}^{\infty} b_i [\psi_i(t) - \sqrt{\lambda_i}\, \phi_i(t)] + g(t).$$

The result now follows since $(\phi_i, g) = (\psi_i, g) = 0$, $i = 1, 2, \ldots$ for $Dg = Bg = 0$, and $(\phi_i, \psi_j) = \sqrt{\lambda_i}\, \delta_{ij}$, $i, j = 1, 2, \ldots$. □

We now prove the main result.

Theorem 3.17.

$$d_n(G; L^2(\mathbb{R})) = \left[\frac{\varepsilon^2 + \eta^2 + 2\varepsilon\eta\sqrt{\lambda_{n+1}}}{1 - \lambda_{n+1}}\right]^{1/2}, \qquad n = 0, 1, 2, \ldots.$$

Furthermore, every set of the form $\{\alpha_i \phi_i(t) + \beta_i \psi_i(t)\}_{i=1}^n$, $|\alpha_i|^2 + |\beta_i|^2 > 0$, $\operatorname{Re} \alpha_i \bar{\beta}_i \geqq 0$, *spans an optimal subspace for* $d_n(G; L^2(\mathbb{R}))$.

Proof. We first prove the upper bound. Let $\zeta_i(t) = \alpha_i \phi_i(t) + \beta_i \psi_i(t)$, $i = 1, \ldots, n$, where $|\alpha_i|^2 + |\beta_i|^2 > 0$, and $\operatorname{Re} \alpha_i \bar{\beta}_i \geqq 0$. Let P_n denote the orthogonal projection of $L^2(\mathbb{R})$ onto $\operatorname{span}\{\zeta_1, \ldots, \zeta_n\}$. We shall make use of the fact that $(\phi_i, \phi_j) = (\psi_i, \psi_j) = \delta_{ij}$, $(\phi_i, \psi_j) = \sqrt{\lambda_i}\, \delta_{ij}$, $i, j = 1, 2, \ldots$, and $(\phi_i, g) = (\psi_i, g) = 0$, $i = 1, 2, \ldots$. The above equations imply that the ζ_i are orthogonal, whence

$$P_n f = \sum_{i=1}^n \frac{(f, \zeta_i)}{(\zeta_i, \zeta_i)} \zeta_i.$$

We wish to determine an upper bound on $\| f - P_n f \|$ for $f \in G$. Thus by Lemma 3.16, let us assume that

$$f(t) = \sum_{i=1}^\infty (d_i \phi_i(t) + b_i \psi_i(t)) + g(t),$$

and $\sum_{i=1}^\infty |b_i|^2 (1 - \lambda_i) + \| g \|^2 \leqq \varepsilon^2$, $\sum_{i=1}^\infty |d_i|^2 (1 - \lambda_i) + \| g \|^2 \leqq \eta^2$.

We first obtain an upper bound on

$$\left\| d_i \phi_i + b_i \psi_i - \frac{(f, \zeta_i)}{(\zeta_i, \zeta_i)} \zeta_i \right\|^2, \qquad i = 1, 2, \ldots, n,$$

for f of the above form. Because of the various orthogonality relationships, it suffices to consider the above quantity for $f(t) = d_i \phi_i(t) + b_i \psi_i(t)$. Thus

$$\begin{aligned} &\left\| d_i \phi_i + b_i \psi_i - \frac{(f, \zeta_i)}{(\zeta_i, \zeta_i)} \zeta_i \right\|^2 \\ &\quad = |d_i|^2 + |b_i|^2 + 2\sqrt{\lambda_i}\, \operatorname{Re} b_i \bar{d}_i - \frac{|(f, \zeta_i)|^2}{(\zeta_i, \zeta_i)} \\ &\quad = \left[|b_i|^2 + |d_i|^2 - \frac{|\bar{\alpha}_i d_i + \bar{\beta}_i b_i|^2 + 2\sqrt{\lambda_i}(|b_i|^2 + |d_i|^2)\operatorname{Re} \alpha_i \bar{\beta}_i}{|\alpha_i|^2 + |\beta_i|^2 + 2\sqrt{\lambda_i}\, \operatorname{Re} \alpha_i \bar{\beta}_i} \right] (1 - \lambda_i). \end{aligned}$$

Since $\operatorname{Re} \alpha_i \bar{\beta}_i \geqq 0$, we obtain

$$\left\| d_i \phi_i + b_i \psi_i - \frac{(f, \zeta_i)}{(\zeta_i, \zeta_i)} \zeta_i \right\|^2 \leqq (|b_i|^2 + |d_i|^2)(1 - \lambda_i).$$

Thus,

$$\begin{aligned} \| f - P_n f \|^2 &= \left\| \sum_{i=1}^\infty (d_i \phi_i + b_i \psi_i + g) - \sum_{i=1}^n \frac{(d_i \phi_i + b_i \psi_i, \zeta_i)}{(\zeta_i, \zeta_i)} \zeta_i \right\|^2 \\ &= \sum_{i=1}^n \left\| d_i \phi_i + b_i \psi_i - \frac{(d_i \phi_i + b_i \psi_i, \zeta_i)}{(\zeta_i, \zeta_i)} \zeta_i \right\|^2 \end{aligned}$$

$$+\sum_{i=n+1}^{\infty}\|d_i\phi_i+b_i\psi_i\|^2+\|g\|^2$$

$$\leqq \sum_{i=1}^{n}(|b_i|^2+|d_i|^2)(1-\lambda_i)$$

$$+\sum_{i=n+1}^{\infty}(|b_i|^2+|d_i|^2+2\sqrt{\lambda_i}\operatorname{Re} b_i\bar{d}_i)+\|g\|^2.$$

Since $1>\lambda_1>\ldots>\lambda_{n+1}\downarrow 0$, it follows that

$$\sum_{i=1}^{n}|b_i|^2(1-\lambda_i)+\sum_{i=n+1}^{\infty}|b_i|^2\leqq\sum_{i=1}^{\infty}|b_i|^2\frac{(1-\lambda_i)}{(1-\lambda_{n+1})},$$

and similarly for the $|d_i|$, and that for $i\geqq n+1$,

$$\sqrt{\lambda_i}\operatorname{Re} b_i\bar{d}_i\leqq\sqrt{\lambda_{n+1}}\,|b_i d_i|\frac{(1-\lambda_i)}{(1-\lambda_{n+1})}.$$

Thus

$$\|f-P_n f\|^2\leqq\frac{\sum_{i=1}^{\infty}|b_i|^2(1-\lambda_i)+\|g\|^2+\sum_{i=1}^{\infty}|d_i|^2(1-\lambda_i)+\|g\|^2}{(1-\lambda_{n+1})}$$

$$+\frac{\sum_{i=1}^{\infty}2\sqrt{\lambda_{n+1}}\,|b_i|\,|d_i|\,(1-\lambda_i)}{(1-\lambda_{n+1})}.$$

Since $f\in G$, we obtain from Lemma 3.16

$$\|f-P_n f\|^2\leqq\frac{\varepsilon^2+\eta^2+2\sqrt{\lambda_{n+1}}\,\varepsilon\eta}{(1-\lambda_{n+1})}.$$

Thus

$$d_n(G;L^2(\mathbb{R}))\leqq\left[\frac{\varepsilon^2+\eta^2+2\sqrt{\lambda_{n+1}}\,\varepsilon\eta}{1-\lambda_{n+1}}\right]^{1/2},\qquad n=0,1,2,\ldots.$$

To prove the lower bound, we shall apply the standard techniques of Proposition 1.6 of Chapter II. Set

$$f_i(t)=\frac{\eta\phi_i(t)+\varepsilon\psi_i(t)}{\sqrt{1-\lambda_i}},\qquad i=1,2,\ldots,$$

and

$$M_{n+1}=\left\{f\colon f(t)=\sum_{i=1}^{n+1}a_i f_i(t),\ \|f\|^2\leqq\frac{\varepsilon^2+\eta^2+2\sqrt{\lambda_{n+1}}\,\varepsilon\eta}{(1-\lambda_{n+1})}\right\}.$$

To prove the lower bound it suffices to show that $M_{n+1}\subseteqq G$.

The $\{f_i\}_{i=1}^{n+1}$ are orthogonal, as are the $\{(I-D)\,f_i\}_{i=1}^{n+1}$ and $\{(I-B)\,f_i\}_{i=1}^{n+1}$. Because $\|(I-D)\,f_i\| = \varepsilon$, $\|(I-B)\,f_i\| = \eta$, it follows that $M_{n+1} \subseteqq G$ if and only if $\sum_{i=1}^{n+1} |a_i|^2 \leqq 1$. A straightforward calculation gives

$$\|f_i\|^2 = \frac{\varepsilon^2 + \eta^2 + 2\sqrt{\lambda_i}\,\varepsilon\eta}{1-\lambda_i} \geqq \frac{\varepsilon^2 + \eta^2 + 2\sqrt{\lambda_{n+1}}\,\varepsilon\eta}{1-\lambda_{n+1}}, \qquad i = 1,\dots,n+1,$$

whence $\sum_{i=1}^{n+1} |a_i|^2 \leqq 1$, and the theorem is proved. □

In Theorem 3.17, G is approximated on the whole real line. One might well be interested in obtaining an approximation only on $(-T, T)$. We will now consider G as a subset of $L^2(-T, T)$.

Theorem 3.18.

$$d_n(G;\, L^2(-T, T)) = \frac{\eta + \varepsilon\sqrt{\lambda_{n+1}}}{\sqrt{1-\lambda_{n+1}}},$$

and $\{\phi_i\}_{i=1}^{n}$ *spans an optimal subspace for* $d_n(G;\, L^2(-T, T))$, $n = 0, 1, 2, \dots$.

Remark. Note that in both Theorems 3.17 and 3.18, the n-widths do *not* tend to zero. This follows from the fact that G is not a compact set.

Proof. From Lemma 3.16, every $f \in G$ has the representation

$$f(t) = \sum_{i=1}^{\infty} (d_i \phi_i(t) + b_i \psi_i(t)) + g(t),$$

where $\sum_{i=1}^{\infty} |b_i|^2(1-\lambda_i) + \|g\|^2 \leqq \varepsilon^2$ and $\sum_{i=1}^{\infty} |d_i|^2(1-\lambda_i) + \|g\|^2 \leqq \eta^2$. Furthermore, on $(-T, T)$, $f(t) = \sum_{i=1}^{\infty} (d_i + b_i\sqrt{\lambda_i})\,\phi_i(t)$ since $Dg = 0$, and $D\psi_i = \sqrt{\lambda_i}\,\phi_i$. Note also that $D\phi_i = \phi_i$ which implies that the support of ϕ_i is contained in $[-T, T]$.

Let P_n denote the orthogonal projection of G onto $\operatorname{span}\{\phi_1, \dots, \phi_n\}$. Thus,

$$\begin{aligned}
\|f - P_n f\|^2_{L^2(-T,T)} &= \left\| \sum_{i=1}^{\infty} (d_i + \sqrt{\lambda_i}\, b_i)\,\phi_i - \sum_{i=1}^{n} (d_i + \sqrt{\lambda_i}\, b_i)\,\phi_i \right\|^2_{L^2(-T,T)} \\
&= \sum_{i=n+1}^{\infty} |d_i + \sqrt{\lambda_i}\, b_i|^2 \\
&= \sum_{i=n+1}^{\infty} (|d_i|^2 + \lambda_i |b_i|^2 + 2\sqrt{\lambda_i}\,\operatorname{Re} b_i \bar{d}_i) \\
&\leqq \sum_{i=n+1}^{\infty} \frac{|d_i|^2(1-\lambda_i)}{(1-\lambda_{n+1})} + \lambda_{n+1} \sum_{i=n+1}^{\infty} \frac{|b_i|^2(1-\lambda_i)}{(1-\lambda_{n+1})}
\end{aligned}$$

$$+ 2\sqrt{\lambda_{n+1}}\left[\sum_{i=n+1}^{\infty} \frac{|b_i|^2(1-\lambda_i)}{(1-\lambda_{n+1})}\right]^{1/2}\left[\sum_{i=n+1}^{\infty} \frac{|d_i|^2(1-\lambda_i)}{(1-\lambda_{n+1})}\right]^{1/2}$$

$$\leqq \frac{\eta^2 + \lambda_{n+1}\varepsilon^2 + 2\sqrt{\lambda_{n+1}}\,\eta\varepsilon}{(1-\lambda_{n+1})} = \frac{(\eta + \sqrt{\lambda_{n+1}}\,\varepsilon)^2}{(1-\lambda_{n+1})}.$$

Hence $d_n(G; L^2(-T, T)) \leqq \dfrac{\eta + \sqrt{\lambda_{n+1}}\,\varepsilon}{\sqrt{1-\lambda_{n+1}}}$.

The proof of the lower bound follows the argument given in proving the lower bound in Theorem 3.17.

Let

$$f_i(t) = \frac{\eta\phi_i(t) + \varepsilon\sqrt{\lambda_i}\,\psi_i(t)}{\sqrt{1-\lambda_i}}$$

and set

$$M_{n+1} = \left\{f\colon f(t) = \sum_{i=1}^{n+1} a_i f_i(t),\ \| f \|_{L^2(-T,T)} \leqq \frac{\eta + \varepsilon\lambda_{n+1}}{\sqrt{1-\lambda_{n+1}}}\right\}.$$

It suffices to prove that $M_{n+1} \subseteqq G$. Since for $f \in M_{n+1}$

$$\| f \|^2_{L^2(-T,T)} = \sum_{i=1}^{n+1} |a_i|^2 \frac{(\eta + \varepsilon\lambda_i)^2}{1-\lambda_i} \leqq \frac{(\eta + \varepsilon\lambda_{n+1})^2}{1-\lambda_{n+1}},$$

and $1 > \lambda_1 > \ldots > \lambda_{n+1} > 0$ which implies $\dfrac{(\eta + \varepsilon\lambda_i)^2}{1-\lambda_i} \geqq \dfrac{(\eta + \varepsilon\lambda_{n+1})^2}{1-\lambda_{n+1}}$, $i = 1, \ldots, n+1$, it follows that $\sum_{i=1}^{n+1} |a_i|^2 \leqq 1$. Hence for $f \in M_{n+1}$

$$\|(I-B) f\|^2 = \sum_{i=1}^{n+1} |a_i|^2 \eta^2 \leqq \eta^2$$

and

$$\|(I-D) f\|^2 = \sum_{i=1}^{n+1} |a_i|^2 \varepsilon^2 \lambda_i \leqq \varepsilon^2.$$

Thus $M_{n+1} \subseteqq G$ and the theorem is proved. □

Theorem 3.18 has certain odd features to it. We considered there the n-widths of the class of functins of the form $\sum_{i=1}^{\infty} (d_i + \sqrt{\lambda_i}\, b_i)\,\phi_i$ with separate restrictions on the $\{d_i\}_1^\infty$ and $\{b_i\}_1^\infty$. To give a certain perspective to this type of result, let us consider the following problem:

Let H be a Hilbert space and $\{\phi_i\}_1^\infty$ an orthonormal basis for H. Let $\varepsilon_1, \ldots, \varepsilon_k$ be fixed non-negative numbers and assume that we are given k sequences of non-increasing positive numbers $\{\lambda_i^j\}_{i=1}^\infty$, $j = 1, \ldots, k$, i.e., $\lambda_1^j \geqq \lambda_2^j \geqq \ldots \geqq 0$, $j = 1, \ldots, k$. Set $G = \Big\{f\colon f = \sum_{i=1}^{\infty} a_i\phi_i,\ a_i \in \mathbb{C}$, there exist $\{b_i^j\}_{i=1}^\infty, j = 1, \ldots, k$ in $\mathbb{C}$

for which $a_i = \sum_{j=1}^{k} b_i^j$, and $\sum_{i=1}^{\infty} \frac{|b_i^j|^2}{\lambda_i^j} \leqq \varepsilon_j^2, j = 1,\ldots,k\Big\}$. G is the class of functions in H for which there exists a decomposition of the coefficients so as to satisfy the above inequalities.

Proposition 3.19. *For H and G as above,*

$$d_n(G;H) = \sum_{j=1}^{k} \sqrt{\lambda_{n+1}^j}\,\varepsilon_j, \qquad n = 0,1,2,3,\ldots.$$

Furthermore, the functions $\{\phi_i\}_{i=1}^n$ span an optimal subspace for $d_n(G;H)$.

Proof. Let P_n denote the orthogonal projection onto span $\{\phi_1,\ldots,\phi_n\}$, and let $f \in G$. Thus we assume that $f = \sum_{i=1}^{\infty} a_i\phi_i$, and $a_i = \sum_{j=1}^{k} b_i^j$, $j = 1,2,3,\ldots$, where b_i^j satisfy the required inequalities. Therefore,

$$\begin{aligned}\|f - P_n f\|_H &= \left(\sum_{i=n+1}^{\infty} |a_i|^2\right)^{1/2} = \left(\sum_{i=n+1}^{\infty} \Big|\sum_{j=1}^{k} b_i^j\Big|^2\right)^{1/2} \\ &\leqq \left(\sum_{i=n+1}^{\infty}\left(\sum_{j=1}^{k} |b_i^j|\right)^2\right)^{1/2} \leqq \left[\sum_{i=n+1}^{\infty}\left[\sum_{j=1}^{k} \frac{|b_i^j|\sqrt{\lambda_{n+1}^j}}{\sqrt{\lambda_i^j}}\right]^2\right]^{1/2} \\ &\leqq \sum_{j=1}^{k} \sqrt{\lambda_{n+1}^j}\left[\sum_{i=n+1}^{\infty} \frac{|b_i^j|^2}{\lambda_i^j}\right]^{1/2} \leqq \sum_{j=1}^{k} \sqrt{\lambda_{n+1}^j}\,\varepsilon_j.\end{aligned}$$

In the above series of inequalities we have used the fact that $\lambda_{n+1}^j \geqq \lambda_i^j > 0$, $i \geqq n+1$, and the inequality

$$\left(\sum_{i=n+1}^{\infty}\left(\sum_{j=1}^{k} |c_{ij} d_j|\right)^2\right)^{1/2} \leqq \sum_{j=1}^{k} |d_j| \left(\sum_{i=n+1}^{\infty} |c_{ij}|^2\right)^{1/2}$$

(see Hardy, Littlewood and Pólya [1952, p. 32]).

The lower bound argument is the usual one given in this chapter. Set $f_i(t) = \left(\sum_{j=1}^{k} \sqrt{\lambda_i^j}\,\varepsilon_j\right)\phi_i(t)$, and $M_{n+1} = \left\{f\colon f = \sum_{i=1}^{n+1} a_i f_i,\ \|f\|_H \leqq \sum_{j=1}^{k} \sqrt{\lambda_{n+1}^j}\,\varepsilon_j\right\}$. We prove that $M_{n+1} \subseteq G$. Choose $f \in M_{n+1}$ and recall that $\lambda_i^j \geqq \lambda_{n+1}^j$, $i = 1,\ldots,n+1$, $j = 1,\ldots,k$. Thus

$$\begin{aligned}\sum_{j=1}^{k} \sqrt{\lambda_{n+1}^j}\,\varepsilon_j \geqq \|f\|_H &= \left(\sum_{i=1}^{n+1} |a_i|^2 \left(\sum_{j=1}^{k} \sqrt{\lambda_i^j}\,\varepsilon_j\right)^2\right)^{1/2} \\ &\geqq \left(\sum_{i=1}^{n+1} |a_i|^2 \left(\sum_{j=1}^{k} \sqrt{\lambda_{n+1}^j}\,\varepsilon_j\right)^2\right)^{1/2} \\ &= \left(\sum_{i=1}^{n+1} |a_i|^2\right)^{1/2} \left(\sum_{j=1}^{k} \sqrt{\lambda_{n+1}^j}\,\varepsilon_j\right),\end{aligned}$$

which implies that $\sum_{i=1}^{n+1} |a_i|^2 \leqq 1$. From this fact it follows that $f \in G$, since

$$f(t) = \sum_{i=1}^{n+1} a_i f_i(t) = \sum_{i=1}^{n+1} a_i \left(\sum_{j=1}^{k} \sqrt{\lambda_i^j}\, \varepsilon_j \right) \phi_i(t) = \sum_{i=1}^{n+1} \left(\sum_{j=1}^{k} a_i \sqrt{\lambda_i^j}\, \varepsilon_j \right) \phi_i(t),$$

and

$$\sum_{i=1}^{n+1} \frac{|a_i|^2 \lambda_i^j \varepsilon_j^2}{\lambda_i^j} = \sum_{i=1}^{n+1} |a_i|^2 \varepsilon_j^2 \leqq \varepsilon_j^2, \quad j = 1, \ldots, k. \quad \square$$

3.4 A Theorem of Ismagilov

In this next example we obtain upper and lower bounds for the n-widths of a general class of functions. These upper and lower bounds differ. In the next section we return to this result and obtain, for a specific class of functions, equality therein.

Set

$$\mathscr{K}_1 = \left\{ f \colon f(x) = \int_0^1 K(x, y)\, h(y)\, dy, \ \|h\|_1 \leqq 1 \right\}$$

where $\|h\|_1 = \int_0^1 |h(y)|\, dy$ is the usual $L^1[0,1]$ norm, and $K(x, y) \in C([0,1] \times [0,1])$. The operator $K'K$, (i.e., the operator induced by the kernel $K'K(x, y) = \int_0^1 K(z, x)\, \overline{K(z, y)}\, dz$) as an operator from $L^2[0,1]$ into $L^2[0,1]$ is compact, self-adjoint, and non-negative, and has eigenvalues $\lambda_1 \geqq \lambda_2 \geqq \ldots \geqq 0$, and associated orthonormal eigenfunctions ϕ_i, $i = 1, 2, \ldots$.

We are interested in determining the n-widths of the set $\mathscr{K}_1$ as a subset of $L^2[0,1]$. As has been noted, see Theorem 1.1 of Chapter II, $d_n(\mathscr{K}_1; L^2) = d_n(\overline{\mathscr{K}}_1; L^2)$, where $\overline{\mathscr{K}}_1$ is the closure of $\mathscr{K}_1$ as a subset of L^2. $\overline{\mathscr{K}}_1$ is easily determined and is given by

$$\overline{\mathscr{K}}_1 = \left\{ f \colon f(x) = \int_0^1 K(x, y)\, d\mu(y), \ \|\mu\|_{T.V.} \leqq 1 \right\}$$

where $\|\mu\|_{T.V.}$ denotes the total variation of the Borel measure $d\mu$.

Theorem 3.20. *For $\mathscr{K}_1$ and $\overline{\mathscr{K}}_1$ as above,*

$$\left(\sum_{j=n+1}^{\infty} \lambda_j \right)^{1/2} \leqq d_n(\mathscr{K}_1; L^2) = d_n(\overline{\mathscr{K}}_1; L^2) \leqq \max_{0 \leqq x \leqq 1} \left(\sum_{j=n+1}^{\infty} \lambda_j |\phi_j(x)|^2 \right)^{1/2}.$$

Proof. Let us first prove the lower bound. Set $\psi_k(x) = \int_0^1 K(x, y)\, \phi_k(y)\, dy$, $k = 1, 2, \ldots$. Thus $(\psi_k, \psi_j) = \lambda_k \delta_{kj}$, $k, j = 1, 2, \ldots$, and the $\{\psi_k\}_{k=1}^{\infty}$ are eigenfunctions of the operator induced by the kernel $KK'(x, y)$. We shall assume, without loss of generality, that $\{\psi_k / \sqrt{\lambda_k}\}_{k=1}^{\infty}$ is a complete orthonormal basis for

$L^2[0,1]$. Let X_n be any n-dimensional subspace of $L^2[0,1]$ and let $\{g_1,\ldots,g_n\}$ be an orthonormal basis for X_n. Complete $\{g_k\}_{k=1}^n$ to $\{g_k\}_{k=1}^\infty$ so that the latter is a complete orthonormal basis for $L^2[0,1]$. Thus $g_k(x) = \sum_{j=1}^\infty \frac{c_{kj}\,\psi_j(x)}{\sqrt{\lambda_j}}$, $k=1,2,\ldots$, where $(c_{kj})_{k,j=1}^\infty$ is a *unitary* matrix. Set

$$E(y_0) = \inf_{g\in X_n} \|K(\cdot, y_0) - g(\cdot)\|.$$

Thus

$$\begin{aligned} E^2(y_0) &= \sum_{k=n+1}^\infty |(K(\cdot, y_0), g_k(\cdot))|^2 \\ &= \sum_{k=n+1}^\infty \left|\sum_{j=1}^\infty \bar{c}_{kj}\left(K(\cdot, y_0), \frac{\psi_j(\cdot)}{\sqrt{\lambda_j}}\right)\right|^2 \\ &= \sum_{k=n+1}^\infty \left|\sum_{j=1}^\infty \bar{c}_{kj}\sqrt{\lambda_j}\,\phi_j(y_0)\right|^2 \end{aligned}$$

and therefore

$$\int_0^1 E^2(y)\,dy = \sum_{k=n+1}^\infty \sum_{j=1}^\infty |c_{kj}|^2\lambda_j,$$

because the $\{\phi_j\}_{j=1}^\infty$ are orthonormal and $(c_{kj})_{k,j=1}^\infty$ is a unitary matrix. Define $q_j = \sum_{k=1}^n |c_{kj}|^2$. Since $\sum_{k=1}^\infty |c_{kj}|^2 = 1$, $j=1,2,\ldots$,

$$\int_0^1 E^2(y)\,dy = \sum_{j=1}^\infty \lambda_j[1-q_j].$$

Now, $0 \leqq q_j \leqq 1$, $\sum_{j=1}^\infty q_j = n$, and $\lambda_1 \geqq \lambda_2 \geqq \ldots > 0$. Therefore, $\sum_{j=1}^\infty \lambda_j[1-q_j] \geqq \sum_{j=n+1}^\infty \lambda_j$, and it follows that

$$\int_0^1 E^2(y)\,dy \geqq \sum_{j=n+1}^\infty \lambda_j,$$

independent of X_n.

Since the extreme points of the set $\{d\mu\colon \|\mu\|_{T.V.} \leqq 1\}$ are simply the point measures with mass one we obtain that for any given X_n,

$$\begin{aligned} &\sup_{\|\mu\|_{T.V.}\leqq 1} \inf_{g\in X_n} \left\|\int_0^1 K(\cdot, y)\,d\mu(y) - g(\cdot)\right\| \\ &= \sup_{0\leqq y\leqq 1} \inf_{g\in X_n} \|K(\cdot, y) - g(\cdot)\| = \sup_{0\leqq y\leqq 1} E(y) \\ &\geqq \left(\int_0^1 E^2(y)\,dy\right)^{1/2} \geqq \left(\sum_{j=n+1}^\infty \lambda_j\right)^{1/2}. \end{aligned}$$

Therefore $d_n(\mathscr{K}_1; L^2) \geqq \left(\sum_{j=n+1}^\infty \lambda_j\right)^{1/2}$.

To prove the upper bound choose $X_n = \text{span}\{\psi_1, \ldots, \psi_n\}$. Then

$$d_n(\mathscr{K}_1; L^2) \leqq \sup_{0 \leqq y \leqq 1} \inf_{g \in X_n} \| K(\cdot, y) - g(\cdot) \|$$
$$= \max_{0 \leqq y \leqq 1} \left[\sum_{k=n+1}^{\infty} \left| \left(K(\cdot, y), \frac{\psi_k(\cdot)}{\sqrt{\lambda_k}} \right) \right|^2 \right]^{1/2}$$
$$= \max_{0 \leqq y \leqq 1} \left(\sum_{k=n+1}^{\infty} \lambda_k \, |\phi_k(y)|^2 \right)^{1/2}. \quad \square$$

Remark. The above theorem does not hold for the Gel'fand and Bernstein n-widths.

4. n-Widths of Compact Periodic Convolution Operators

4.1 n-Widths as Fourier Coefficients

There is one particular class of compact operators whose s-numbers are easily calculable. Let $k(x)$ be a real, square-integrable function on $[0, 2\pi)$. We shall consider $k(x)$ as a 2π-periodic function by the convention that $k(x + 2\pi) = k(x)$ for all x. The (convolution) operator

$$T_k \phi(x) = (k * \phi)(x) = (1/2\pi) \int_0^{2\pi} k(x - y) \, \phi(y) \, dy$$

is a Hilbert-Schmidt operator and hence is compact. The s-numbers of T_k are computed as follows:

Associated with $k(x)$ is its formal Fourier series

$$k(x) \sim \sum_{n=-\infty}^{\infty} a_n e^{inx},$$

where $a_n = \frac{1}{2\pi} \int_0^{2\pi} k(x) \, e^{-inx} \, dx$, $n = 0, \pm 1, \pm 2, \ldots$. Since k is real, $a_{-n} = \bar{a}_n$, $n = 0, \pm 1, \pm 2, \ldots$. The self-adjoint, non-negative, compact operator $T_k' T_k$ is given by

$$T_k' T_k \, \phi(x) = \frac{1}{2\pi} \int_0^{2\pi} k'k(x - y) \, \phi(y) \, dy,$$

where $k'k(x) = \frac{1}{2\pi} \int_0^{2\pi} k(z) \, k(z + x) \, dz$. The Fourier series of $k'k(x)$ is $\sum_{n=-\infty}^{\infty} |a_n|^2 \, e^{inx}$. The following are easily shown to be the eigenvalues and eigenfunctions (eigenvectors) of $T_k' T_k$. The function $\phi_1(x) \equiv 1$ is an eigenfunction with eigenvalue $|a_0|^2$. The functions $\phi_{2n}(x) = e^{inx}$ and $\phi_{2n+1}(x) = e^{-inx}$ are eigenfunctions with corresponding eigenvalue $|a_n|^2$, $n = 1, 2, \ldots$. Thus the n-widths and associated optimal subspaces are easily obtained in this case.

Remark. It should be noted that the above analysis also extends to the case of a multivariate function. Assume that $k(\mathbf{x})$ is a real, square-integrable function on $Q = [0, 2\pi)^N$ (periodically extended in each variable), where $\mathbf{x} = (x_1, \ldots, x_N)$ and consider the map

$$T_k \phi(\mathbf{x}) = \frac{1}{(2\pi)^N} \int_Q \ldots \int k(\mathbf{x} - \mathbf{y}) \phi(\mathbf{y}) \, d\mathbf{y}.$$

If $k(\mathbf{x}) \sim \sum_{n_1=-\infty}^{\infty} \cdots \sum_{n_N=-\infty}^{\infty} a_{\mathbf{n}} e^{i\mathbf{n}\cdot\mathbf{x}}$, where $\mathbf{n} = (n_1, \ldots, n_N)$ and $\mathbf{n} \cdot \mathbf{x} = n_1 x_1 + \ldots + n_N x_N$, then $a_{-\mathbf{n}} = \bar{a}_{\mathbf{n}}$ and the analysis of the eigenvalue-eigenfunction problem is identical to the above.

To ease our notation, and because of the importance of the class, consider $k(x)$ as above whose Fourier coefficients satisfy $|a_n| \geqq |a_{n+1}|$, $n = 0, 1, 2, \ldots$. For ease of exposition we fix our notation and set

$$\tilde{\mathscr{K}}_p = \{(k * \phi)(x) \colon \|\phi\|_p \leqq 1\},$$

where $\|\phi\|_p = \left(\frac{1}{2\pi} \int_0^{2\pi} |\phi(y)|^p \, dy\right)^{1/p}$, $1 \leqq p < \infty$, and $\|\phi\|_\infty = \operatorname{ess\,sup}\{|\phi(y)| \colon 0 \leqq y \leqq 2\pi\}$.

Theorem 4.1. *For k as above, $d_0(\tilde{\mathscr{K}}_2; L^2) = |a_0|$, and $d_{2n-1}(\tilde{\mathscr{K}}_2; L^2) = d_{2n}(\tilde{\mathscr{K}}_2; L^2) = |a_n|$, $n = 1, 2, \ldots$. Furthermore, $T_{n-1} = \operatorname{span}\{1, \sin x, \cos x, \ldots, \sin(n-1)x, \cos(n-1)x\}$ is an optimal $(2n-1)$-dimensional subspace for $d_{2n-1}(\tilde{\mathscr{K}}_2; L^2)$ (and hence also optimal for $d_{2n}(\tilde{\mathscr{K}}_2; L^2)$).*

One other important class is the following:

Let $G(x)$ be a real, square-integrable function on $[0, 2\pi)$. We consider $G(x)$ as a 2π-periodic function by the convention that $G(x + 2\pi) = G(x)$ for all x. Set

$$\tilde{\mathscr{B}}_p = \{a + (G * \phi)(x) \colon \|\phi\|_p \leqq 1, \ \phi \perp 1, \ a \in \mathbb{R}\}$$

where $\phi \perp 1$ means that $\int_0^{2\pi} \phi(y) \, dy = 0$. Associated with G is its formal Fourier series

$$G(x) \sim \sum_{n=-\infty}^{\infty} a_n e^{inx}.$$

In all that follows, we assume that $|a_n| \geqq |a_{n+1}|$, $n = 1, 2, \ldots$. (Note that there is here no condition on a_0.)

From Theorem 2.2 and the results of Sections 3.1 and 3.2, we have

Theorem 4.2. *For G as above, $d_0(\tilde{\mathscr{B}}_2; L^2) = \infty$, and $d_{2n-1}(\tilde{\mathscr{B}}_2; L^2) = d_{2n}(\tilde{\mathscr{B}}_2; L^2) = |a_n|$, $n = 1, 2, \ldots$. Furthermore, T_{n-1} is an optimal $(2n-1)$-dimensional subspace for $d_{2n-1}(\tilde{\mathscr{B}}_2; L^2)$ (and hence also for $d_{2n}(\tilde{\mathscr{B}}_2; L^2)$).*

We now apply Theorem 4.2 to the example originally considered by Kolmogorov, and discussed in Chapter I.

Example 4.1. The Sobolev space of functions $\tilde{W}_p^{(r)}$ is defined by

$$\tilde{W}_p^{(r)} = \tilde{W}_p^{(r)}[0, 2\pi] = \{f : f \ 2\pi\text{-periodic}, f^{(r-1)} \text{ abs. cont.}, f^{(r)} \in L^p[0, 2\pi)\}.$$

Set $\tilde{B}_p^{(r)} = \{f : f \in \tilde{W}_p^{(r)}, \ \|f^{(r)}\|_p \leq 1\}$, and let us calculate $d_n(\tilde{B}_2^{(r)}; L^2)$.

Every $f \in \tilde{W}_2^{(r)}$ may be written in the form

$$f(x) = a_0 + \frac{1}{\pi} \int_0^{2\pi} D_r(x-t) f^{(r)}(t)\, dt,$$

with the stipulation that $\|f^{(r)}\|_2 < \infty$, $\int_0^{2\pi} f^{(r)}(t)\, dt = 0$, (i.e., $f^{(r)} \perp 1$) and $a_0 = 1/2\pi \int_0^{2\pi} f(t)\, dt$.

The function (kernel) D_r is known both as the Dirichlet kernel (this is not the usual function referred to as the Dirichlet kernel) and as the Bernouilli monospline. The function D_r may be written in the form

$$D_r(x) = \sum_{k=1}^{\infty} \frac{\cos(kx - (r\pi/2))}{k^r}, \qquad r = 1, 2, \ldots.$$

However it is not difficult to establish that $D_r(x)$ is, for $x \in [0, 2\pi)$, an algebraic polynomial of degree r. Essentially (there is a problem of uniform convergence of the above infinite series for $r = 1$), if we set $D_0(x) = -1/2$, then $D_r'(x) = D_{r-1}(x)$, $D_r(x) = (-1)^r D_r(2\pi - x)$, $r = 1, 2, \ldots$, and $D_r(0) = 0$, $r = 3, 5, 7, \ldots$. Thus D_r on the interval $[0, 2\pi)$ is what is generally referred to as the Bernouilli polynomial B_r, suitably normalized. D_r is taken as the 2π-periodic extension of B_r and is therefore known as the Bernouilli monospline. This latter form shall prove useful (see Section 6). For our present purposes, however, the infinite series is ideal since we immediately obtain the Fourier coefficients of D_r. The absolute value of the nth Fourier coefficient of $2D_r(x)$ (the constant 2 being introduced because of the form of $f(x)$) is n^{-r}, $n = 1, 2, \ldots$.

Corollary 4.3. *For $\tilde{B}_2^{(r)}$ as above,*

$$d_0(\tilde{B}_2^{(r)}; L^2) = \infty,$$
$$d_{2n-1}(\tilde{B}_2^{(r)}; L^2) = d_{2n}(\tilde{B}_2^{(r)}; L^2) = n^{-r}, \qquad n \geqq 1.$$

Furthermore, T_{n-1} is an optimal $(2n-1)$-dimensional subspace for $d_{2n-1}(\tilde{B}_2^{(r)}; L^2)$ (and thus also for $d_{2n}(\tilde{B}_2^{(r)}; L^2)$).

An application of Proposition 3.19 gives us the following:

Let $\{k_j\}_{j=1}^m$ be real, 2π-periodic, square integrable functions with Fourier series

$$k_j(x) \sim \sum_{n=-\infty}^{\infty} a_n^j e^{inx},$$

where $|a_n^j| \geqq |a_{n+1}^j|$, $n = 0, 1, 2, \ldots; j = 1, \ldots, m$. Set

$$\mathscr{C} = \left\{ \sum_{j=1}^{m} (k_j * \phi_j)(x) \colon \|\phi_j\|_2 \leqq \varepsilon_j,\ j = 1, \ldots, m \right\}.$$

Then

Proposition 4.4. *For $\mathscr{C}$ as above, $d_0(\mathscr{C}; L^2) = \sum_{j=1}^{m} |a_0^j| \, \varepsilon_j$, and*

$$d_{2n-1}(\mathscr{C}; L^2) = d_{2n}(\mathscr{C}; L^2) = \sum_{j=1}^{m} |a_n^j| \, \varepsilon_j, \ n = 1, 2, \ldots.$$

Furthermore, T_{n-1} is an optimal $(2n-1)$-dimensional subspace for $d_{2n-1}(\mathscr{C}; L^2)$ (and hence for $d_{2n}(\mathscr{C}; L^2)$).

A similar result holds where G_j replaces k_j, a free constant term is introduced, and the ϕ_j are restricted to be orthogonal to the constant function (see the definition of $\tilde{\mathscr{B}}_2$).

4.2 A Return to Ismagilov's Theorem

Consider once again the problem discussed in Section 3.4 (Theorem 3.20). Assume that $k(x)$ is continuous and 2π-periodic, and $|a_n| \geqq |a_{n+1}|$, $n = 0, 1, 2, \ldots$. The closure of $\tilde{\mathscr{K}}_1$ as a subset of $L^2[0, 2\pi)$ is

$$\bar{\tilde{\mathscr{K}}}_1 = \left\{ \frac{1}{2\pi} \int_0^{2\pi} k(x-y)\, d\mu(y) \colon \|\mu\|_{T.V.} \leqq 1 \right\},$$

where $\|\mu\|_{T.V.}$ is the total variation of a regular Borel measure $d\mu$ (suitably normalized because of the factor $1/2\pi$ in the L^1-norm). Since the orthonormal eigenfunctions $\phi_i(x)$ of $T_k' T_k$ may be chosen so that $|\phi_i(x)| = 1$ for all $x \in [0, 2\pi)$ and all i we obtain, as a consequence of Theorem 3.20,

Proposition 4.5. *For $\bar{\tilde{\mathscr{K}}}_1$ as above,*

$$d_{2n-1}(\bar{\tilde{\mathscr{K}}}_1; L^2) = \Big(\sum_{|k| \geqq n} |a_k|^2\Big)^{1/2}$$

and

$$d_{2n}(\bar{\tilde{\mathscr{K}}}_1; L^2) = \Big(\sum_{|k| \geqq n+1} |a_k|^2 + |a_n|^2\Big)^{1/2}.$$

Furthermore, T_{n-1} is an optimal subspace for $d_{2n-1}(\bar{\tilde{\mathscr{K}}}_1; L^2)$, and span$\{T_{n-1}$, $\cos(nx+\alpha)\}$ *is an optimal subspace for $d_{2n}(\bar{\tilde{\mathscr{K}}}_1; L^2)$ for any real α.*

Note that unlike the result of Theorem 4.1, $d_{2n-1} \neq d_{2n}$.

Remark. It easily follows (see Proposition 5.2 of Chapter II) that $\delta_n(\bar{\tilde{\mathscr{K}}}_1; L^2) = d_n(\bar{\tilde{\mathscr{K}}}_1; L^2)$ for all n. However, in general the Gel'fand n-width $d^n(\bar{\tilde{\mathscr{K}}}_1; L^2)$ is strictly less than this quantity.

Example 4.2. Let

$$P(r,\theta) = \frac{1-r^2}{1-2r\cos\theta + r^2}$$

be the Poisson kernel and set

$$u(z) = u(re^{i\theta}) = \frac{1}{2\pi}\int_0^{2\pi} P(r,\theta-t)\, d\mu(t)$$

where $\mu(t)$ is a real measure of bounded variation on $[0, 2\pi)$. Each such function u is a harmonic function in $|z| < 1$. The above integral is known as the Poisson-Stieltjes integral and the class of Poisson-Stieltjes integrals is identical with the class h^1 of real-valued harmonic functions in $|z| < 1$ for which

$$M_1(r,u) = \frac{1}{2\pi}\int_0^{2\pi} |u(re^{i\theta})|\, d\theta$$

remains bounded as $r \uparrow 1$ (see Duren [1970; p. 2]). The Poisson kernel may be rewritten as

$$P(r,\theta) = 1 + 2\sum_{k=1}^{\infty} r^k \cos k\theta = \sum_{k=-\infty}^{\infty} r^{|k|} e^{ik\theta},$$

a form more suitable to our purposes. Set

$$P_r = \left\{u(re^{i\theta}) \colon u(re^{i\theta}) = \frac{1}{2\pi}\int_0^{2\pi} P(r,\theta-t)\, d\mu(t),\ \|\mu\|_{T.V.} \leqq 1\right\}.$$

Assume that we wish to approximate such functions in L^2 on a circle of radius r, $0 < r < 1$, and in this manner $d_n(P_r; L^2)$ should be understood. From Proposition 4.5 we obtain

Proposition 4.6. *For P_r as above, r fixed, $0 < r < 1$,*

$$d_{2n}(P_r; L^2) = r^n \left(\frac{1+r^2}{1-r^2}\right)^{1/2}, \quad d_{2n+1}(P_r; L^2) = r^{n+1}\left(\frac{2}{1-r^2}\right)^{1/2}, \qquad n = 0, 1, 2, \ldots$$

and T_n is an optimal subspace for $d_{2n+1}(P_r; L^2)$, while span $\{T_{n-1}, \cos(n\theta+\alpha)\}$ *is an optimal subspace for $d_{2n}(P_r; L^2)$, for any real α.*

Note that these subspaces are independent of r.

Remark. We have shown that $d_{2n-1}(\tilde{\mathscr{K}}_2; L^2) = d_{2n}(\tilde{\mathscr{K}}_2; L^2) = |a_n|$, while the quantities $d_{2n-1}(\tilde{\mathscr{K}}_1; L^2)$ and $d_{2n}(\tilde{\mathscr{K}}_1; L^2)$ are given in Proposition 4.5. What are the n-widths of $d_n(\tilde{\mathscr{K}}_p; L^2)$ when p lies between 1 and 2? One might suppose, on the basis of the above results, that we would obtain

$$d_{2n-1}(\tilde{\mathscr{K}}_p; L^2) = \left(\sum_{|k|\geqq n}^{\infty} |a_k|^q\right)^{1/2} \quad \text{and} \quad d_{2n}(\tilde{\mathscr{K}}_p; L^2) = \left(\sum_{|k|\geqq n+1}^{\infty} |a_k|^q + |a_n|^q\right)^{1/2},$$

where $1/q = 1/p - 1/2$. This is in agreement with our known results for $p = 1, 2$.

Unfortunately, we are unable to prove the above equalities. (It is probable that equality does not generally hold.) However, it is true that the above quantities are upper bounds on the n-widths. We prove that

$$d_{2n-1}(\tilde{\mathcal{K}}_p; L^2) \leqq \Big(\sum_{|k| \geqq n} |a_k|^q\Big)^{1/q}$$

for $p \in (1,2)$, $1/q = 1/p - 1/2$. The upper bound for $d_{2n}(\tilde{\mathcal{K}}_p; L^2)$ is similarly proven.

The key to the proof of the upper bound is the Hausdorff-Young inequality which says that if $1 \leqq p \leqq 2$, and $b_k = \frac{1}{2\pi} \int_0^{2\pi} \phi(y)\, e^{-iky}\, dy$, $k = 0, \pm 1, \pm 2, \ldots$, then

$$\left(\sum_{k=-\infty}^{\infty} |b_k|^{p'}\right)^{1/p'} \leqq \|\phi\|_p,$$

where $1/p + 1/p' = 1$. Now,

$$d_{2n-1}(\tilde{\mathcal{K}}_p; L^2) \leqq \sup_{\|\phi\|_p \leqq 1} \inf_{g \in T_{n-1}} \left\| \frac{1}{2\pi} \int_0^{2\pi} k(\cdot - y)\, \phi(y)\, dy - g(\cdot) \right\|_2$$

$$= \sup_{\|\phi\|_p \leqq 1} \Big(\sum_{|k| \geqq n} |a_k b_k|^2\Big)^{1/2},$$

where the a_k and b_k are the Fourier coefficients of k and ϕ, respectively. Since $1/q = 1/p - 1/2 = 1/2 - 1/p'$, we have by Hölder's inequality

$$\Big(\sum_{|k| \geqq n} |a_k b_k|^2\Big)^{1/2} \leqq \Big(\sum_{|k| \geqq n} |a_k|^q\Big)^{1/q} \Big(\sum_{|k| \geqq n} |b_k|^{p'}\Big)^{1/p'}.$$

The result now follows from the Hausdorff-Young inequality.

Let us now consider $d_{2n-1}(\tilde{\mathcal{B}}_1; L^2)$. (The evaluation of $d_{2n}(\tilde{\mathcal{B}}_1; L^2)$ may be similarly treated.) Unfortunately, the orthogonality condition $\phi \perp 1$ presents problems. The following is a result of the method of proof of Theorem 3.20.

Theorem 4.7. *For $\tilde{\mathcal{B}}_1$ as defined in Section* 4.1,

$$\sup_{0 \leqq y \leqq 2\pi} \left(\sum_{\substack{k=1 \\ k \neq i_1(y), \ldots, i_{n-1}(y)}}^{\infty} |a_k|^2 (1 - \cos ky) \right)^{1/2}$$

$$\leqq d_{2n-1}(\tilde{\mathcal{B}}_1; L^2) \leqq \sup_{0 \leqq y \leqq 2\pi} \left(\sum_{k=n}^{\infty} |a_k|^2 (1 - \cos ky) \right)^{1/2}$$

where $\{i_1(y), \ldots, i_{n-1}(y)\}$ *are the* $n - 1$ *largest terms of the sequence* $\{|a_k|^2 (1 - \cos ky)\}_{k=1}^{\infty}$.

The upper bound is the exact error in the best approximation by T_{n-1}. For the special case $\tilde{\mathcal{B}}_1 = \tilde{B}_1^{(r)}$, we can obtain a lower bound which is somewhat simpler to state. Whether it is, in fact, much worse than the lower bound given above is not known.

Proposition 4.8. *For* $\tilde{B}_1^{(r)}$ *as defined in Example* 4.1,

$$\sup_{0 \leqq y \leqq \frac{\pi}{n-1}} \left(\sum_{k=n}^{\infty} k^{-2r}(1 - \cos k y) \right)^{1/2}$$
$$\leqq d_{2n-1}(\tilde{B}_1^{(r)}; L^2) \leqq \sup_{0 \leqq y \leqq 2\pi} \left(\sum_{k=n}^{\infty} k^{-2r}(1 - \cos k y) \right)^{1/2}.$$

Proof. For $\tilde{\mathscr{B}}_1 = \tilde{B}_1^{(r)}$ we immediately obtain the upper bound from the identification $|a_k| = k^{-r}$. We also have the lower bound

$$d_{2n-1}(\tilde{B}_1^{(r)}; L^2) \geqq \sup_{0 \leqq y \leqq 2\pi} \left(\sum_{\substack{k=1 \\ k \neq i_1(y), \ldots, i_{n-1}(y)}}^{\infty} k^{-2r}(1 - \cos k y) \right)^{1/2}$$

where $\{i_1(y), \ldots, i_{n-1}(y)\}$ are the $n-1$ largest terms of the sequence $\{k^{-2r}(1 - \cos k y)\}_{k=1}^{\infty}$. It therefore remains to prove that for all $y \in [0, \pi/(n-1)]$, $j^{-2r}(1 - \cos j y) \geqq k^{-2r}(1 - \cos k y)$ for all $j \leqq n - 1 < k$, and all $r = 1, 2, \ldots$. It suffices to prove the inequality for $r = 1$. A simple change of variable shows that it suffices to prove that $x^{-2}(1 - \cos x) \geqq y^{-2}(1 - \cos y)$ for all $x \leqq y$, $x \in [0, \pi]$. Since $1 - \cos x = 2(\sin(x/2))^2$, this is equivalent to showing that

$$(\sin x)/x \geqq |(\sin y)/y|$$

for all $x \leqq y$, and $x \in [0, \pi/2]$. Now $(\sin x)/x$ is a decreasing positive function of x on $(0, \pi)$, and for $y \geqq \pi$,

$$|(\sin y)/y| \leqq 1/\pi < 2/\pi = (\sin(\pi/2))/(\pi/2) = \min_{0 \leqq x \leqq \pi/2} (\sin x)/x. \quad \square$$

4.3 Bounded mth Modulus of Continuity

$k(x)$ is again taken to be a fixed, 2π-periodic, real, square-integrable function with formal Fourier series

$$k(x) \sim \sum_{n=-\infty}^{\infty} a_n e^{inx}.$$

For $\phi \in L^2[0, 2\pi)$, 2π-periodic, let $\Delta_m(\phi; t)$ denote the L^2-norm of the mth forward difference operator with jumps t, i.e.,

$$\Delta_m(\phi; t) = \left\| \sum_{k=0}^{m} (-1)^k \binom{m}{k} \phi(\cdot + k t) \right\|$$
$$= \left(\frac{1}{2\pi} \int_0^{2\pi} \left| \sum_{k=0}^{m} (-1)^k \binom{m}{k} \phi(x + k t) \right|^2 dx \right)^{1/2}.$$

The mth modulus of continuity of ϕ in L^2 is defined by

$$\omega_m(\phi;t) = \sup_{0 \leqq s \leqq t} \Delta_m(\phi;s).$$

Set

$$\tilde{K}(m;p;\alpha) = \left\{(k*\phi)(x): \left(\int_0^\alpha \omega_m^p(\phi;t)\,dt\right)^{1/p} \leqq 1\right\}.$$

We shall explicitly calculate the n-widths of $\tilde{K}(m;p;\alpha)$ as a subset of $L^2[0,2\pi)$ for certain choices of k, p, and α.

Theorem 4.9. *Assume that k is as above,* $0 < p \leqq 2$, *m a positive integer,* $n\alpha \leqq \pi$, $a_0 \neq 0$, *and* $|a_j|\,j^{1/p} \geqq |a_{j+1}|\,(j+1)^{1/p}$ *for all* $j \geqq 1$. *Then*

$$d_0(\tilde{K}(m;p;\alpha);L^2) = \infty$$

and

$$d_{2n-1}(\tilde{K}(m;p;\alpha);L^2) = d_{2n}(\tilde{K}(m;p;\alpha);L^2) = \frac{|a_n|}{2^m\left(\int_0^\alpha \left(\sin\frac{nx}{2}\right)^{mp} dx\right)^{1/p}}$$

for $n = 1,2,3,\ldots$. *Furthermore,* T_{n-1} *is an optimal* $(2n-1)$*-dimensional subspace for* $d_{2n-1}(\tilde{K}(m;p;\alpha);L^2)$ *(and thus for* $d_{2n}(\tilde{K}(m;p;\alpha);L^2)$*).*

Proof. We first prove the upper bound. For ease of notation, set

$$C_n = \frac{|a_n|}{2^m\left(\int_0^\alpha \left(\sin\frac{nx}{2}\right)^{mp} dx\right)^{1/p}}$$

for fixed α, m and p. The upper bound is thus equivalent to showing that for any $f(x) = \frac{1}{2\pi}\int_0^{2\pi} k(x-y)\,\phi(y)\,dy$,

$$\inf_{t \in T_{n-1}} \|f - t\| \leqq C_n \left(\int_0^\alpha \omega_m^p(\phi;t)\,dt\right)^{1/p}.$$

If ϕ has the formal Fourier series $\phi(y) \sim \sum_{k=-\infty}^{\infty} b_k e^{iky}$, then $f(x) \sim \sum_{k=-\infty}^{\infty} a_k b_k e^{ikx}$ and the best approximation to f in L^2 from T_{n-1} is given by the partial sum $S_{n-1}(f) = \sum_{k=-n+1}^{n-1} a_k b_k e^{ikx}$. Furthermore, it is easily seen that

$$\omega_m(\phi - S_{n-1}(\phi);t) \leqq \omega_m(\phi;t).$$

Thus it suffices to prove that

$$\|f\| \leqq C_n \left(\int_0^\alpha \omega_m^p(\phi;t)\,dt\right)^{1/p}$$

for every $f(x) = \frac{1}{2\pi} \int_0^{2\pi} k(x-y)\,\phi(y)\,dy$ which is orthogonal to T_{n-1}. Since $|a_k| \neq 0$, $k = 0, \pm 1, \ldots, \pm n-1$, (if not it follows from the assumptions that the theorem is trivially true), this is equivalent to proving the above inequalities for ϕ orthogonal to T_{n-1}. Thus we shall assume that

$$\phi(y) \sim \sum_{|k| \geqq n} b_k e^{iky},$$

and

$$f(x) \sim \sum_{|k| \geqq n} a_k b_k e^{ikx}.$$

For ϕ of the above form, a simple calculation shows that

$$\Delta_m(\phi;t) = \left(\sum_{|k| \geqq n} |b_k|^2\, 2^{2m} \left(\sin \frac{kt}{2}\right)^{2m}\right)^{1/2}.$$

Since $\Delta_m(\phi;t) \leqq \omega_m(\phi;t)$, it therefore suffices to prove that

$$\|f\| = \Big(\sum_{|k| \geqq n} |a_k b_k|^2\Big)^{1/2} \leqq C_n \left(\int_0^{\alpha} \left(\sum_{|k| \geqq n} |b_k|^2\, 2^{2m} \left(\sin \frac{kx}{2}\right)^{2m}\right)^{p/2} dx\right)^{1/p}.$$

Because $|a_j|\, j^{1/p}$ is a decreasing sequence

$$\Big(\sum_{|k| \geqq n} |a_k b_k|^2\Big)^{1/2} \leqq \left(\sum_{|k| \geqq n} |b_k|^2 \left(\frac{n}{k}\right)^{2/p}\right)^{1/2} |a_n|.$$

Also for $k \geqq n$,

$$\int_0^{\alpha} \left|\sin \frac{kx}{2}\right|^{pm} dx = \frac{n}{k} \int_0^{\frac{k}{n}\alpha} \left|\sin \frac{nx}{2}\right|^{pm} dx \geqq \frac{n}{k} \int_0^{\alpha} \left|\sin \frac{nx}{2}\right|^{pm} dx.$$

Thus

$$\left(\frac{n}{k}\right)^{2/p} \leqq \frac{\left(\int_0^{\alpha} \left|\sin \frac{kx}{2}\right|^{pm} dx\right)^{2/p}}{\left(\int_0^{\alpha} \left|\sin \frac{nx}{2}\right|^{pm} dx\right)^{2/p}},$$

and

$$\left(\sum_{|k| \geqq n} |b_k|^2 \left(\frac{n}{k}\right)^{2/p}\right)^{1/2} \leqq \frac{\left(\sum_{|k| \geqq n} |b_k|^2 \left(\int_0^{\alpha} \left|\sin \frac{kx}{2}\right|^{pm} dx\right)^{2/p}\right)^{1/2}}{\left(\int_0^{\alpha} \left|\sin \frac{nx}{2}\right|^{pm} dx\right)^{1/p}}.$$

$$= \frac{\left(\sum_{|k| \geqq n} \left(\int_0^{\alpha} |b_k|^p \left|\sin \frac{kx}{2}\right|^{pm} dx\right)^{2/p}\right)^{1/2}}{\left(\int_0^{\alpha} \left|\sin \frac{nx}{2}\right|^{pm} dx\right)^{1/p}}.$$

The desired inequality thereby reduces to showing that

$$\left(\sum_{|k|\geqq n}\left(\int_0^\alpha |b_k|^p \left|\sin\frac{kx}{2}\right|^{pm} dx\right)^{2/p}\right)^{1/2} \leqq \left(\int_0^\alpha \left(\sum_{|k|\geqq n} |b_k|^2 \left|\sin\frac{kx}{2}\right|^{2m}\right)^{p/2} dx\right)^{1/p}.$$

Set $f_k(x) = b_k\left(\sin\frac{kx}{2}\right)^m$. We rewrite the above inequality in the form

$$\left(\sum_{|k|\geqq n}\left(\int_0^\alpha |f_k(x)|^p\, dx\right)^{2/p}\right)^{1/2} \leqq \left(\int_0^\alpha \left(\sum_{|k|\geqq n} |f_k(x)|^2\right)^{p/2} dx\right)^{1/p}.$$

This, however, is simply a version of Minkowski's inequality which is valid for $0 < p \leqq 2$ (see Hardy, Littlewood and Pólya [1952, p. 32]). Thus the upper bound holds.

To prove the lower bound, let

$$M_{2n+1} = \{t_n \colon \|t_n\| \leqq C_n\},$$

where C_n is as above, and t_n is any trigonometric polynomial of degree less than or equal to n, i.e., $t_n \in T_n$. Since M_{2n+1} is a ball of radius C_n in a $(2n+1)$-dimensional subspace it suffices to prove that $M_{2n+1} \subseteqq \tilde{K}(m;p;\alpha)$.

Let $t_n(x) = \sum_{k=-n}^{n} c_k e^{ikx}$. Since $a_k \neq 0$, $k = -n, \ldots, n$, the function $\phi(y) = \sum_{k=-n}^{n} (c_k/a_k)\, e^{iky}$ satisfies

$$t_n(x) = \frac{1}{2\pi}\int_0^{2\pi} k(x-y)\,\phi(y)\,dy.$$

We assume that $\|t_n\| = \left(\sum_{k=-n}^{n} |c_k|^2\right)^{1/2} \leqq C_n$ and must prove that $\left(\int_0^\alpha \omega_m^p(\phi;t)\,dt\right)^{1/p} \leqq 1$. Now

$$\Delta_m(\phi;s) = \left(\sum_{k=-n}^{n}\left|\frac{c_k}{a_k}\right|^2 2^{2m}\left(\sin\frac{ks}{2}\right)^{2m}\right)^{1/2}.$$

One of the assumptions of our theorem is that $n\alpha \leqq \pi$. The function $\left|\sin\frac{ks}{2}\right|$ is an increasing function in $|k|$ and $|s|$ for $|k| \leqq n$ and $|s| \leqq |t| \leqq \alpha$ when $n\alpha \leqq \pi$. Thus

$$\omega_m(\phi;t) \leqq 2^m\left(\sin\frac{nt}{2}\right)^m\left(\sum_{k=-n}^{n}\left|\frac{c_k}{a_k}\right|^2\right)^{1/2}.$$

Since $|a_j|\, j^{1/p}$ is decreasing, $j = 1, 2, \ldots, n$, the $|a_j|, j = 1, 2, \ldots, n$ is also a decreasing sequence and therefore

$$\omega_m(\phi;t) \leqq \frac{2^m\left(\sin\frac{nt}{2}\right)^m}{|a_n|}\, C_n.$$

Hence

$$\left(\int_0^\alpha \omega_m^p(\phi;t)\,dt\right)^{1/p} \leqq \frac{C_n 2^m}{|a_n|}\left(\int_0^\alpha \left(\sin\frac{nt}{2}\right)^{mp} dt\right)^{1/p} = 1.$$

This proves the lower bound. □

We now return to the important example of the Sobolev space $\tilde{W}_2^{(r)}$. Set

$$\tilde{B}_2^{(r)}(m;p;\alpha) = \left\{f\colon f \in \tilde{W}_2^{(r)}, \left(\int_0^\alpha \omega_m^p(f^{(r)};t)\,dt\right)^{1/p} \leqq 1\right\}.$$

Then we have

Corollary 4.10. *If $n\alpha \leqq \pi$, m a non-negative integer, and $2 \geqq p \geqq \frac{1}{r}$, then*

$$d_0(\tilde{B}_2^{(r)}(m;p;\alpha);L^2) = \infty,$$

and

$$d_{2n-1}(\tilde{B}_2^{(r)}(m;p;\alpha);L^2) = d_{2n}(\tilde{B}_2^{(r)}(m;p;\alpha);L^2) = n^{-r}2^{-m}\left(\int_0^\alpha \left(\sin\frac{nx}{2}\right)^{mp} dx\right)^{-1/p},$$

$n = 1,2,\ldots$. Furthermore, T_{n-1} is an optimal $(2n-1)$-dimensional subspace for $d_{2n-1}(\tilde{B}_2^{(r)}(m;p;\alpha);L^2)$.

Since $|a_n| = n^{-r}$, the condition $p \geqq \frac{1}{r}$ ensures that $|a_j|\,j^{1/p} \geqq |a_{j+1}|\,(j+1)^{1/p}$ for $j = 1,2,\ldots$. Corollary 4.10 follows from Theorem 4.9 where we note that the condition $a_0 \neq 0$ used in the proof of the lower bound in Theorem 4.9 is unnecessary here due to the existence of the arbitrary constant term.

It may be the case that we are not interested in a bound of the form $\left(\int_0^\alpha \omega_m^p(\phi;t)\,dt\right)^{1/p} \leqq 1$, for some fixed α, $n\alpha \leqq \pi$. For which fixed increasing functions $\Psi(\alpha)$ can we determine the n-widths of the set of functions

$$\tilde{K}(m;p;\Psi) = \left\{(k * \phi)(x)\colon \left(\int_0^\alpha \omega_m^p(\phi;t)\,dt\right)^{1/p} \leqq \Psi(\alpha),\ \text{for all } \alpha \in [0,2\pi)\right\}.$$

The following result is a simple consequence of Theorem 4.9.

Theorem 4.11. *Assume that $\Psi(\alpha) \in C[0,2\pi]$, and that*

$$\inf_{0 \leqq \alpha \leqq 2\pi} \frac{\Psi(\alpha)}{\left[\int_0^{\min(\alpha,\frac{\pi}{n})} \left(\sin\frac{nx}{2}\right)^{mp} dx + \frac{(n\alpha-\pi)_+}{n}\right]^{1/p}} = D$$

is attained for some $\alpha^* \in \left[0,\frac{\pi}{n}\right]$. Then under the assumptions of Theorem 4.9 *(without the restriction $n\alpha \leqq \pi$), we have*

$$d_{2n-1}(\tilde{K}(m;p;\Psi);L^2) = d_{2n}(\tilde{K}(m;p;\Psi);L^2) = \frac{|a_n|}{2^m} D, \qquad n = 1,2,\ldots,$$

and T_{n-1} is an optimal $(2n-1)$-dimensional subspace.

Note that the condition on Ψ is independent of k.

Proof. From Theorem 4.9, we immediately see that

$$d_{2n-1}(\tilde{K}(m;p;\Psi);L^2) \leqq \frac{|a_n|}{2^m} \frac{\Psi(\alpha)}{\left(\int\limits_0^\alpha \left(\sin\frac{nx}{2}\right)^{mp} dx\right)^{1/p}}$$

for every $\alpha \in \left[0, \frac{\pi}{n}\right]$. Thus,

$$d_{2n-1}(\tilde{K}(m;p;\Psi);L^2) \leqq \frac{|a_n|}{2^m} D.$$

To prove the lower bound, we again set

$$M_{2n+1} = \left\{t_n\colon\ t_n \in T_n,\ \|t_n\| \leqq \frac{|a_n|}{2^m} D\right\}.$$

We shall show that $M_{2n+1} \subseteq \tilde{K}(m;p;\Psi)$. We first assume that $\alpha \leqq \frac{\pi}{n}$. From the analysis of Theorem 4.9, it follows that

$$\omega_m(\phi;x) \leqq \frac{2^m \left(\sin\frac{nx}{2}\right)^{mp}}{|a_n|} \|t_n\|$$

which implies

$$\left(\int\limits_0^\alpha \omega_m^p(\phi;x)\,dx\right)^{1/p} \leqq \frac{2^m}{|a_n|} \|t_n\| \left(\int\limits_0^\alpha \left(\sin\frac{nx}{2}\right)^{mp} dx\right)^{1/p}.$$

By our choice of D, we then obtain $\left(\int\limits_0^\alpha \omega_m^p(\phi;x)\,dx\right)^{1/p} \leqq \Psi(\alpha)$. Now let us assume that $\alpha > \frac{\pi}{n}$. It is easily proven that

$$\omega_m(\phi;x) = \begin{cases} \dfrac{2^m}{|a_n|} \|t_n\| \left(\sin\dfrac{nx}{2}\right)^m, & x \leqq \dfrac{\pi}{n} \\[2ex] \dfrac{2^m}{|a_n|} \|t_n\|, & x > \dfrac{\pi}{n}. \end{cases}$$

Thus

$$\left(\int_0^\alpha \omega_m^p(\phi;x)\,dx\right)^{1/p} = \left(\int_0^{\frac{\pi}{n}} \omega_m^p(\phi;x)\,dx + \int_{\frac{\pi}{n}}^{\alpha} \omega_m^p(\phi;x)\,dx\right)^{1/p}$$
$$\leqq \frac{2^m}{|a_n|}\,\|t_n\| \left[\int_0^{\frac{\pi}{n}} \left(\sin\frac{nx}{2}\right)^{mp} dx + \left(\alpha - \frac{\pi}{n}\right)\right]^{1/p}.$$

The result follows as above. □

For which functions $\Psi(\alpha)$ do the conditions of Theorem 4.11 hold? Let us set $\Psi(\alpha) = \alpha^a$, $a \geqq 0$, and determine values of a for which the condition given in Theorem 4.11 is satisfied. The condition will certainly be valid if

$$\frac{d}{d\alpha}\, \frac{\alpha^a}{\left[\int_0^{\frac{\pi}{n}} \left(\sin\frac{nx}{2}\right)^{mp} dx + \left(\alpha - \frac{\pi}{n}\right)\right]^{1/p}} \geqq 0$$

for all $\alpha \in \left[\frac{\pi}{n}, 2\pi\right]$. Differentiating we see that the above inequality is equivalent to

$$ap \left[\int_0^{\frac{\pi}{n}} \left(\sin\frac{nx}{2}\right)^{mp} dx - \frac{\pi}{n}\right] \geqq \alpha(1-ap),$$

for all $\alpha \in \left[\frac{\pi}{n}, 2\pi\right]$. Since $\int_0^{\frac{\pi}{n}} \left(\sin\frac{nx}{2}\right)^{mp} dx - \frac{\pi}{n} < 0$, it is necessary that $1 - ap < 0$, i.e., $a > 1/p$. The left hand side is independent of α, and since $(1-ap) < 0$ and $\max\{\alpha(1-ap)\colon \alpha \in [\pi/n, 2\pi]\} = \frac{\pi}{n}(1-ap)$ we obtain the condition

$$a \geqq \frac{1}{p}\left[\frac{n}{\pi}\int_0^{\frac{\pi}{n}} \left(\sin\frac{nx}{2}\right)^{mp} dx\right]^{-1} = \frac{1}{p}\left[\frac{2}{\pi}\int_0^{\frac{\pi}{2}} (\sin y)^{mp}\, dy\right]^{-1},$$

which is independent of n.

5. n-Widths of Totally Positive Operators in L^2

In the previous sections of this chapter we calculated the n-widths from an appropriate eigenvalue problem and obtained an optimal subspace by choosing specific eigenfunctions of this problem. In this section we shall construct, for a large class of compact operators induced by totally positive kernels, certain additional optimal subspaces which are quite different in nature from the eigenfunctions, and which are often more useful.

Set

$$\mathscr{K}_2 = \{Kh(x)\colon \|h\| \leqq 1\},$$

where $\|\cdot\|$ is the usual $L^2[0,1]$ norm, $Kh(x) = \int_0^1 K(x,y)\,h(y)\,dy$, and $K(x,y)$ is a Hilbert-Schmidt kernel on $[0,1] \times [0,1]$. Set $K'K(x,y) = \int_0^1 K(z,x)\,\overline{K(z,y)}\,dz$. As we are by now well aware, the operator induced by the kernel $K'K(x,y)$ is self-adjoint, non-negative, and compact, has eigenvalues $\lambda_1 \geqq \lambda_2 \geqq \ldots \geqq 0$, and associated eigenfunctions $\phi_1, \phi_2, \ldots$. The operator induced by the kernel $KK'(x,y) = \int_0^1 K(x,z)\,\overline{K(y,z)}\,dz$ has these same eigenvalues, and associated eigenfunctions $\psi_1, \psi_2, \ldots$ given by $\psi_i = K\phi_i$, $i = 1, 2, \ldots$. From Theorem 2.2 we have that $d_n(\mathscr{K}_2; L^2) = \lambda_{n+1}^{1/2}$, while span $\{\psi_1, \ldots, \psi_n\}$ is an optimal n-dimensional subspace for $d_n(\mathscr{K}_2; L^2)$.

One method of proof of the above result, as well as a proof of the analogous result for $d^n(\mathscr{K}_2; L^2)$ and $\delta_n(\mathscr{K}_2; L^2)$, depends on the min-max characterization of eigenvalues of self-adjoint, non-negative, compact operators. This is an immediate consequence of the easily proven

$$\text{(A)} \qquad d_n(\mathscr{K}_2; L^2) = \inf_{X_n} \sup_{\|h\| \leqq 1} \inf_{g \in X_n} \|Kh - g\|$$

$$\text{(B)} \qquad = \left[\inf_{X_n} \sup_{(f, X_n) = 0} \frac{(KK'f, f)}{(f, f)} \right]^{1/2}$$

where the X_n which is optimal in (A) is optimal in (B) and vice-versa. In any discussion of the above problem one is naturally led to the question of the uniqueness of the optimal subspaces X_n in (B) (and thus in (A)). One choice of an optimal subspace is $X_n = \text{span}\{\psi_1, \ldots, \psi_n\}$. This choice is referred to as the *classical* subspace. Other choices, that is *nonclassical* subspaces, may and do in general exist. The fact that the classical subspace is not unique has often been overlooked. For example, Kolmogorov [1936] erroneously claimed the uniqueness of the optimal subspace. Criteria under which a subspace is a nonclassical subspace are given in the literature (see especially Weinstein and Stenger [1972], and Karlovitz [1973], [1976]). We shall not deal with this general question here but shall concern ourselves with the construction of explicit nonclassical extremal subspaces for certain classes of functions.

5.1 The Main Theorem

The basic class of functions considered will be those induced by nondegenerate totally positive kernels. (For a definition and discussion of total positivity, see Chapter III.)

Definition 5.1. We say that the kernel $K(x,y)$ is *nondegenerate totally positive* (*NTP*) on $[0,1] \times [0,1]$ if $K(x,y)$ is totally positive (*TP*) and each of the sets $\{K(x_1,\cdot), K(x_2,\cdot), \ldots, K(x_n,\cdot)\}$ and $\{K(\cdot,y_1), K(\cdot,y_2), \ldots, K(\cdot,y_n)\}$ span a space of dimension n for every choice of $0 < x_1 < \ldots < x_n < 1$, $0 < y_1 < \ldots < y_n < 1$, and all $n = 1, 2, \ldots$.

A motivation behind the above definition is contained in the following result.

Proposition 5.1. *Assume that $K(x, y)$ is an NTP kernel. Then $KK'(x, y)$ is a totally positive kernel and*

$$KK'\begin{pmatrix} x_1, \ldots, x_n \\ x_1, \ldots, x_n \end{pmatrix} > 0 \quad \textit{for all } 0 < x_1 < \ldots < x_n < 1,$$

and $n = 1, 2, \ldots$.

Proof. It follows from the basic composition formula (see Karlin [1968, p. 17]) that

$$KK'\begin{pmatrix} x_1, \ldots, x_n \\ y_1, \ldots, y_n \end{pmatrix} = \int_{0 < s_1 < \ldots < s_n < 1} \!\!\ldots\!\! \int K\begin{pmatrix} x_1, \ldots, x_n \\ s_1, \ldots, s_n \end{pmatrix} K\begin{pmatrix} y_1, \ldots, y_n \\ s_1, \ldots, s_n \end{pmatrix} ds_1 \ldots ds_n.$$

Since K is *TP*, we see that KK' is also *TP* (and of course symmetric). Now

$$KK'\begin{pmatrix} x_1, \ldots, x_n \\ x_1, \ldots, x_n \end{pmatrix} = \int_{0 < s_1 < \ldots < s_n < 1} \!\!\ldots\!\! \int \left[K\begin{pmatrix} x_1, \ldots, x_n \\ s_1, \ldots, s_n \end{pmatrix} \right]^2 ds_1 \ldots ds_n.$$

The assumption that K is *NTP* implies that $KK'\begin{pmatrix} x_1, \ldots, x_n \\ x_1, \ldots, x_n \end{pmatrix} > 0$. □

This same result also holds for the kernel $K'K(x, y)$.

We recall information concerning the spectral and oscillation properties of $KK'(x, y)$ and $K'K(x, y)$. (See Theorem 3.4 of Chapter III and the remark thereafter.)

Theorem 5.2. *Let $L(x, y)$ be a TP symmetric kernel with $L\begin{pmatrix} x_1, \ldots, x_n \\ x_1, \ldots, x_n \end{pmatrix} > 0$ for all $0 < x_1 < \ldots < x_n < 1$, and $n = 1, 2, \ldots$. Then the eigenvalue problem*

$$L\phi(x) = \int_0^1 L(x, y)\, \phi(y)\, dy = \lambda \phi(x)$$

possesses an infinite number of eigenvalues, all of which are positive and simple. We denote these eigenvalues by $\lambda_1 > \lambda_2 > \ldots > 0$.

Let ϕ_i denote an eigenfunction (unique up to multiplication by a nonzero constant) associated with the eigenvalue λ_i. On $(0, 1)$ the functions $\{\phi_i\}_{i=1}^{\infty}$ are orthogonal, ϕ_i has exactly $(i - 1)$ sign changes and no other zeros, and $\{\phi_i\}_{i=1}^{n}$ is a T-system.

Since KK' (and $K'K$) satisfies the conditions of Theorem 5.2 for K *NTP*,

$$\lambda_1 > \lambda_2 > \ldots > 0.$$

Let ϕ_i, $i = 1, 2, \ldots$, and ψ_i, $i = 1, 2, \ldots$, denote eigenfunctions of the operators induced by the kernels $K'K$ and KK', respectively, associated with the eigenvalue

λ_i, and assume that $K\phi_i = \psi_i$, $i = 1, 2, \ldots$ (this is simply a normalization). For fixed n, let

$$0 < \xi_1 < \ldots < \xi_n < 1, \quad \text{and} \quad 0 < \eta_1 < \ldots < \eta_n < 1$$

denote the n sign changes (zeros) of ϕ_{n+1} and ψ_{n+1} in $(0,1)$, respectively. We may now state the main theorem of this section.

Theorem 5.3. *Let $K(x, y)$ be a nondegenerate totally positive kernel on $[0,1] \times [0,1]$. Then*

$$d_n(\mathscr{K}_2; L^2) = \lambda_{n+1}^{1/2}, \qquad n = 0, 1, 2, \ldots.$$

Furthermore, the following three subspaces are optimal for $d_n(\mathscr{K}_2; L^2)$.

(1) $X_n^1 = \operatorname{span}\{\psi_1, \ldots, \psi_n\}$.
(2) $X_n^2 = \operatorname{span}\{K(\cdot, \xi_1), \ldots, K(\cdot, \xi_n)\}$.
(3) $X_n^3 = \operatorname{span}\{KK'(\cdot, \eta_1), \ldots, KK'(\cdot, \eta_n)\}$.

Proof. We have shown that $d_n(\mathscr{K}_2; L^2) = \lambda_{n+1}^{1/2}$ and also that X_n^1 is an optimal subspace. It remains to prove the optimality of X_n^2 and X_n^3.

To prove that X_n^2 is an optimal subspace we must show that

$$\sup_{\|h\| \leqq 1} \inf_{a_1, \ldots, a_n} \left\| Kh(\cdot) - \sum_{i=1}^{n} a_i K(\cdot, \xi_i) \right\| = \lambda_{n+1}^{1/2}.$$

Let P denote the orthogonal projection of $L^2[0,1]$ onto X_n^2. Thus

$$\begin{aligned}
\sup_{\|h\| \leqq 1} \inf_{a_1, \ldots, a_n} \| Kh(\cdot) - \sum_{i=1}^{n} a_i K(\cdot, \xi_i) \| &= \sup_{\|h\| \leqq 1} \| Kh - PKh \| \\
&= [\sup_{\|h\| \leqq 1} ((K - PK)h, (K - PK)h)]^{1/2} \\
&= [\sup_{\|h\| \leqq 1} ((I - P)Kh, (I - P)Kh)]^{1/2} \\
&= [\sup_{\|h\| \leqq 1} (K'(I - P)Kh, h)]^{1/2}.
\end{aligned}$$

Setting $L = K'(I - P)K$, it follows that our problem is equivalent to calculating the largest eigenvalue of the self-adjoint, non-negative, compact operator with kernel $L(x, y)$. We first claim that

$$L(x, y) = \frac{K'K\begin{pmatrix} x, \xi_1, \ldots, \xi_n \\ y, \xi_1, \ldots, \xi_n \end{pmatrix}}{K'K\begin{pmatrix} \xi_1, \ldots, \xi_n \\ \xi_1, \ldots, \xi_n \end{pmatrix}}.$$

If P is the orthogonal projection onto span $\{u_1,\ldots,u_n\}$ for any linearly independent $u_1,\ldots,u_n$, then

$$(I-P)\,h(x)=\frac{\begin{vmatrix} h(x) & u_1(x) & \ldots & u_n(x) \\ (u_1,h) & (u_1,u_1) & \ldots & (u_1,u_n) \\ \vdots & \vdots & & \vdots \\ (u_n,h) & (u_n,u_1) & \ldots & (u_n,u_n) \end{vmatrix}}{G(u_1,\ldots,u_n)}$$

where $G(u_1,\ldots,u_n)$ is the Grammian of $\{u_1,\ldots,u_n\}$. Thus for $u_i(x)=K(x,\xi_i)$, $i=1,\ldots,n$,

$$(I-P)\,K(x,y)=\frac{\begin{vmatrix} K(x,y) & K(x,\xi_1) & \ldots & K(x,\xi_n) \\ K'K(\xi_1,y) & K'K(\xi_1,\xi_1) & \ldots & K'K(\xi_1,\xi_n) \\ \vdots & \vdots & & \vdots \\ K'K(\xi_n,y) & K'K(\xi_n,\xi_1) & \ldots & K'K(\xi_n,\xi_n) \end{vmatrix}}{K'K\begin{pmatrix} \xi_1,\ldots,\xi_n \\ \xi_1,\ldots,\xi_n \end{pmatrix}}$$

and

$$L(x,y)=(K'(I-P)K)(x,y)=\int_0^1 K(z,x)\,[(I-P)K](z,y)\,dz$$

$$=\frac{\begin{vmatrix} K'K(x,y) & K'K(x,\xi_1) & \ldots & K'K(x,\xi_n) \\ K'K(\xi_1,y) & K'K(\xi_1,\xi_1) & \ldots & K'K(\xi_1,\xi_n) \\ \vdots & \vdots & & \vdots \\ K'K(\xi_n,y) & K'K(\xi_n,\xi_1) & \ldots & K'K(\xi_n,\xi_n) \end{vmatrix}}{K'K\begin{pmatrix} \xi_1,\ldots,\xi_n \\ \xi_1,\ldots,\xi_n \end{pmatrix}}$$

$$=\frac{K'K\begin{pmatrix} x,\xi_1,\ldots,\xi_n \\ y,\xi_1,\ldots,\xi_n \end{pmatrix}}{K'K\begin{pmatrix} \xi_1,\ldots,\xi_n \\ \xi_1,\ldots,\xi_n \end{pmatrix}}.$$

Let us now show that ϕ_{n+1} is an eigenfunction of $L(x,y)$ with associated eigenvalue λ_{n+1}, i.e., $\int_0^1 L(x,y)\,\phi_{n+1}(y)\,dy=\lambda_{n+1}\phi_{n+1}(x)$. Because of the form of $L(x,y)$ we see that for any $\phi\in L^2[0,1]$, $\int_0^1 L(x,y)\,\phi(y)\,dy=\int_0^1 K'K(x,y)\,\phi(y)\,dy+\sum_{i=1}^n u_i(x)\int_0^1 K'K(\xi_i,y)\,\phi(y)\,dy$, for some fixed functions $u_i(x)$, $i=1,\ldots,n$, independent of ϕ. Since

$$\int_0^1 K'K(\xi_i,y)\,\phi_{n+1}(y)\,dy=\lambda_{n+1}\phi_{n+1}(\xi_i)=0,\qquad i=1,\ldots,n,$$

it follows that

$$\int_0^1 L(x,y)\,\phi_{n+1}(y)\,dy = \int_0^1 K'K(x,y)\,\phi_{n+1}(y)\,dy = \lambda_{n+1}\,\phi_{n+1}(x).$$

The symmetric kernel $L(x,y)$ is not non-negative. However $L(x,y)$ has the rather simple sign pattern given by

$$|L(x,y)| = (\operatorname{sgn}\phi_{n+1}(x))\,L(x,y)\,(\operatorname{sgn}\phi_{n+1}(y))$$

for $x, y \in (0,1)$. Furthermore, associated with the kernel $L_0(x,y) = |L(x,y)|$ is the non-negative eigenfunction $|\phi_{n+1}(x)|$ and eigenvalue λ_{n+1}. A modification of Jentzsch's Theorem (Section 3 of Chapter III) shows that λ_{n+1} is the largest eigenvalue of $L_0(x,y)$. Moreover, every eigenvalue λ of $L(x,y)$, with corresponding eigenfunction $\phi(x)$, is an eigenvalue of $L_0(x,y)$ with corresponding eigenfunction $\phi(x)(\operatorname{sgn}\phi_{n+1}(x))$. Thus λ_{n+1} is the largest eigenvalue of $L(x,y)$.

This completes the proof of the fact that X_n^2 is an optimal subspace for $d_n(\mathscr{K}_2; L^2)$.

Let us now prove this same result for X_n^3. The proof of this fact is a rather simple consequence of the above analysis. We proved that

$$\sup_{\|h\| \leq 1} \|(K - PK)h\| = \lambda_{n+1}^{1/2},$$

where P was the orthogonal projection onto X_n^2. Replacing K by K' it follows that

$$\sup_{\|h\| \leq 1} \|(K' - QK')h\| = \lambda_{n+1}^{1/2},$$

where Q takes the role of P, but with respect to K', and is therefore the orthogonal projection onto $\operatorname{span}\{K(\eta_1,\cdot),\ldots,K(\eta_n,\cdot)\}$. Moreover,

$$\sup_{\|h\| \leq 1} \|(K' - QK')h\| = \sup_{\|h\| \leq 1} \|(K - KQ)h\|$$

which implies that the image of KQ, being n-dimensional, is also an optimal subspace for $d_n(\mathscr{K}_2; L^2)$. The image of KQ is explicitly

$$X_n^3 = \operatorname{span}\{KK'(\cdot,\eta_1),\ldots,KK'(\cdot,\eta_n)\}. \quad \square$$

In the proof of the optimality of X_n^3 we did not, for each $h \in L^2[0,1]$, construct the best approximation to Kh from X_n^3. This is equivalent to saying that KQ is not the orthogonal projection onto X_n^3. It is not necessary that we choose the best approximation to every element in the set. It is only necessary that we never do worse than the value of the associated n-width. Returning to the optimality of X_n^3, one may easily show that the operator KQ simply maps $h \in L^2[0,1]$ into the interpolant (from X_n^3) to $Kh(x)$ at the points $\eta_1, \ldots, \eta_n$. The existence of such an interpolant follows from the fact that $KK'\begin{pmatrix}\eta_1, \ldots, \eta_n \\ \eta_1, \ldots, \eta_n\end{pmatrix} > 0$ (see Proposition 5.1). To prove that we are interpolating at $\eta_1, \ldots, \eta_n$ note that $(Kh - KQh)(\eta_i) = (h - Qh, K(\eta_i, \cdot))$. Since Q is the orthogonal projection onto span$\{K(\eta_1, \cdot), \ldots, K(\eta_n, \cdot)\}$, $((I - Q)h, K(\eta_i, \cdot)) = 0$, $i = 1, \ldots, n$. Thus $Kh(\eta_i) = (KQh)(\eta_i)$, $i = 1, \ldots, n$.

We proved that X_n^2 was an optimal subspace by constructing the best approximant to each function in $\mathscr{K}_2$. We shall now show that interpolation to $Kh(x)$ at $\eta_1, \ldots, \eta_n$ from X_n^2 is also an optimal method. In the proof of this fact we employ the following lemma.

Lemma 5.4. *Let K be an NTP kernel. Assume that $\phi \in L^1[0,1]$ has n strict sign changes in $(0,1)$ at $\tau_1 < \ldots < \tau_n$, by which we mean that*

$$\varepsilon \phi(x)(-1)^i > 0 \quad \textit{a.e. for } x \in (\tau_{i-1}, \tau_i), \quad i = 1, \ldots, n+1,$$

where $\tau_0 = 0$, $\tau_{n+1} = 1$, and $\varepsilon = +1$ or -1, fixed. Assume that $\psi(x) = \int_0^1 K(x,y)\phi(y)\,dy$ has n zeros at $\zeta_1 < \ldots < \zeta_n$ in $(0,1)$. Then

$$K\begin{pmatrix}\zeta_1, \ldots, \zeta_n \\ \tau_1, \ldots, \tau_n\end{pmatrix} > 0.$$

Applying the lemma with $\phi = \phi_{n+1}$ and $\psi = \psi_{n+1}$, we obtain $K\begin{pmatrix}\eta_1, \ldots, \eta_n \\ \xi_1, \ldots, \xi_n\end{pmatrix} > 0$ where the $\{\xi_i\}_{i=1}^n$ and $\{\eta_i\}_{i=1}^n$ are as previously defined.

Proof. If the relevant determinant is zero there then exist constants $\alpha_1, \ldots, \alpha_n$, not all zero, for which the function

$$f(x) = \sum_{i=1}^{n} \alpha_i K(x, \tau_i)$$

vanishes at $\zeta_1, \ldots, \zeta_n$. Since $\{K(x,\tau_1), \ldots, K(x,\tau_n)\}$ is a set of n linearly independent functions on $(0,1)$ there exists a $\zeta_{n+1} \in (0,1) \backslash (\zeta_1, \ldots, \zeta_n)$ for which $f(\zeta_{n+1}) \neq 0$. Choose a constant c so that $\psi(\zeta_j) - cf(\zeta_j) = 0$, for $j = 1, \ldots, n+1$. Since the functions $K(\zeta_1, y), \ldots, K(\zeta_n, y)$, $K(\zeta_{n+1}, y)$ form a weak Tchebycheff system of $n+1$ linearly independent functions on $[0,1]$, there exists a non-zero function

$$g(x) = \sum_{j=1}^{n+1} \beta_j K(\zeta_j, x)$$

for which $\varepsilon g(x)(-1)^i \geqq 0$, $\tau_{i-1} \leqq x \leqq \tau_i$, $i = 1, \ldots, n+1$, where $\tau_0 = 0$, $\tau_{n+1} = 1$, (see Proposition 1.2 of Chapter III). Now

$$0 = \sum_{j=1}^{n+1} \beta_j(\psi(\zeta_j) - cf(\zeta_j))$$

$$= \int_0^1 g(x)\,\phi(x)\,dx - c\sum_{j=1}^{n} \alpha_j g(\tau_j)$$

$$= \int_0^1 |g(x)|\,|\phi(x)|\,dx > 0.$$

This contradiction proves the lemma. □

Let R denote the interpolation operator from $L^2[0,1]$ to X_n^2 given by $Rh(\eta_i) = Kh(\eta_i)$, $i = 1,\ldots,n$. By the above lemma such an operator exists.

Theorem 5.5. *Let K be an NTP kernel. Then*

$$\sup_{\|h\| \leqq 1} \|Kh - Rh\| = d_n(\mathscr{K}_2; L^2) = \lambda_{n+1}^{1/2}.$$

Proof. It is easily seen that

$$\sup_{\|h\| \leqq 1} \|Kh - Rh\| = \sup_{\|h\| \leqq 1} \|Mh\|$$

where

$$M(x,y) = \frac{K\begin{pmatrix} x, \eta_1, \ldots, \eta_n \\ y, \xi_1, \ldots, \xi_n \end{pmatrix}}{K\begin{pmatrix} \eta_1, \ldots, \eta_n \\ \xi_1, \ldots, \xi_n \end{pmatrix}}.$$

The proof of this theorem now follows the analysis of the proof of Theorem 5.3. We simply show that the largest eigenvalue of $M'M(x,y)$ is λ_{n+1}. Since

$$M'M(x,y) = \int_0^1 \frac{K\begin{pmatrix} z, \eta_1, \ldots, \eta_n \\ x, \xi_1, \ldots, \xi_n \end{pmatrix} K\begin{pmatrix} z, \eta_1, \ldots, \eta_n \\ y, \xi_1, \ldots, \xi_n \end{pmatrix}}{\left[K\begin{pmatrix} \eta_1, \ldots, \eta_n \\ \xi_1, \ldots, \xi_n \end{pmatrix}\right]^2}\,dz,$$

let us first prove that $M'M\phi_{n+1} = \lambda_{n+1}\phi_{n+1}$.

$$M\phi_{n+1}(z) = \int_0^1 \frac{K\begin{pmatrix} z, \eta_1, \ldots, \eta_n \\ y, \xi_1, \ldots, \xi_n \end{pmatrix}}{K\begin{pmatrix} \eta_1, \ldots, \eta_n \\ \xi_1, \ldots, \xi_n \end{pmatrix}}\,\phi_{n+1}(y)\,dy,$$

and $\int_0^1 K(\eta_i, y)\, \phi_{n+1}(y)\, dy = \psi_{n+1}(\eta_i) = 0$, $i = 1, \ldots, n$. Thus

$$M\phi_{n+1}(z) = \int_0^1 K(z, y)\, \phi_{n+1}(y)\, dy = \psi_{n+1}(z).$$

The function ψ_{n+1} satisfies

$$\int_0^1 K(z, x)\, \psi_{n+1}(z)\, dz = \lambda_{n+1}\, \phi_{n+1}(x),$$

and $\phi_{n+1}(\xi_i) = 0$, $i = 1, \ldots, n$. Therefore $M'M\phi_{n+1} = M'\psi_{n+1} = \lambda_{n+1}\phi_{n+1}$. In addition, it follows from the total positivity of K that

$$(\operatorname{sgn} \phi_{n+1}(x))\, M'M(x, y)\, (\operatorname{sgn} \phi_{n+1}(y)) \geqq 0$$

for all $x, y \in [0, 1]$. We conclude, as in Theorem 5.3, that λ_{n+1} is the largest eigenvalue associated with the kernel $M'M(x, y)$. □

Remark. The problem of the calculation of the n-widths $d_n(\mathscr{K}_2; L^2)$ and their optimal subspaces, independent of whether the kernel is totally positive or not, is an exceedingly difficult if not impossible problem since one must obtain the corresponding eigenvalues and eigenfunctions. For this very reason the importance of the subspaces X_n^2 and X_n^3 is further enhanced. To exactly determine X_n^2 and X_n^3 one must of course calculate the points $\{\xi_i\}_{i=1}^n$ and $\{\eta_i\}_{i=1}^n$ and this problem is certainly no easier than the former. However it is often possible, in specific cases, to obtain an estimate of how well a subspace of the form span $\{K(\cdot, \zeta_i)\}_{i=1}^n$, for fixed ζ_i, approximates Kh with respect to how well span $\{K(\cdot, \tau_i)\}_{i=1}^n$ approximates Kh for an arbitrary choice of $\{\tau_i\}_{i=1}^n$.

Here is, moreover, one example where X_n^1, X_n^2 and X_n^3 are explicitly obtained.

Example 5.1. Given any positive real numbers $t_1, \ldots, t_m$, we define the polynomial

$$q_{2m}(x) = \prod_{j=1}^m (x^2 - t_j^2),$$

and set

$$\mathscr{D}_2 = \{f : f \in W_2^{(2m)},\, f^{(2k)}(0) = f^{(2k)}(1) = 0,\, k = 0, 1, \ldots, m-1,\, \|q_{2m}(D) f\| \leqq 1\},$$

where $D = \dfrac{d}{dx}$. $\mathscr{D}_2$ is a subset of the Sobolev space $W_2^{(2m)}[0, 1]$. We may express $\mathscr{D}_2$ as $\mathscr{K}_2$ where $K(x, y)$ is the Green's function of the differential operator

$$q_{2m}(D) f = \prod_{j=1}^m \left(\frac{d^2}{dx^2} - t_j^2\right) f = 0$$

with boundary conditions

$$f^{(2k)}(0) = f^{(2k)}(1) = 0, \qquad k = 0, 1, \ldots, m-1.$$

It may be verified that

$$(*)\ K(x, y) = 2 \sum_{k=1}^{\infty} \frac{\sin k\pi x \sin k\pi y}{q_{2m}(ik\pi)}$$

and $(-1)^m K(x, y)$ is a nondegenerate totally positive kernel on $[0, 1] \times [0, 1]$. To prove this latter fact, we may apply the methods of Chapter III, Section 4. We can also prove it directly by observing that the differential operator given above is the product of the second order differential operators

$$\begin{aligned} f''(x) - \alpha^2 f(x) &= 0, \qquad \alpha = t_1, \ldots, t_m \\ f(0) = f(1) &= 0. \end{aligned}$$

If we denote the Green's function of this differential operator by $K(x, y; \alpha)$ and define

$$K_j(x, y) = \int_0^1 K(x, z; t_j)\, K_{j-1}(z, y)\, dz,$$

where $K_1(x, y) = K(x, y; t_1)$, then $K(x, y) = K_m(x, y)$. Hence it suffices to prove that $-K(x, y; \alpha)$ is totally positive for any $\alpha > 0$ in order to obtain the total positivity of $(-1)^m K(x, y)$. A direct calculation shows that

$$K(x, y; \alpha) = \begin{cases} -\dfrac{1}{\alpha} \sinh \alpha x \sinh \alpha(1-y), & x \leqq y \\ -\dfrac{1}{\alpha} \sinh \alpha(1-x) \sinh \alpha y, & x \geqq y. \end{cases}$$

It is a known fact that $L(x, y)$ of the form

$$L(x, y) = \begin{cases} \phi(x)\, \psi(y), & a < x \leqq y < b \\ \phi(y)\, \psi(x), & a < y \leqq x < b, \end{cases}$$

is totally positive on (a, b) if $\phi(x) \cdot \psi(x) > 0$ and continuous on (a, b) and $\phi(x)/\psi(x)$ is increasing (see Karlin [1968, Cor. 3.1, p. 112]). We therefore conclude that $-K(x, y; \alpha)$ is totally positive for $\alpha > 0$. The nondegeneracy also follows.

It is easily verified that the kernel $K(x, y)$ is symmetric with eigenvalues $q_{2m}^{-1}(ik\pi)$, $k = 1, 2, \ldots$, and corresponding eigenfunctions $\sin k\pi x$, $k = 1, 2, \ldots$. (This provides a proof of the fact that $K(x, y)$ is of the form $(*)$.) From Theorem 5.3, the subspaces

$$\begin{aligned} X_n^1 &= \operatorname{span}\{\sin \pi x, \ldots, \sin n\pi x\} \\ X_n^2 &= \operatorname{span}\left\{K\left(\cdot, \frac{1}{n+1}\right), \ldots, K\left(\cdot, \frac{n}{n+1}\right)\right\} \end{aligned}$$

and

$$X_n^3 = \operatorname{span}\left\{G\left(\cdot, \frac{1}{n+1}\right), \ldots, G\left(\cdot, \frac{n}{n+1}\right)\right\}$$

are optimal subspaces for $\mathscr{D}_2$, where $G(x, y)$ is of the form

$$G(x, y) = 2 \sum_{k=1}^{\infty} \frac{\sin k\pi x \sin k\pi y}{[q_{2m}(ik\pi)]^2}.$$

Furthermore it follows from Theorem 5.5 and from the proof of Theorem 5.3 that interpolation from either of the subspaces X_n^2 and X_n^3 to the functions at the points $i/(n+1)$, $i = 1, \ldots, n$, is an optimal procedure.

Remark. Having obtained the optimal subspaces X_n^2 and X_n^3 for $d_n(\mathscr{K}_2; L^2)$, it is natural to ask for additional optimal subspaces and operators for the linear, Gel'fand, and Bernstein n-widths of $\mathscr{K}_2$ in L^2. As is easily seen (considering that we are in a Hilbert space setting), the rank n operators PK and KQ of the proof of Theorem 5.3, as well as R of Theorem 5.5, are optimal rank n linear operators for $\delta_n(\mathscr{K}_2; L^2)$. The following two subspaces of codimension n are also optimal for $d^n(\mathscr{K}_2; L^2)$.

$$L_2^n = \{h\colon Kh(\eta_i) = 0, \quad i = 1, \ldots, n\}$$

$$L_3^n = \{h\colon K'Kh(\xi_i) = 0, \ i = 1, \ldots, n\}.$$

No analogue of X_n^2 and X_n^3 has been found for the Bernstein n-width.

A discussion of the matrix analogue of the results of this section is deferred to Chapter VI.

5.2 Restricted Approximating Subspaces

We shall now extend the above analysis to the class of functions

$$\mathscr{K}_2^r = \left\{\sum_{i=1}^{r} a_i k_i(x) + \int_0^1 K(x, y)\, h(y)\, dy\colon \|h\| \leqq 1, \ a_i \in \mathbb{R}\right\},$$

where $K(x, y)$ together with the $\{k_i\}_{i=1}^r$ satisfy certain non-degenerate total positivity requirements which we shall soon delineate. The motivation behind the consideration of the n-widths of the class $\mathscr{K}_2^r$ comes first and foremost from the class of disconjugate rth order differential equations. The $\{k_i\}_{i=1}^r$ may be regarded as the fundamental solutions of the differential equation and K as the Green's function of this same equation with Cauchy boundary conditions (see Section 4 of Chapter III).

The assumptions which we shall heretofore assume to hold are the following:

Property A.

1. $\{k_1,\ldots,k_r\}$ *is a* T^+*-system on* $(0,1)$, *i.e., for all* $0 < x_1 < \ldots < x_r < 1$,
$$K\begin{pmatrix} x_1,\ldots,x_r \\ 1,\ldots,r \end{pmatrix} = \det(k_j(x_i))_{i,j=1}^r > 0.$$

2. *For every choice of points* $0 \leqq y_1 < \ldots < y_m \leqq 1$ *and* $0 \leqq x_1 < \ldots < x_{r+m} \leqq 1$, $m \geqq 0$, *the determinant*

$$K\begin{pmatrix} x_1,\ldots,x_r, & x_{r+1},\ldots,x_{r+m} \\ 1,\ldots,r, & y_1,\ldots,y_m \end{pmatrix}$$
$$= \begin{vmatrix} k_1(x_1) & \ldots & k_r(x_1) & K(x_1,y_1) & \ldots & K(x_1,y_m) \\ \vdots & & \vdots & \vdots & & \vdots \\ k_1(x_{r+m}) & \ldots & k_r(x_{r+m}) & K(x_{r+m},y_1) & \ldots & K(x_{r+m},y_m) \end{vmatrix}$$

is nonnegative. Furthermore for any given $0 < y_1 < \ldots < y_m < 1$ *the above determinant is not identically zero, and for any given* $0 < x_1 < \ldots < x_{r+m} < 1$, *the above determinant is also not identically zero.*

In Section 3.1 we considered sets of the form $\mathscr{K}_2^r$. From Theorem 3.1 we obtain:

Proposition 5.6. *Let* Q_r *denote the orthogonal projection of* $L^2[0,1]$ *onto* $Y_r = \operatorname{span}\{k_1,\ldots,k_r\}$. *Set* $K_r(x,y) = K(x,y) - Q_r K(x,y)$, *and*

$$\hat{\mathscr{K}}_2^r = \left\{\int_0^1 K_r(x,y)\,h(y)\,dy\colon \|h\| \leqq 1\right\}.$$

Then,

$$d_n(\mathscr{K}_2^r; L^2) = \begin{cases} \infty, & n < r \\ d_{n-r}(\hat{\mathscr{K}}_2^r; L^2), & n \leqq r. \end{cases}$$

Furthermore, for $n \geqq r$, $\operatorname{span}\{k_1,\ldots,k_r, u_1,\ldots,u_{n-r}\}$ *is an optimal subspace for* $d_n(\mathscr{K}_2^r; L^2)$ *whenever* $\operatorname{span}\{u_1,\ldots,u_{n-r}\}$ *is an optimal subspace for* $d_{n-r}(\hat{\mathscr{K}}_2^r; L^2)$.

Thus $K_r(x,y)$ is the kernel which we must consider. Firstly we note that $K_r(x,y)$ itself is *not* a totally positive kernel. This follows from the fact that for any $h \in L^2$ it is easily seen that $K_r h(x)$ is orthogonal to each element of Y_r. Since Y_r is a T-system this in turn implies that $K_r h$ has at least r sign changes in $(0,1)$. However, this fact contradicts the variation diminishing property of a totally positive kernel. Hence $K_r(x,y)$ is not a totally positive kernel.

However a reading of the proof of Theorem 5.3 shows that it is only necessary that $K_r' K_r(x,y)$ be totally positive. Let us now prove this fact. (By reasoning, similar to the above, it can also be shown that $K_r K_r'(x,y)$ is *not* totally positive.)

Lemma 5.7. *The kernel* $K_r' K_r(x,y)$ *is totally positive and*

$$K_r' K_r\begin{pmatrix} x_1,\ldots,x_n \\ x_1,\ldots,x_n \end{pmatrix} > 0 \quad \textit{for all } 0 < x_1 < \ldots < x_n < 1.$$

Proof. As was shown in Section 3.1,

$$(K_r' K_r h)(x) = \frac{\begin{vmatrix} (k_1,k_1) \dots (k_1,k_r) & (k_1,Kh) \\ \vdots \quad\quad \vdots & \vdots \\ (k_r,k_1) \dots (k_r,k_r) & (k_r,Kh) \\ K'k_1(x) \dots K'k_r(x) & (K'Kh)(x) \end{vmatrix}}{G(k_1,\dots,k_r)}$$

where $G(k_1,\dots,k_r)$ is the Grammian of $\{k_1,\dots,k_r\}$. Thus

$$K_r' K_r(x,y) = \frac{\begin{vmatrix} (k_1,k_1) & \dots & (k_1,k_r) & (k_1, K(\cdot,y)) \\ \vdots & & \vdots & \vdots \\ (k_r,k_1) & \dots & (k_r,k_r) & (k_r, K(\cdot,y)) \\ (K(\cdot,x),k_1) & \dots & (K(\cdot,x),k_r) & (K(\cdot,x), K(\cdot,y)) \end{vmatrix}}{G(k_1,\dots,k_r)}.$$

This unwieldy form may be simplified by the basic composition formula to

$$K_r' K_r(x,y) = \frac{\int\limits_{0<s_1<\dots<s_{r+1}<1} \!\!\!\dots\! \int K\begin{pmatrix} s_1,\dots,s_{r+1} \\ 1,\dots,r,x \end{pmatrix} K\begin{pmatrix} s_1,\dots,s_{r+1} \\ 1,\dots,r,y \end{pmatrix} ds_1 \dots ds_{r+1}}{G(k_1,\dots,k_r)}.$$

Finally, from Sylvester's determinant identity (see Section 2 of Chapter III)

$$K_r' K_r\begin{pmatrix} x_1,\dots,x_n \\ y_1,\dots,y_n \end{pmatrix} = \frac{\int\limits_{0<s_1<\dots<s_{r+n}<1} \!\!\!\dots\! \int K\begin{pmatrix} s_1,\dots,s_{r+n} \\ 1,\dots,r,x_1,\dots,x_n \end{pmatrix} K\begin{pmatrix} s_1,\dots,s_{r+n} \\ 1,\dots,r,y_1,\dots,y_n \end{pmatrix} ds_1 \dots ds_{r+n}}{G(k_1,\dots,k_r)}.$$

Since $G(k_1,\dots,k_r) > 0$, the results of the lemma follow from Property A. □

Let $\lambda_1 > \lambda_2 > \dots$ denote the eigenvalues induced by the kernel $K_r' K_r(x,y)$ and $\phi_1,\phi_2,\dots,$ their respective eigenfunctions. Set $K_r \phi_i = \psi_i$, $i = 1,2,\dots$. Thus $K_r K_r' \psi_i = \lambda_i \psi_i$, $i = 1,2,\dots$.

Lemma 5.8. *The set of functions* $\{k_1,\dots,k_r,\psi_1,\dots,\psi_n\}$ *form a T-system on* $(0,1)$ *for all* $n = 0,1,2,\dots$.

Proof. One way of proving that a set is a T-system is simply to prove that the associated determinant maintains a fixed sign. Thus let $0 < x_1 < \dots < x_{n+r} < 1$

and consider the determinant

$$\begin{vmatrix} k_1(x_1) \ldots k_r(x_1) & (K_r\phi_1)(x_1) \ldots (K_r\phi_n)(x_1) \\ \vdots \quad\quad \vdots & \vdots \quad\quad \vdots \\ k_1(x_{n+r}) \ldots k_r(x_{n+r}) & (K_r\phi_1)(x_{n+r}) \ldots (K_r\phi_n)(x_{n+r}) \end{vmatrix}.$$

An application of the basic composition formula implies that the above determinant is equal to

$$\int_{0<\sigma_1<\ldots<\sigma_n<1} \!\!\!\ldots \int \Phi\begin{pmatrix}\sigma_1,\ldots,\sigma_n\\ 1,\ldots,n\end{pmatrix} K\begin{pmatrix}x_1,\ldots,x_r,\ldots,x_{r+n}\\ 1,\ldots,r,\sigma_1,\ldots,\sigma_n\end{pmatrix} d\sigma_1\ldots d\sigma_n,$$

where $\Phi\begin{pmatrix}\sigma_1,\ldots,\sigma_n\\ 1,\ldots,n\end{pmatrix} = \det(\phi_j(\sigma_i))_{i,j=1}^n$. Since $\{\phi_1,\ldots,\phi_n\}$ is a T-system on $(0,1)$ (see Theorem 5.2), the determinant $\Phi\begin{pmatrix}\sigma_1,\ldots,\sigma_n\\ 1,\ldots,n\end{pmatrix}$ is of one fixed sign for each choice of $0<\sigma_1<\ldots<\sigma_n<1$. The desired result now follows from Property A. □

Lemma 5.9. *For $n \geqq 1$, the function ψ_n has exactly $n+r-1$ zeros (all sign changes) in* $(0,1)$.

Proof. Since $(I-Q_r)k_i = 0$, $i=1,\ldots,r$, it follows that $K'_r k_i = 0$, for $i=1,\ldots,r$. Now,

$$(\psi_n, k_i) = (K_r\phi_n, k_i) = (\phi_n, K'_r k_i) = 0.$$

Therefore, $(\psi_n, k_i) = 0$, $i=1,\ldots,r$. Furthermore, $(\psi_n,\psi_i) = 0$, $i \neq n$, since the $\{\psi_i\}_{i=1}^\infty$ are eigenfunctions of $K_r K'_r \psi = \lambda\psi$ associated with distinct eigenvalues. Thus ψ_n is orthogonal to each of the elements of the T-system $\{k_1,\ldots,k_r,\psi_1,\ldots,\psi_{n-1}\}$ of order $n+r-1$. By Proposition 1.4 of Chapter III, ψ_n must exhibit at least $n+r-1$ sign changes in $(0,1)$. On the other hand, ψ_n is an element of the T-system $\{k_1,\ldots,k_r,\psi_1,\ldots,\psi_n\}$ and therefore has at most $n+r-1$ zeros in $(0,1)$. □

Let $\{\xi_i\}_{i=1}^{n-r}$ and $\{\eta_i\}_{i=1}^n$ denote the zeros (sign changes) of ϕ_{n-r+1} and ψ_{n-r+1} in $(0,1)$, respectively. Thus

$$\begin{aligned} \phi_{n-r+1}(\xi_i) &= 0, \quad 0<\xi_1<\ldots<\xi_{n-r}<1,\\ \psi_{n-r+1}(\eta_i) &= 0, \quad 0<\eta_1<\ldots<\eta_n<1. \end{aligned}$$

We divide the proof of the analogue of Theorem 5.3 into two parts.

Theorem 5.10.

$$d_n(\mathscr{K}_2^r; L^2) = \begin{cases} \infty, & n<r\\ \lambda_{n-r+1}^{1/2}, & n\geqq r, \end{cases}$$

and both

$$X_n^1 = \operatorname{span}\{k_1,\ldots,k_r,\psi_1,\ldots,\psi_{n-r}\}$$

and

$$X_n^2 = \text{span}\{k_1,\ldots,k_r,\, K_r(\cdot,\xi_1),\ldots,K_r(\cdot,\xi_{n-r})\}$$

are optimal subspaces for the n-widths of $\mathscr{K}_2^r$ in L^2 for $n \geqq r$.

Remark. We can also write $X_n^1 = \text{span}\{k_1,\ldots,k_r, K\phi_1,\ldots,K\phi_{n-r}\}$ and $X_n^2 = \text{span}\{k_1,\ldots,k_r, K(\cdot,\xi_1),\ldots,K(\cdot,\xi_{n-r})\}$ since $K_r(\cdot,\xi) - K(\cdot,\xi)$ and $K_r\phi - K\phi$ are both functions in $\text{span}\{k_1,\ldots,k_r\}$.

Proof. The fact that $d_n(\mathscr{K}_2^r; L^2)$ assumes the above value and that X_n^1 is an optimal subspace for $n \geqq r$ is a consequence of Proposition 5.6. From this proposition it suffices to prove that $\text{span}\{K_r(\cdot,\xi_1),\ldots,K_r(\cdot,\xi_{n-r})\}$ is an optimal subspace for $d_{n-r}(\hat{\mathscr{K}}_2^r; L^2)$. The proof of this fact uses the arguments of Theorem 5.3. Note that the proof therein really only required the consequences of Lemma 5.7 for $K_r'K_r(x,y)$ and did not require that $K_r(x,y)$ itself be a nondegenerate totally positive kernel. Theorem 5.10 is proved. □

The fact that $K_rK_r'(x,y)$ is *not* totally positive, and more importantly, the fact that ψ_{n-r+1} has n rather than $n-r$ zeros complicates the problem of determining an analogue to X_n^3 of Theorem 5.3 which is also optimal for $d_n(\mathscr{K}_2^r; L^2)$.

We shall show that the n-dimensional subspace spanned by $k_1,\ldots,k_r$ and the $(n-r)$-dimensional subspace of $\{KK'(\cdot,\eta_1),\ldots,KK'(\cdot,\eta_n)\}$ whose elements vanish identically at $\eta_{i_1},\ldots,\eta_{i_r}$, for $1 \leqq i_1 < \ldots < i_r \leqq n$, arbitrarily chosen, is an optimal subspace for $d_n(\mathscr{K}_2^r; L^2)$. To ease our notation we shall, without loss of generality, choose $\{i_1, i_2,\ldots,i_r\} = \{1,2,,\ldots,r\}$.

Set $\bar{K}_r = (I - J_r)K$, where J_r is the interpolation operator onto $\text{span}\{k_1,\ldots,k_r\}$ defined by $\bar{K}_r(\eta_i, y) = 0$, $i = 1,\ldots,r$. Thus,

$$\bar{K}_r(x,y) = \frac{K\begin{pmatrix}\eta_1,\ldots,\eta_r, x\\ 1,\ldots,r, y\end{pmatrix}}{K\begin{pmatrix}\eta_1,\ldots,\eta_r\\ 1,\ldots,r\end{pmatrix}}.$$

Theorem 5.11. *For $n \geqq r$, the subspace*

$$X_n^3 = \text{span}\{k_1,\ldots,k_r,\, \bar{K}_r\bar{K}_r'(\cdot,\eta_{r+1}),\ldots,\bar{K}_r\bar{K}_r'(\cdot,\eta_n)\}$$

is an optimal subspace for $d_n(\mathscr{K}_2^r; L^2)$.

Proof. To prove the optimality of X_n^3, we must show that

$$\sup_{\|h\|\leqq 1}\ \inf_{a_1,\ldots,a_n} \left\| Kh - \sum_{i=1}^{r} a_i k_i - \sum_{i=r+1}^{n} a_i \bar{K}_r\bar{K}_r'(\cdot,\eta_i)\right\| \leqq \lambda_{n-r+1}^{1/2}.$$

Since $\bar{K}_r(x,y) = K(x,y) + \sum_{i=1}^{r} k_i(x) M_i(y)$, for some $M_i(y)$, $i = 1,\ldots,r$, it suffices to prove that

$$\sup_{\|h\|\leqq 1}\ \inf_{a_{r+1},\ldots,a_n} \left\| \bar{K}_r h - \sum_{i=r+1}^{n} a_i \bar{K}_r\bar{K}_r'(\cdot,\eta_i)\right\| \leqq \lambda_{n-r+1}^{1/2}.$$

Let Q denote the orthogonal projection onto the subspace spanned by $\{\bar{K}_r(\eta_{r+1},\cdot),\ldots,\bar{K}_r(\eta_n,\cdot)\}$. Thus we must prove that

$$\sup_{\|h\|\leqq 1} \|\bar{K}_r h - \bar{K}_r Q h\| \leqq \lambda_{n-r+1}^{1/2}.$$

We have reduced our problem to showing that $\lambda_{n-r+1}^{1/2}$ is the largest singular value (s-number) of the operator induced by the kernel $\bar{K}_r(I-Q)$. This is equivalent to showing that λ_{n-r+1} is the largest eigenvalue associated with the kernel $L_r = \bar{K}_r(I-Q)\bar{K}_r'$. The kernel L_r is given by

$$L_r(x,y) = \frac{\bar{K}_r\bar{K}_r'\begin{pmatrix} x, \eta_{r+1}, \ldots, \eta_n \\ y, \eta_{r+1}, \ldots, \eta_n \end{pmatrix}}{\bar{K}_r\bar{K}_r'\begin{pmatrix} \eta_{r+1}, \ldots, \eta_n \\ \eta_{r+1}, \ldots, \eta_n \end{pmatrix}}.$$

(See the proof of Theorem 5.3 for the analogue of the derivation of this formula.) The denominator in the above formula is positive because it is the Grammian of $n-r$ linearly independent functions.

Let us first show that ψ_{n-r+1} is an eigenfunction and λ_{n-r+1} a corresponding eigenvalue associated with the kernel L_r. It suffices, since $\psi_{n-r+1}(\eta_i)=0$, $i=r+1,\ldots,n$, to prove this same result for the kernel $\bar{K}_r\bar{K}_r'$. From the definition of $\bar{K}_r$ it is easily seen that

$$\bar{K}_r\bar{K}_r'\psi_{n-r+1}(x) = KK'\psi_{n-r+1}(x) + \sum_{i=1}^{r} b_i k_i(x)$$

for some choice of $\{b_i\}_{i=1}^r$. Since $\bar{K}_r = (I-J_r)K$, and $(I-J_r)k_i = 0$, $i=1,\ldots,r$, the constants $\{b_i\}_{i=1}^r$ are uniquely determined by the property that

$$\left(KK'\psi_{n-r+1} + \sum_{i=1}^{r} b_i k_i, k_j\right) = 0, \qquad j=1,\ldots,r.$$

The eigenfunctions ψ_i satisfy $K_r K_r' \psi_i = \lambda_i \psi_i$, where $K_r = K - Q_r K$, Q_r being the orthogonal projection onto span $\{k_1,\ldots,k_r\}$. Thus

$$KK'\psi_{n-r+1} = K_r K_r' \psi_{n-r+1} + \sum_{i=1}^{r} a_i k_i = \lambda_{n-r+1}\psi_{n-r+1} + \sum_{i=1}^{r} a_i k_i,$$

and $\left(\lambda_{n-r+1}\psi_{n-r+1} + \sum_{i=1}^{r}(a_i+b_i)k_i, k_j\right) = 0, j=1,\ldots,r$. Since $(\psi_{n-r+1},k_j)=0$, $j=1,\ldots,r$ (see Lemma 5.9), we obtain $a_i+b_i=0$, $i=1,\ldots,r$, and $\bar{K}_r\bar{K}_r'\psi_{n-r+1} = \lambda_{n-r+1}\psi_{n-r+1}$. Thus ψ_{n-r+1} is an eigenfunction of the kernel L_r with corresponding eigenvalue λ_{n-r+1}. It remains to prove that $(\operatorname{sgn}\psi_{n-r+1}(x))$ $L_r(x,y)(\operatorname{sgn}\psi_{n-r+1}(y)) \geqq 0$ for all $x, y \in [0,1]$. From an application of Sylvester's determinant identity

$$\bar{K}_r \bar{K}_r' \begin{pmatrix} x, \eta_{r+1}, \ldots, \eta_n \\ y, \eta_{r+1}, \ldots, \eta_n \end{pmatrix}$$
$$= \left[K \begin{pmatrix} \eta_1, \ldots, \eta_r \\ 1, \ldots, r \end{pmatrix} \right]^{-2} \int_{0<\sigma_1<\ldots<\sigma_{n-r+1}<1} \!\!\!\cdots\!\int K \begin{pmatrix} \eta_1, \ldots, \eta_n, x \\ 1, \ldots, r, \sigma_1, \ldots, \sigma_{n-r+1} \end{pmatrix}$$
$$\cdot K \begin{pmatrix} \eta_1, \ldots, \eta_n, y \\ 1, \ldots, r, \sigma_1, \ldots, \sigma_{n-r+1} \end{pmatrix} d\sigma_1 \ldots d\sigma_{n-r+1}.$$

The desired result now follow from Property A. □

Analogous to Theorem 5.5, we prove that interpolation to each function in $\mathscr{K}_2^r$ at $\eta_1, \ldots, \eta_n$ from X_n^2 is an optimal linear method of approximation. To prove such a result we must show that interpolation is indeed possible. We state, without proof, the following analogue of Lemma 5.4.

Lemma 5.12. *For* $\{\xi_i\}_{i=1}^{n-r}$ *and* $\{\eta_i\}_{i=1}^{n}$ *as above,* $K \begin{pmatrix} \eta_1, \ldots, \eta_n \\ 1, \ldots, r, \xi_1, \ldots, \xi_{n-r} \end{pmatrix} > 0.$

On the basis of the above lemma, we may define R_n, the interpolation operator from $L^2[0,1]$ to X_n^2 given by

$$R_n h(\eta_i) = K h(\eta_i), \qquad i = 1, \ldots, n.$$

Theorem 5.13. *Let* K *and* $\{k_i\}_{i=1}^r$ *be as above. Then,*

$$\sup_{\|h\| \leqq 1} \|Kh - R_n h\| = d_n(\mathscr{K}_2^r; L^2).$$

Proof. The proof of Theorem 5.13 follows the reasoning of the proofs of the previous theorems. We have

$$\sup_{\|h\| \leqq 1} \|Kh - R_n h\| = \sup_{\|h\| \leqq 1} \|M_r h\|$$

where

$$M_r(x, y) = \frac{K \begin{pmatrix} \eta_1, \ldots, \eta_n, x \\ 1, \ldots, r, \xi_1, \ldots, \xi_{n-r}, y \end{pmatrix}}{K \begin{pmatrix} \eta_1, \ldots, \eta_n \\ 1, \ldots, r, \xi_1, \ldots, \xi_{n-r} \end{pmatrix}}.$$

It therefore suffices to show that $M_r M_r' \psi_{n-r+1} = \lambda_{n-r+1} \psi_{n-r+1}$, and that $(\operatorname{sgn} \psi_{n-r+1}(x))\, M_r M_r'(x, y) (\operatorname{sgn} \psi_{n-r+1}(y)) \geqq 0$ for all $x, y \in [0,1]$. This latter fact follows from the formula for M_r given above.

Let us therefore prove that $M_r' \psi_{n-r+1} = \lambda_{n-r+1} \phi_{n-r+1}$, and $M_r \phi_{n-r+1} = \psi_{n-r+1}$. Recall that $K_r = (I - Q_r) K$ and $K_r \phi_{n-r+1} = \psi_{n-r+1}$, $K_r' \psi_{n-r+1} = \lambda_{n-r+1} \phi_{n-r+1}$. This last equation may be further simplified since $K_r' = K'(I - Q_r)$, and ψ_{n-r+1} is orthogonal to each k_i, $i = 1, \ldots, r$. Thus $K_r' \psi_{n-r+1} =$

$K'\psi_{n-r+1} = \lambda_{n-r+1}\phi_{n-r+1}$. Because $\phi_{n-r+1}(\xi_i) = 0$, $i = 1,\ldots,n-r$, it is easily seen that $M_r'\psi_{n-r+1} = \lambda_{n-r+1}\phi_{n-r+1}$.

To prove that $M_r\phi_{n-r+1} = \psi_{n-r+1}$, we write

$$M_r\phi_{n-r+1}(x) = \frac{\begin{vmatrix} k_1(\eta_1)\ldots k_r(\eta_1) & K(\eta_1,\xi_1)\ldots K(\eta_1,\xi_{n-r}) & K\phi_{n-r+1}(\eta_1) \\ \vdots \quad\quad \vdots & \vdots \quad\quad \vdots & \vdots \\ k_1(\eta_n)\ldots k_r(\eta_n) & K(\eta_n,\xi_1)\ldots K(\eta_n,\xi_{n-r}) & K\phi_{n-r+1}(\eta_n) \\ k_1(x)\ldots k_r(x) & K(x,\xi_1)\ldots K(x,\xi_{n-r}) & K\phi_{n-r+1}(x) \end{vmatrix}}{K\begin{pmatrix} \eta_1,\ldots,\eta_n \\ 1,\ldots,r,\ \xi_1,\ldots,\xi_{n-r} \end{pmatrix}}.$$

Since $\psi_{n-r+1} = K_r\phi_{n-r+1} = K\phi_{n-r+1} - Q_rK\phi_{n-r+1}$, and $\psi_{n-r+1}(\eta_i) = 0$, $i = 1,\ldots,n$, the last column of the numerator satisfies

$$\begin{pmatrix} K\phi_{n-r+1}(\eta_1) \\ \vdots \\ K\phi_{n-r+1}(\eta_n) \\ K\phi_{n-r+1}(x) \end{pmatrix} = \begin{pmatrix} Q_rK\phi_{n-r+1}(\eta_1) \\ \vdots \\ Q_rK\phi_{n-r+1}(\eta_n) \\ Q_rK\phi_{n-r+1}(x) \end{pmatrix} + \begin{pmatrix} 0 \\ \vdots \\ 0 \\ \psi_{n-r+1}(x) \end{pmatrix}.$$

The first of the two columns on the right side is a linear combination of the first r columns of the matrix in the numerator given above for $M_r\phi_{n-r+1}(x)$. Thus $M_r\phi_{n-r+1}(x) = \psi_{n-r+1}(x)$. □

Example 5.2. The Sobolev space of functions $W_2^{(r)}[0,1]$ was defined in Example 3.1. Recall that each $f \in W_2^{(r)}[0,1]$ may be written in the form

$$f(x) = \sum_{i=0}^{r-1} \frac{f^{(i)}(0)}{i!}x^i + \frac{1}{(r-1)!}\int_0^1 (x-y)_+^{r-1} f^{(r)}(y)\,dy$$

and that

$$B_2^{(r)} = \{f\colon f \in W_2^{(r)}[0,1],\ \|f^{(r)}\| \leqq 1\}.$$

In Corollary 3.5 we obtained the value of the n-widths $d_n(B_2^{(r)}; L^2)$ as well as an optimal subspace of the form X_n^1. A major motivation behind the preceding analysis is that with the choice $k_i(x) = x^{i-1}$, $i = 1,\ldots,r$, and $K(x,y) = \dfrac{(x-y)_+^{r-1}}{(r-1)!}$, Property A is satisfied. Thus it follows from Theorems 5.10, 5.11 and 5.13 that there exist "knots" $\{\xi_1,\ldots,\xi_{n-r}\}$ and $\{\eta_1,\ldots,\eta_n\}$ for which the subspace

$$X_n^2 = \operatorname{span}\{1, x, \ldots, x^{r-1}, (x-\xi_1)_+^{r-1}, \ldots, (x-\xi_{n-r})_+^{r-1}\}$$

is optimal, and interpolation to $f \in B_2^{(r)}$ at $\eta_1,\ldots,\eta_n$ from X_n^2 is also an optimal method.

Furthermore, the space of "natural splines" X_n^3 given by functions s in

$$\text{span}\,\{1, x, \ldots, x^{2r-1}, (x-\eta_1)_+^{2r-1}, \ldots, (x-\eta_n)_+^{2r-1}\}$$

which satisfy $s^{(i)}(0) = s^{(i)}(1) = 0$, $i = r, \ldots, 2r-1$, is also an optimal subspace, and interpolation at $\eta_1, \ldots, \eta_n$ an optimal method of approximation.

(The natural spline s, which satisfies the above boundary conditions and which interpolates a function f at $\eta_1, \ldots, \eta_n$, possesses the important minimal property that among all functions $g \in W_2^{(r)}$ for which $g(\eta_i) = f(\eta_i)$, $i = 1, \ldots, n$, s minimizes $\|g^{(r)}\|$. This fact does not depend on the particular choice of $\eta_1, \ldots, \eta_n$ and holds for any given n fixed knots which are also taken to be the interpolation points (see de Boor [1963]).)

Remark. We might just as well have considered the class

$$\mathscr{K}_2^r(\zeta) = \left\{ \sum_{i=1}^{r} a_i K(x, \zeta_i) + \int_0^1 K(x, y)\, h(y)\, dy \colon \|h\| \leq 1,\ a_i \in \mathbb{R} \right\},$$

where $\{\zeta_i\}_{i=1}^r$ are fixed points in a set E containing $[0, 1]$, under the assumption that $K(x, y)$ is nondegenerate totally positive on $[0, 1] \times E$. The class $\mathscr{K}_2^r$ may be thought of as corresponding to the case where the $\{\zeta_i\}_{i=1}^r$ all lie to the left of zero. However it is also possible to consider the n-widths of the class $\mathscr{K}_2^r(\zeta)$ as a subset of $L^2[0, 1]$ for some $\zeta_i \in (0, 1)$. To deal with this case, it is necessary to assume not only that $K(x, y)$ is nondegenerate totally positive as given by Definition 5.1, but also that $K(x, y)$ satisfy certain linear independence conditions with respect to $\dfrac{\partial K(x, \zeta_i)}{\partial y}$ for $\zeta_i \in (0, 1)$. Under conditions of the above form we would then obtain

$$X_n^{2,r} = \text{span}\,\{K(\cdot, \zeta_1), \ldots, K(\cdot, \zeta_r), K(\cdot, \xi_1), \ldots, K(\cdot, \xi_{n-r})\}$$

as an optimal n-dimensional subspace for $d_n(\mathscr{K}_2^r(\zeta); L^2)$ for some choice of $0 < \xi_1 < \ldots < \xi_{n-r} < 1$, with the understanding that if $\xi_j = \zeta_i$, and such a case can and does arise, then we write $\dfrac{\partial K(\cdot, \zeta_i)}{\partial y}$ in place of $K(\cdot, \xi_j)$. Aside from the problem wherein $\xi_j = \zeta_i$, the analysis of the n-widths of $\mathscr{K}_2^r$ and $\mathscr{K}_2^r(\zeta)$ as subsets of L^2 is fundamentally the same.

6. Certain Classes of Periodic Functions

In this section we prove analogues of the results of Section 5.1 for two specific classes of periodic functions. The first of these is:

$$\tilde{\mathscr{K}}_2 = \{(k * \phi)(x) \colon \|\phi\| \leq 1\},$$

where $(k * \phi)(x) = 1/2\pi \int_0^{2\pi} k(x-y)\,\phi(y)\,dy$, $\|\phi\| = \left(1/2\pi \int_0^{2\pi} |\phi(y)|^2\,dy\right)^{1/2}$, and the kernel k is cyclic variation diminishing (CVD) (with an added condition (Definition 6.1)). The second class is:

$$\tilde{\mathscr{B}}_2 = \{a + (G * \phi)(x) \colon \|\phi\| \leqq 1,\ \phi \perp 1,\ a \in \mathbb{R}\}$$

where by $\phi \perp 1$ we mean $\int_0^{2\pi} \phi(y)\,dy = 0$, and where G satisfies certain positivity conditions which we state as Property B.

6.1 n-Widths of Cyclic Variation Diminishing Operators

Let k be a real, square-integrable function on $[0, 2\pi)$. We assume that k is 2π-periodic by the convention $k(x+2\pi) = k(x)$ for all x. In Section 4 we calculated the n-widths and exhibited optimal subspaces for $d_n(\tilde{\mathscr{K}}_2; L^2[0, 2\pi])$. One problem in obtaining analogues to Theorems 5.3 and 5.5 is that the kernel $K(x,y) = k(x-y)$ cannot possibly be totally positive. It can only satisfy half the requirements of total positivity. These, together with the simple form of the associated eigenvalue-eigenfunction problem, enable us to obtain analogues of X_n^2 and X_n^3 of Section 5.1 for even n.

Definition 6.1. We say that k is *nondegenerate cyclic variation diminishing* ($NCVD$) if k is 2π-periodic, continuous, non-negative CVD and

$$\dim \operatorname{span}\{k(x_1 - \cdot), \ldots, k(x_n - \cdot)\} = (\dim \operatorname{span}\{k(\cdot - y_1), \ldots, k(\cdot - y_n)\} =)\, n$$

for every choice of $0 \leqq x_1 < \ldots < x_n < 2\pi$, $(0 \leqq y_1 < \ldots < y_n < 2\pi)$ and for all n.

(See Section 4 of Chapter III for a definition of cyclic variation diminishing (CVD).)

We prove the following theorem.

Theorem 6.1. *Let k be an $NCVD$ kernel. Associated with k is its formal Fourier series*

$$k(x) \sim \sum_{-\infty}^{\infty} a_n e^{inx}$$

where $a_n = 1/2\pi \int_0^{2\pi} k(x) e^{-inx}\,dx$. Then

$$d_0(\tilde{\mathscr{K}}_2; L^2) = a_0, \quad \textit{and} \quad d_{2n-1}(\tilde{\mathscr{K}}_2; L^2) = d_{2n}(\tilde{\mathscr{K}}_2; L^2) = |a_n|, \qquad n = 1, 2, \ldots.$$

Furthermore,

(1) T_{n-1} *is an optimal subspace for* $d_{2n-1}(\tilde{\mathscr{K}}_2; L^2)$,
(2) $X_{2n}^2 = \operatorname{span}\{k(\cdot), k(\cdot - \pi/n), \ldots, k(\cdot - (2n-1)\pi/n)\}$ *is an optimal $2n$-dimensional subspace for* $d_{2n}(\tilde{\mathscr{K}}_2; L^2)$,

(3) $X^3_{2n} = \operatorname{span}\{k'k(\cdot), k'k(\cdot - \pi/n), \ldots, k'k(\cdot - (2n-1)\pi/n)\}$ *is an optimal $2n$-dimensional subspace for* $d_{2n}(\tilde{\mathscr{K}}_2; L^2)$.

Proof. The values of the n-widths and the optimality of T_{n-1} for $d_{2n-1}(\tilde{\mathscr{K}}_2; L^2)$ are direct consequences of Theorem 4.1, and of the fact that $a_0 \geqq |a_1| \geqq |a_2| \geqq \ldots$, since k is *CVD* (see Theorem 4.6 of Chapter III). Recall that $\phi_1(x) = 1$ is an eigenfunction associated with the kernel $k'k(x-y)$ with eigenvalue a_0^2, and $\phi_{2n}(x) = e^{inx}$, $\phi_{2n-1}(x) = e^{-inx}$ are eigenfunctions with corresponding eigenvalue $|a_n|^2$.

The *NCVD* property of k enters into the analysis in proving that X^2_{2n} and X^3_{2n} are also optimal subspaces.

The proof is totally analogous to the proof of Theorem 5.3 except that we cannot appeal to results concerning eigenfunctions and eigenvalues of totally positive kernels. This is hardly necessary here since we have explicitly exhibited them.

Let P denote the orthogonal projection of $L^2[0, 2\pi]$ onto X^2_{2n}. To prove the optimality of X^2_{2n}, we must show that

$$\sup\{\|k * \phi - P(k * \phi)\| : \|\phi\| \leqq 1\} = |a_n|.$$

Thus we are once again led to the problem of calculating the largest eigenvalue of the compact self-adjoint, non-negative operator generated by the kernel $L(x, y)$ given by

$$L(x,y) = \frac{k'k\begin{pmatrix} x, 0, \pi/n, \ldots, (2n-1)\pi/n \\ y, 0, \pi/n, \ldots, (2n-1)\pi/n \end{pmatrix}}{k'k\begin{pmatrix} 0, \pi/n, \ldots, (2n-1)\pi/n \\ 0, \pi/n, \ldots, (2n-1)\pi/n \end{pmatrix}}.$$

Note that the denominator is positive since it is simply the Grammian of the linearly independent functions $\{k(\cdot), k(\cdot - \pi/n), \ldots, k(\cdot - (2n-1)\pi/n)\}$.

The function $\sin nx$ is an eigenfunction associated with the kernel $k'k(x-y)$ and the eigenvalue $|a_n|^2$. Furthermore, $\sin nx$ vanishes at $i\pi/n$, $i = 0, 1, \ldots, 2n-1$. Thus

$$1/2\pi \int_0^{2\pi} L(x,y) \sin ny \, dy = |a_n|^2 \sin nx.$$

Moreover $(\sin nx)\, L(x, y)\,(\sin ny) \geqq 0$ for all $x, y \in [0, 2\pi]$. Thus $|a_n|^2$ is the largest eigenvalue of the operator induced by the kernel $L(x, y)$ which proves that X^2_{2n} is an optimal subspace for $d_{2n}(\tilde{\mathscr{K}}_2; L^2)$. The optimality of X^3_{2n} follows from the method of proof given in Theorem 5.3, which is easily adapted to our problem. □

Let $\psi_{2n}(x) = 1/2\pi \int_0^{2\pi} k(x-y) \sin ny \, dy$. We can explicitly calculate this integral to see that ψ_{2n} is a linear combination of $\sin nx$ and $\cos nx$. Thus ψ_{2n} has exactly $2n$ zeros (sign changes) in $[0, 2\pi)$ at $\{\beta + i\pi/n\}_{i=0}^{2n-1}$, $0 \leqq \beta < \pi/n$.

Analogous to Lemma 5.4, we obtain:

Lemma 6.2. *For k and β as above,*

$$k\begin{pmatrix}\beta, \beta+\pi/n, \ldots, \beta+(2n-1)\,\pi/n \\ 0, \pi/n, \ldots, (2n-1)\,\pi/n\end{pmatrix} \neq 0.$$

Let R denote the interpolation operator from $L^2[0,2\pi)$ to X_{2n}^2 given by

$$R\phi(\beta+i\pi/n) = (k*\phi)(\beta+i\pi/n), \qquad i=0,1,\ldots,2n-1.$$

The above lemma implies the existence of such an operator and the next theorem is then proven in exactly the same manner as was Theorem 5.5.

Theorem 6.3. *Let k be an NCVD kernel. Then*

$$\sup_{\|\phi\|\leqq 1} \|k*\phi - R\phi\| = d_{2n}(\tilde{\mathscr{K}}_2; L^2).$$

Example 6.1. We consider a class of functions which may be characterized by convolutions against a *CVD* kernel.

Let $S_\beta = \{z\colon z\in\mathbb{C}, |\operatorname{Im} z| < \beta\}$, and let $\tilde{H}_\beta$ denote the class of functions f which are analytic in S_β, real and 2π-periodic on the x-axis, and satisfy $1/2\pi \int_0^{2\pi} |\operatorname{Re} f(t+i\beta)|^2\,dt \leqq 1$. (The boundary values are taken in the correct sense.) A function f is in $\tilde{H}_\beta$ if and only if it is of the form

$$f(z) = 1/2\pi \int_0^{2\pi} K_\beta(z-t)\operatorname{Re}[f(t+i\beta)]\,dt$$

for $z\in S_\beta$, where $K_\beta(z) = 1 + 2\sum_{k=1}^{\infty} (\cos kz)/(\cosh k\beta)$.

One derives the above formula as follows. Since f is 2π-periodic on the real axis, $w=e^{iz}$ maps S_β onto the annulus $e^{-\beta} < |w| < e^{\beta}$ and $g(w) = g(e^{iz}) = f(z)$ is analytic thereon. Thus $g(w) = \sum_{k=-\infty}^{\infty} \gamma_k w^k = \sum_{k=-\infty}^{\infty} \gamma_k e^{ikz} = f(z)$. We evaluate the γ_k by contour integration about the circles of radii e^{β} and $e^{-\beta}$, and use the fact that $f(z) = \overline{f(\bar{z})}$ (since f is real on the real axis) to obtain the desired formula. In Section 4 of Chapter III, we showed that the kernel K_β is *CVD* on $[0,2\pi)$ in a strong sense which certainly implies that it is *NCVD*. We may therefore apply Theorems 6.1 and 6.3 to the class $\tilde{H}_\beta$, where we are approximating functions in the class on the interval $[0,2\pi)$ in the L^2-norm.

6.2 n-Widths for Kernels Satisfying Property B

The set $\tilde{\mathscr{B}}_2$ was defined above and also in Section 4.1. We shall consider $\tilde{\mathscr{B}}_2$ for functions G satisfying Property B.

Definition 6.2. A real, 2π-periodic, continuous function G satisfies *Property B* if for every choice of $0 \leqq y_1 < \ldots < y_m < 2\pi$, and each m, the subspace

$$X_m = \left\{ b + \sum_{i=1}^{m} b_i G(\cdot - y_i) \colon \sum_{i=1}^{m} b_i = 0 \right\}$$

is of dimension m, and is a weak Tchebycheff (WT-) system for m odd.

This definition parallels Definition 6.1. A function k is $NCVD$ if and only if for every choice of $0 \leqq y_1 < \ldots < y_m < 2\pi$, and each m, the subspace

$$Y_m = \operatorname{span}\{k(\cdot - y_1), \ldots, k(\cdot - y_m)\}$$

is of dimension m, and is a WT-system for m odd.

Before entering into a discussion of the n-widths of $\tilde{\mathscr{B}}_2$ in L^2, we present various properties of functions satisfying Property B. The first result will only be used in Section 4 of Chapter V.

It is known (see Section 4 of Chapter III) that for ϕ a piecewise continuous, 2π-periodic function, and k $NCVD$,

$$S_c(k * \phi) \leqq S_c(\phi)$$

where S_c is the (cyclic) sign change count for piecewise continuous, 2π-periodic functions as defined in Chapter III.

We prove a similar result for G satisfying Property B.

Proposition 6.4. *Assume G satisfies Property B. Let ϕ be a piecewise continuous, 2π-periodic function satisfying $\phi \perp 1$, and set*

$$\psi(x) = a + (G * \phi)(x).$$

Then $S_c(\psi) \leqq S_c(\phi)$.

Proof. We prove that for ϕ as above satisfying meas $\{x \colon \phi(x) = 0\} = 0$, ψ has at most $S_c(\phi)$ zeros in $[0, 2\pi)$. Since we are only interested, in the statement of the proposition, in the number of sign changes of ψ this fact, by continuity considerations, is sufficient to establish our result.

Let $S_c(\phi) = 2m$, and $0 \leqq y_1 < \ldots < y_{2m} < 2\pi$ satisfy

$$\varepsilon \phi(y)(-1)^i > 0, \quad y \in (y_i, y_{i+1}),$$

$i = 0, 1, \ldots, 2m$, where $y_0 = 0$, $y_{2m+1} = 2\pi$, and $\varepsilon \in \{-1, 1\}$. Assume ψ has $2m + 1$ zeros at $0 \leqq x_1 < \ldots < x_{2m+1} < 2\pi$. The space

$$X_{2m+1} = \left\{ b + \sum_{i=1}^{2m+1} b_i G(x_i - \cdot) \colon \sum_{i=1}^{2m+1} b_i = 0 \right\}$$

is a WT-system of dimension $2m+1$. As such there exists a nontrivial function

$$g(y) = b + \sum_{i=1}^{2m+1} b_i G(x_i - y)$$

for which $\sum_{i=1}^{2m+1} b_i = 0$, and

$$\varepsilon g(y)(-1)^i \geqq 0, \quad y \in (y_i, y_{i+1}), \qquad i = 0, 1, \ldots, 2m.$$

Thus

$$\begin{aligned}
0 &< 1/2\pi \int_0^{2\pi} g(y)\,\phi(y)\,dy \\
&= 1/2\pi \int_0^{2\pi} \left[b + \sum_{i=1}^{2m+1} b_i G(x_i - y) \right] \phi(y)\,dy \\
&= \sum_{i=1}^{2m+1} b_i \left[1/2\pi \int_0^{2\pi} G(x_i - y)\,\phi(y)\,dy \right] \\
&= \sum_{i=1}^{2m+1} b_i \left[a + 1/2\pi \int_0^{2\pi} G(x_i - y)\,\phi(y)\,dy \right] \\
&= \sum_{i=1}^{2m+1} b_i \psi(x_i) \\
&= 0.
\end{aligned}$$

$\left(\text{We used both the fact that } \phi \perp 1, \text{ and } \sum_{i=1}^{2m+1} b_i = 0.\right)$ This contradiction proves the proposition. □

Our next result is an analogue of Lemma 5.4.

Proposition 6.5. *Let G satisfy Property B. Let $\phi \in L^1[0, 2\pi]$ be such that $\phi \perp 1$, $\operatorname{meas}\{x: \phi(x) = 0\} = 0$, and ϕ has $2m$ sign changes on $[0, 2\pi)$ (considered cyclically). By this we mean that there exist points $0 \leqq y_1 < \ldots < y_{2m} < 2\pi$ for which*

$$\varepsilon \phi(y)(-1)^i > 0, \quad \text{a.e.} \quad y \in (y_i, y_{i+1}), \qquad i = 0, 1, \ldots, 2m,$$

where $y_0 = 0$, $y_{2m+1} = 2\pi$, and $\varepsilon \in \{-1, 1\}$. Assume that

$$\psi(x) = a + (G * \phi)(x)$$

has $2m$ zeros at $0 \leqq x_1 < \ldots < x_{2m} < 2\pi$. Then for any given data $\{c_i\}_{i=1}^{2m}$, there exists a unique function f in

$$X_{2m} = \left\{ b + \sum_{i=1}^{2m} b_i G(\cdot - y_i) : \sum_{i=1}^{2m} b_i = 0 \right\}$$

such that $f(x_i) = c_i$, $i = 1, \ldots, 2m$.

Proof. Assume that this is not the case. There then exist constants d, $\{d_i\}_{i=1}^{2m}$ not all zero, such that $\sum_{i=1}^{2m} d_i = 0$, and

$$h(x) = d + \sum_{i=1}^{2m} d_i G(x - y_i)$$

vanishes at $x = x_1, \ldots, x_{2m}$. By Property B, h is not identically zero. Thus there exists a $\gamma \in [0, 2\pi)/\{x_1, \ldots, x_{2m}\}$ for which $h(\gamma) \neq 0$. Choose c so that

$$\psi(\gamma) - c h(\gamma) = 0.$$

By Property B, the space

$$X_{2m+1} = \left\{ b + \sum_{i=1}^{2m} b_i G(x_i - \cdot) + b_{2m+1} G(\gamma - \cdot) \colon \sum_{i=1}^{2m+1} b_i = 0 \right\}$$

is a *WT*-system of dimension $2m+1$. There therefore exist coefficients α, $\{\alpha_j\}_{j=1}^{2m+1}$, not all zero, such that $\sum_{j=1}^{2m+1} \alpha_j = 0$, and

$$g(y) = \alpha + \sum_{j=1}^{2m} \alpha_j G(x_j - y) + \alpha_{2m+1} G(\gamma - y)$$

satisfies $\varepsilon g(y)(-1)^i \geqq 0$, $y \in (y_i, y_{i+1})$, $i = 0, 1, \ldots, 2m$. Thus

$$\begin{aligned}
0 &= \sum_{j=1}^{2m} \alpha_j [\psi(x_j) - c h(x_j)] + \alpha_{2m+1} [\psi(\gamma) - c h(\gamma)] \\
&= \sum_{j=1}^{2m} \alpha_j \left[a + 1/2\pi \int_0^{2\pi} G(x_j - y)\, \phi(y)\, dy \right] - c \sum_{j=1}^{2m} \alpha_j \left[d + \sum_{i=1}^{2m} d_i G(x_j - y_i) \right] \\
&\quad + \alpha_{2m+1} \left[a + 1/2\pi \int_0^{2\pi} G(\gamma - y)\, \phi(y)\, dy \right] - c \alpha_{2m+1} \left[d + \sum_{i=1}^{2m} d_i G(\gamma - y_i) \right].
\end{aligned}$$

Since $\sum_{j=1}^{2m+1} \alpha_j = \sum_{i=1}^{2m} d_i = 0$ and $\phi \perp 1$, this reduces to

$$\begin{aligned}
0 &= 1/2\pi \int_0^{2\pi} \left[\alpha + \sum_{j=1}^{2m} \alpha_j G(x_j - y) + \alpha_{2m+1} G(\gamma - y) \right] \phi(y)\, dy \\
&\quad - c \sum_{i=1}^{2m} d_i \left[\alpha + \sum_{j=1}^{2m} \alpha_j G(x_j - y_i) + \alpha_{2m+1} G(\gamma - y_i) \right] \\
&= 1/2\pi \int_0^{2\pi} g(y)\, \phi(y)\, dy - c \sum_{i=1}^{2m} d_i g(y_i) \\
&= 1/2\pi \int_0^{2\pi} |g(y)|\, |\phi(y)|\, dy \\
&> 0. \quad \square
\end{aligned}$$

From a function satisfying Property B, it is a very simple matter to generate a class of functions which satisfy Property B.

Proposition 6.6. *Assume G satisfies Property B. Set*

$$\tilde{G}(x) = \int_0^x (G(y) - a_0)\, dy + c$$

where $a_0 = 1/2\pi \int_0^{2\pi} G(y)\, dy$, and $c \in \mathbb{R}$. Then $\tilde{G}$ satisfies Property B.

Proof. By the choice of a_0, it follows (since $\tilde{G}(0) = \tilde{G}(2\pi) = c$) that $\tilde{G}$ is a real, continuous, 2π-periodic function. It is also easy to prove that X_m, as in Definition 6.2 but with $\tilde{G}$ replacing G, is of dimension m. It remains to show that X_{2m-1} is a WT-system for each m.

If f and f' are real, continuous and 2π-periodic, then by Rolle's Theorem,

$$S_c(f) \leqq S_c(f').$$

For $0 \leqq y_1 < \ldots < y_{2m-1} < 2\pi$, and $\sum_{i=1}^{2m-1} b_i = 0$,

$$\begin{aligned} S_c\left(b + \sum_{i=1}^{2m-1} b_i \tilde{G}(\cdot - y_i)\right) &\leqq S_c\left(\sum_{i=1}^{2m-1} b_i \tilde{G}'(\cdot - y_i)\right) \\ &= S_c\left(\sum_{i=1}^{2m-1} b_i [G(\cdot - y_i) - a_0]\right) \\ &= S_c\left(\sum_{i=1}^{2m-1} b_i G(\cdot - y_i)\right) \\ &\leqq 2m - 2. \end{aligned}$$

The last inequality follows from the WT property of G. Thus X_{2m-1} is a WT-system. □

We now delineate two particular classes of functions which satisfy Property B.

Proposition 6.7. *Let k be an NCVD kernel. Set*

$$G(x) = \int_0^x (k(y) - a_0)\, dy + c$$

where $a_0 = 1/2\pi \int_0^{2\pi} k(y)\, dy$ and $c \in \mathbb{R}$. Then G satisfies Property B.

The proof of this result is a repeat of the proof of Proposition 6.6.

The Sobolev space of periodic functions in $L^2[0, 2\pi]$ was introduced, and discussed in some detail in Example 4.1.Recall that $\tilde{W}_2^{(r)}$ is defined as the space of 2π-periodic functions whose $(r-1)$st derivative is absolutely continuous and whose rth derivative is in $L^2[0,2\pi]$. Let

$$\tilde{B}_2^{(r)} = \{f\colon f \in \tilde{W}_2^{(r)},\ \|f^{(r)}\| \leqq 1\}.$$

Every $f \in \tilde{W}_2^{(r)}$ has the representation

$$f(x) = a_0 + 1/\pi \int_0^{2\pi} D_r(x-y)\,\phi(y)\,dy,$$

where $a_0 = 1/2\pi \int_0^{2\pi} f(x)\,dx$, and $\phi(y) = f^{(r)}(y)$ satisfies $\phi \perp 1$. Thus $\tilde{B}_2^{(r)}$ is of the form $\tilde{\mathscr{B}}_2$ where $G(x) = 2D_r(x)$. The function D_r was defined in two equivalent ways and is referred to as the Dirichlet kernel or the Bernouilli monospline. We now prove:

Proposition 6.8. *For each $r \geqq 2$, D_r satisfies Property B.*

Proof. Since $D_r'(x) = D_{r-1}(x)$, it suffices, by Proposition 6.6 to establish the result for $r=2$. It is a simple matter to prove that $\dim X_m = m$, where X_m is as defined in Definition 6.2. It remains to show that X_{2m-1} is a WT-system for $m = 1, 2, \ldots$.

Various methods exist of proving this fact. We prove it directly. $D_2(x-y) = -(x-y)^2/4 - \pi(x-y)/2 - \pi^2/6 + \pi(x-y)_+^1/2$. Set $f(x) = b + \sum_{i=1}^{2m-1} b_i D_2(x-y_i)$, where $\sum_{i=1}^{2m-1} b_i = 0$, and for convenience, $0 < y_1 < \ldots < y_{2m-1} < 2\pi$. Thus after simplification,

$$f(x) = c + dx + \pi/2 \sum_{i=1}^{2m-1} b_i(x-y_i)_+^1,$$

with $f(0) = f(2\pi)$. We must prove that f has at most $2m-2$ sign changes (counted cyclically) on $[0,2\pi)$. Since f is linear on $[y_i, y_{i+1}]$, $i = 0, 1, \ldots, 2m-1$, $y_0 = 0$, $y_{2m} = 2\pi$, and $f(0) = f(2\pi)$, f has at most $2m$ sign changes on $[0,2\pi)$. f has exactly $2m$ sign changes on $[0,2\pi)$ if and only if f changes sign in (y_i, y_{i+1}) for each i. However, since $2m-1$ is odd, it is therefore necessary that f be strictly monotone in opposite directions on $[y_0, y_1]$ and on $[y_{2m-1}, y_{2m}]$. But $f'(x) = d$ for $x \in (y_0, y_1) \cup (y_{2m-1}, y_{2m})$. This contradiction proves the proposition. □

Remark. The function $D_1(x)$ also satisfies all the conditions of Property B except that it has a jump discontinuity at $x = 0$ (identifying 0 with 2π). For this technical reason it is excluded from the above analysis. However the results of this section also apply, mutatis mutandis, to $\tilde{B}_2^{(1)}$.

We now prove:

Theorem 6.9. *Let G satisfy Property B. Associated with G is its formal Fourier series*

$$G(x) \sim \sum_{-\infty}^{\infty} a_n e^{inx}.$$

Set

$$\tilde{\mathscr{B}}_2 = \{a + (G * \phi)(x) \colon \|\phi\| \leqq 1,\ \phi \perp 1,\ a \in \mathbb{R}\}.$$

Then,

$$d_0(\tilde{\mathscr{B}}_2; L^2) = \infty, \quad \text{and} \quad d_{2n-1}(\tilde{\mathscr{B}}_2; L^2) = d_{2n}(\tilde{\mathscr{B}}_2; L^2) = |a_n|, \qquad n = 1, 2, \ldots.$$

Furthermore,

(1) T_{n-1} *is an optimal subspace for* $d_{2n-1}(\tilde{\mathscr{B}}_2; L^2)$,

(2) $X_{2n}^2 = \left\{ b + \sum_{i=0}^{2n-1} b_i G(\cdot - i\pi/n) \colon \sum_{i=0}^{2n-1} b_i = 0 \right\}$ *is an optimal* $2n$*-dimensional subspace for* $d_{2n}(\tilde{\mathscr{B}}_2; L^2)$.

Remark. Implied in the statement of Theorem 6.9 is the fact that $|a_n| \geqq |a_{n+1}|$, $n = 1, 2, \ldots$. This might well be proven directly. However it also follows as a consequence of the proof of Theorem 6.9. (Note that there is no condition on a_0.)

Proof. Since G is real, $|a_{-k}| = |a_k|$ for $k = 1, 2, \ldots$. Let $\{|a_{k_j}|\}_{j=1}^{\infty}$ denote a non-increasing rearrangement of $\{|a_k|\}_{k=1}^{\infty}$. Thus $|a_{k_1}| \geqq |a_{k_2}| \geqq \ldots$, where $|a_{k_1}| = \max\{|a_k| \colon k = 1, 2, \ldots\}$, $|a_{k_2}| = \max\{|a_k| \colon k = 1, 2, \ldots, k \neq k_1\}$, etc.....

On the basis of the method of proof of Theorem 4.2, it may be shown that $d_0(\tilde{\mathscr{B}}_2; L^2) = \infty$, $d_{2n-1}(\tilde{\mathscr{B}}_2; L^2) = d_{2n}(\tilde{\mathscr{B}}_2; L^2) = |a_{k_n}|$, and

$$\text{span}\{1, \sin k_1 x, \cos k_1 x, \ldots, \sin k_{n-1} x, \cos k_{n-1} x\}$$

is an optimal subspace for $d_{2n-1}(\tilde{\mathscr{B}}_2; L^2)$.

It therefore remains to prove that we may choose $k_j = j$ for all j, and that X_{2n}^2 is optimal for $d_{2n}(\tilde{\mathscr{B}}_2; L^2)$. We prove the former result from the latter. We shall not prove the optimality of X_{2n}^2 as in Theorem 6.1, but rather by the method of Theorem 6.3.

Set $\psi_{2n}(x) = 1/2\pi \int_0^{2\pi} G(x-y) \sin ny\, dy$. Explicitly, $\psi_{2n}(x) = a \sin nx + b \cos nx$, where $a_n = a + ib$. Thus ψ_{2n} has $2n$ zeros at $\{\beta + i\pi/n\}_{i=0}^{2n-1}$ for some $\beta \in [0, \pi/n)$. For $f \in \tilde{\mathscr{B}}_2$, let R_{2n} denote the interpolation operator from X_{2n}^2 which interpolates to f at $\{\beta + i\pi/n\}_{i=0}^{2n-1}$. By Proposition 6.5, such an operator is well-defined.

One basis for X_{2n}^2 is the set of functions 1, $\{G(x - i\pi/n) - G(x)\}_{i=1}^{2n-1}$. To somewhat ease the notational complexity, set

$$M(x, y) = \begin{vmatrix} 1 & 1 \\ G(\beta - y) - G(\beta) & G(x - y) - G(x) \end{vmatrix}.$$

By Sylvester's determinant identity (see Section 2 of Chapter III), the determinant

$$M\begin{pmatrix} \beta + \pi/n, \ldots, \beta + (2n-1)\pi/n \\ \pi/n, \ldots, (2n-1)\pi/n \end{pmatrix}$$

is, up to a fixed interchange of rows and columns, the determinant of the matrix with columns obtained by evaluating the vector

$$\{1, G(x-\pi/n)-G(x), \ldots, G(x-(2n-1)\pi/n)-G(x)\}$$

at $x=\beta, \beta+\pi/n, \ldots, \beta+(2n-1)\pi/n$, and is therefore not zero by Proposition 6.5.

For ϕ satisfying $\phi \perp 1$,

$$1/2\pi \int_0^{2\pi} M(x,y)\,\phi(y)\,dy = 1/2\pi \int_0^{2\pi} G(x-y)\,\phi(y)\,dy + c,$$

where c is some constant. For $\{b_i\}_{i=0}^{2n-1}$ satisfying $\sum_{i=0}^{2n-1} b_i = 0$,

$$\sum_{i=1}^{2n-1} b_i M(x, i\pi/n) = \sum_{i=0}^{2n-1} b_i G(x-i\pi/n) + a,$$

where a is some constant. Thus

$$\begin{aligned} &1/2\pi \int_0^{2\pi} M(x,y)\,\phi(y)\,dy - \sum_{i=1}^{2n-1} b_i M(x, i\pi/n) \\ &\quad = 1/2\pi \int_0^{2\pi} G(x-y)\,\phi(y)\,dy - \sum_{i=0}^{2n-1} b_i G(x-i\pi/n) - b \end{aligned}$$

if $\phi \perp 1$ and $\sum_{i=0}^{2n-1} b_i = 0$.

Set

$$\bar{M}(x,y) = M\begin{pmatrix} x, \beta+\pi/n, \ldots, \beta+(2n-1)\pi/n \\ y, \pi/n, \ldots, (2n-1)\pi/n \end{pmatrix} \Big/ M\begin{pmatrix} \beta+\pi/n, \ldots, \beta+(2n-1)\pi/n \\ \pi/n, \ldots, (2n-1)\pi/n \end{pmatrix}.$$

Therefore,

$$\begin{aligned} &\sup\{\|a + G*\phi - R_{2n}\phi\| : \|\phi\| \leqq 1,\ \phi \perp 1,\ a \in \mathbb{R}\} \\ &\quad = \sup\left\{\left\| 1/2\pi \int_0^{2\pi} \bar{M}(\cdot,y)\,\phi(y)\,dy \right\| : \|\phi\| \leqq 1,\ \phi \perp 1\right\} \\ &\quad \leqq \sup\left\{\left\| 1/2\pi \int_0^{2\pi} \bar{M}(\cdot,y)\,\phi(y)\,dy \right\| : \|\phi\| \leqq 1\right\}. \end{aligned}$$

Again, by Sylvester's determinant identity, it follows from Property B that

$$(\sin n(x-\beta))\,\bar{M}(x,y)\,(\sin ny)$$

does not change sign on $[0,2\pi] \times [0,2\pi]$. Furthermore,

$$1/2\pi \int_0^{2\pi} \bar{M}(x,y) \sin ny\, dy = 1/2\pi \int_0^{2\pi} G(x-y) \sin ny\, dy,$$

since $\int_0^{2\pi} M(\beta + i\pi/n, y) \sin ny\, dy = 0,\ i = 1,\ldots,2n-1.$

Following the pattern of proof given in Theorem 6.1, we can now prove that $|a_n|^2$ is the largest eigenvalue of the integral operator with kernel $\bar{M}'\bar{M}(x,y)$. This implies that $E(\tilde{\mathscr{B}}_2; X_{2n}^2) \leqq |a_n|$ and therefore $d_{2n}(\tilde{\mathscr{B}}_2; L^2) \leqq |a_n|$.

Setting $n = 1$, we obtain $|a_{k_1}| = d_2(\tilde{\mathscr{B}}_2; L^2) \leqq |a_1|$. From the definition of k_1, equality holds and we may choose $k_1 = 1$. Assume $k_j = j, j = 1,\ldots,n-1$. Then $|a_{k_n}| = d_{2n}(\tilde{\mathscr{B}}_2; L^2) \leqq |a_n|$. Again from the definition of k_n, we may choose $k_n = n$. This proves the theorem. □

From the proof of Theorem 6.9 we have also shown:

Theorem 6.10. *Let G satisfy Property B, and let R_{2n} be as defined above. Then*

$$\sup\{\|a + G*\phi - R_{2n}\phi\| : \|\phi\| \leqq 1,\ \phi \perp 1,\ a \in \mathbb{R}\} = d_{2n}(\tilde{\mathscr{B}}_2; L^2).$$

Notes and References

Section 2. Theorem 2.1 is to be found in various references, see e.g. Riesz, Sz. Nagy [1955], Gohberg, Krein [1969], or Weinstein, Stenger [1972]. Theorem 2.2 is one of those known results whose authorship is unclear. The equality $\delta_n(T(H); H) = s_{n+1}(T)$ for $T \in K(H,H)$ is due to Allahverdiev [1957], and is sometimes used as an alternative definition of s-numbers (see e.g. Pietsch [1974]). It is generally accepted that $d_n(T(H); H) = s_{n+1}(T)$ goes back to Kolmogorov [1936], although it is not therein explicitly stated as such. Examples 2.3 and 2.4 are due to Jagerman. Example 2.3 is to be found in Jagerman [1969], while Example 2.4 is in Jagerman [1970].

Section 3. Theorem 3.10 and Example 3.4 are due to Melkman, Micchelli [1979], although the analysis of Example 3.4, as given there, was somewhat lacking. Example 3.5 (i.e., Theorems 3.17 and 3.18) is due to Melkman [1977] (see also Melkman, Micchelli [1979]). For other applications of n-widths to signal representation, see Parks [1974], and Parks, Meier [1971]. Theorem 3.20 is, of course, a result of Ismagilov [1968].

Section 4. Proposition 4.5 is due to Ismagilov [1968]. The statement of Proposition 4.8 may be found in Taikov [1980]. The results of Subsection 4.3 are generalizations of work of Taikov [1976], [1977a], [1979].

Section 5. The material of this section is taken from Melkman, Micchelli [1978].

Section 6. Theorems 6.1 and 6.3 are due to Melkman, Micchelli [1978]. The results of Subsection 6.2 are all new. The method of proof of Theorem 6.9 may be applied, in place of Theorem 4.6 of Chapter III, to prove that if k is CVD, then $|a_n| \geqq |a_{n+1}|$, $n = 0, 1, \ldots$, where $k(x) \sim \sum_{-\infty}^{\infty} a_n e^{inx}$.

Chapter V. Exact n-Widths of Integral Operators

1. Introduction

Let $K(x, y) \in C([0,1] \times [0,1])$ and set

$$\mathscr{K}_p = \left\{Kh(x)\colon Kh(x) = \int_0^1 K(x, y)\, h(y)\, dy,\ \|h\|_p \leqq 1\right\}$$

where

$$\|h\|_p = \begin{cases} \left(\int_0^1 |h(y)|^p\, dy\right)^{1/p}, & 1 \leqq p < \infty \\ \operatorname{ess\,sup}\{|h(y)|\colon 0 \leqq y \leqq 1\}, & p = \infty \end{cases}$$

In Section 2 of this chapter we determine the n-widths (d_n, d^n and δ_n) of $\mathscr{K}_p$ in L^q for $p = \infty$, $1 \leqq q \leqq \infty$, and $1 \leqq p \leqq \infty$, $q = 1$, where K is a nondegenerate totally positive kernel. (Such kernels were defined in Section 5 of Chapter IV and are redefined in Section 2.) We prove that all three n-widths considered are equal (the common value depending on p and q) and that there exists a set of n distinct points $\{\xi_i\}_{i=1}^n$ in $(0,1)$ for which $X_n = \operatorname{span}\{K(\cdot, \xi_i)\}_{i=1}^n$ is optimal for $d_n(\mathscr{K}_p; L^q)$, with the $\{\xi_i\}_{i=1}^n$ dependent on p, q. Furthermore, there exists an additional set $\{\eta_i\}_{i=1}^n$ of n distinct points in $(0,1)$ (again dependent on p, q) for which interpolation from X_n to $Kh(x)$ at the $\{\eta_i\}_{i=1}^n$ is optimal for $\delta_n(\mathscr{K}_p; L^q)$. (An analogous statement holds for $d^n(\mathscr{K}_p; L^q)$.)

In Section 3 we extend the analysis to

$$\mathscr{K}_p^r = \left\{\sum_{i=1}^r a_i k_i(x) + Kh(x)\colon \|h\|_p \leqq 1,\ a_i \in \mathbb{R}\right\}$$

under the assumption that Property A of Section 5.2 of Chapter IV holds. We obtain results totally analogous to those proven in Section 2.

The results of Sections 2 and 3 were motivated by the important paper of Tichomirov [1969], wherein is determined $d_n(B_\infty^{(r)}; L^\infty)$.

On the basis of these results, as well as on those of Section 5 of Chapter IV, it is conjectured that similar theorems hold for $d_n(\mathscr{K}_p; L^q)$ (and for d^n and δ_n), where $1 \leqq q \leqq p \leqq \infty$. In other words we conjecture that for given $1 \leqq q \leqq p \leqq \infty$ all three n-widths are equal, and that there exist two sets of n distinct points in $(0,1)$, $\{\xi_i\}_{i=1}^n$ and $\{\eta_i\}_{i=1}^n$, such that $X_n = \operatorname{span}\{K(\cdot, \xi_i)\}_{i=1}^n$ is optimal for $d_n(\mathscr{K}_p; L^q)$, and interpolation from X_n to $Kh(x)$ at the $\{\eta_i\}_{i=1}^n$ is optimal for $\delta_n(\mathscr{K}_p; L^q)$ (with an analogous statement for $d^n(\mathscr{K}_p; L^q)$ and for the n-widths of $\mathscr{K}_p^r$ in L^q)[1]. When dealing with the specific example of $d_n(B_p^{(r)}; L^q)$, this would translate into the statement that there exists, for $n \geqq r$, an n-dimensional optimal subspace of splines of degree $r-1$ with $n-r$ fixed distinct knots.

In Section 4 we study periodic analogues of the results of Section 2. Two sets of functions are considered. They are

$$\tilde{\mathscr{K}}_p = \left\{(k * h)(x) \colon (k * h)(x) = 1/2\pi \int_0^{2\pi} k(x-y)\, h(y)\, dy,\ \|h\|_p \leqq 1\right\}$$

where k is a nondegenerate cyclic variation diminishing kernel (defined in Section 6.1 of Chapter IV and redefined in Section 4) and

$$\tilde{\mathscr{B}}_p = \{a + (G * h)(x) \colon h \perp 1,\ \|h\|_p \leqq 1,\ a \in \mathbb{R}\},$$

where G satisfies Property B $\left(\text{and } h \perp 1 \text{ means } \int_0^{2\pi} h(y)\, dy = 0\right)$, as defined in Section 6.2 of Chapter IV. We prove that $X_{2n} = \operatorname{span}\{k(\cdot - \pi i/n)\}_{i=0}^{2n-1}$ is an optimal subspace for $d_{2n}(\tilde{\mathscr{K}}_p; L^q)$ when $p = \infty$ or $q = 1$, and we determine an optimal operator and subspace for $\delta_{2n}(\tilde{\mathscr{K}}_p; L^q)$ and $d^{2n}(\tilde{\mathscr{K}}_p; L^q)$, respectively. When $p = q = \infty$ and $1 \leqq p \leqq \infty$, $q = 1$, we prove that T_{n-1}, the space of trigonometric polynomials of degree $\leqq n-1$, is optimal for $d_{2n-1}(\tilde{\mathscr{K}}_p; L^q)$ (and $d_{2n-1}(\tilde{\mathscr{K}}_p; L^q) = d_{2n}(\tilde{\mathscr{K}}_p; L^q)$). Results analogous to these are proven for the set $\tilde{\mathscr{B}}_p$.

Very much motivated by the conjectures concerning $d_n(\mathscr{K}_p; L^q)$, we again consider, in Section 5, the n-widths of $\mathscr{K}_p$ in L^q for all $p, q \in [1, \infty]$. However this time we assume that K is simply a rank $n+1$ kernel, i.e., $K(x, y) = \sum_{i=1}^{n+1} u_i(x)\, v_i(y)$, for some linearly independent subsets $\{u_i\}_{i=1}^{n+1}$, $\{v_i\}_{i=1}^{n+1}$ of $C[0,1]$. Based on Theorem 2.1 of Chapter II we are able to determine $d_n(\mathscr{K}_p; L^q)$. If, in addition, K enjoys properties very much like those of total positivity, then we also prove that there exist $\{\xi_i\}_{i=1}^n$ and $\{\eta_i\}_{i=1}^n$, two sets of distinct points in $(0, 1)$ (dependent on p, q) for which $X_n = \operatorname{span}\{K(\cdot, \xi_i)\}_{i=1}^n$ is optimal for $d_n(\mathscr{K}_p; L^q)$ and interpolation from X_n to $Kh(x)$ at the $\{\eta_i\}_{i=1}^n$ is optimal for $\delta_n(\mathscr{K}_p; L^q)$.

2. Exact n-Widths of $\mathscr{K}_\infty$ in L^q and $\mathscr{K}_p$ in L^1

In Section 5 of Chapter IV we defined the set of nondegenerate totally positive (NTP) kernels. For convenience we now redefine this class.

Definition 2.1. We say that a kernel K is *nondegenerate totally positive* (NTP) on $[0,1] \times [0,1]$ if K is totally positive (TP) thereon and each of the sets $\operatorname{span}\{K(x_1, \cdot), \ldots, K(x_n, \cdot)\}$ and $\operatorname{span}\{K(\cdot, y_1), \ldots, K(\cdot, y_n)\}$ has dimension n for every choice of $0 < x_1 < \ldots < x_n < 1$, $0 < y_1 < \ldots < y_n < 1$, and all $n = 1, 2, \ldots$.

---

1 The author has recently proven these conjectures for $p = q$, see A. Pinkus, n-Widths of Sobolev spaces in L^p, Constructive Approximation **1** (1984).

The symmetry of this property implies that $K(x, y)$ is NTP if and only if $K'(x, y)$ $(= K(y, x))$ is NTP.

We now introduce the following notation. Let Λ_m denote the m-dimensional open simplex on $([0, 1])^m$ defined by

$$\Lambda_m = \{\mathbf{t}\colon \mathbf{t} = (t_1, \ldots, t_m),\ 0 < t_1 < \ldots < t_m < 1\}.$$

The closure of Λ_m, $\bar{\Lambda}_m$, is given by

$$\bar{\Lambda}_m = \{\mathbf{t}\colon \mathbf{t} = (t_1, \ldots, t_m),\ 0 \leqq t_1 \leqq \ldots \leqq t_m \leqq 1\}.$$

For each $\mathbf{t} \in \bar{\Lambda}_m$, set

$$h_{\mathbf{t}}(x) = (-1)^i, \quad x \in [t_i, t_{i+1}), \qquad i = 0, 1, \ldots, m,$$

where $t_0 = 0$, $t_{m+1} = 1$. It therefore follows that if $\mathbf{t} \in \bar{\Lambda}_m$, then $h_{\mathbf{t}} = \pm h_{\mathbf{s}}$ for some $\mathbf{s} \in \Lambda_k$ (the open simplex) and some $k \leqq m$.

The following simple lemma will prove useful.

Lemma 2.1. (i) *If* $\mathbf{t} \in \Lambda_m$ *and* $\mathbf{s} \in \Lambda_k$, *then* $S(h_{\mathbf{t}} \pm h_{\mathbf{s}}) \leqq \min\{m, k\}$.
(ii) *If* $\mathbf{t}, \mathbf{s} \in \Lambda_m$, *then* $S(h_{\mathbf{t}} - h_{\mathbf{s}}) \leqq m - 1$.

Proof. To prove (i), assume without loss of generality that $m \leqq k$ and $\mathbf{t} = (t_1, \ldots, t_m)$. Then $(h_{\mathbf{t}} \pm h_{\mathbf{s}})(x)(-1)^i \geqq 0$, $x \in [t_i, t_{i+1})$, $i = 0, 1, \ldots, m$, and therefore $S(h_{\mathbf{t}} \pm h_{\mathbf{s}}) \leqq m$.

To prove (ii), let $\mathbf{t} = (t_1, \ldots, t_m)$, $\mathbf{s} = (s_1, \ldots, s_m)$ and assume that $t_1 \leqq s_1$. Now, $(h_{\mathbf{t}} - h_{\mathbf{s}})(x)(-1)^i \geqq 0$, $x \in [t_i, t_{i+1})$, $i = 1, \ldots, m$ and $(h_{\mathbf{t}} - h_{\mathbf{s}})(x) = 0$ for $x \in [0, t_1)$. Thus $S(h_{\mathbf{t}} - h_{\mathbf{s}}) \leqq m - 1$. □

Since $S(h_{\mathbf{t}}) = m$ for $\mathbf{t} \in \Lambda_m$, it follows (from Theorem 3.3 of Chapter III) that for any TP kernel K, $S(K h_{\mathbf{t}}) \leqq m$, where

$$K h_{\mathbf{t}}(x) = \int_0^1 K(x, y)\, h_{\mathbf{t}}(y)\, dy.$$

One reason for considering the class of NTP kernels is to strengthen this result.

Proposition 2.2. *Let K be an NTP kernel.*

(i) *Let f be any integrable function on $[0, 1]$ for which* $\operatorname{sgn} f(y) = \operatorname{sgn} h_{\mathbf{t}}(y)$ *a.e. on $(0, 1)$, where $\mathbf{t} \in \Lambda_m$. If Kf has m zeros in $(0, 1)$ at $\mathbf{s} \in \Lambda_m$, then* $\operatorname{sgn} Kf(x) = h_{\mathbf{s}}(x)$ *for all $x \in (0, 1)$.*
(ii) *For $\mathbf{t} = (t_1, \ldots, t_m)$ and $\mathbf{s} = (s_1, \ldots, s_m)$ as in* (i)

$$K\begin{pmatrix} s_1, \ldots, s_m \\ t_1, \ldots, t_m \end{pmatrix} > 0.$$

(iii) *Given $\mathbf{s} = (s_1, \ldots, s_m) \in \Lambda_m$, there exists a $\mathbf{t} \in \Lambda_m$ for which $K h_{\mathbf{t}}(s_i) = 0$, $i = 1, \ldots, m$.*

Proof. Part (ii) of the proposition is simply a restatement of Lemma 5.4 of Chapter IV.

Part (i) of the proposition says that if Kf has m zeros in $(0, 1)$, then these zeros are necessarily strict sign changes and furthermore that the orientation of the sign of Kf is uniquely determined. To prove (i), we show that under the above assumptions, $(-1)^i Kf(x) > 0$ for $x \in (s_i, s_{i+1})$, $i = 0, 1, \ldots, m$.

Since K is NTP, the functions $\{K(s_1, \cdot), \ldots, K(s_m, \cdot), K(x, \cdot)\}$ are linearly independent on $[0, 1]$ and since $x \in (s_i, s_{i+1})$, the set $\{K(s_1, \cdot), \ldots, K(s_i, \cdot), K(x, \cdot), K(s_{i+1}, \cdot), \ldots, K(s_m, \cdot)\}$ is a weak Descartes system of order $m+1$ (with associated determinants nonnegative). Thus there exist coefficients $\alpha_1, \ldots, \alpha_{m+1}$, $\alpha_j(-1)^{j+1} \geqq 0$, $j = 1, \ldots, m+1$, and a not identically zero function

$$u(y) = \alpha_1 K(s_1, y) + \ldots + \alpha_i K(s_i, y) + \alpha_{i+1} K(x, y) + \alpha_{i+2} K(s_{i+1}, y) + \ldots + \alpha_{m+1} K(s_m, y)$$

for which $(-1)^i u(y) \geqq 0$, $y \in [t_i, t_{i+1}]$, $i = 0, 1, \ldots, m$. Hence since $Kf(s_i) = 0$, $i = 1, \ldots, m$,

$$\alpha_{i+1}(Kf)(x) = \int_0^1 u(y) f(y)\, dy = \int_0^1 |u(y)|\, |f(y)|\, dy > 0.$$

This implies that $\alpha_{i+1}(-1)^i > 0$ and $Kf(x)(-1)^i > 0$ for $x \in (s_i, s_{i+1})$.

The proof of part (iii) of the proposition is based on a result known as the Hobby-Rice Theorem. Since we shall make frequent use of this result, we present it here (with a slight addition).

Theorem 2.3 (Hobby, Rice [1965]). *Let $\{u_i\}_{i=1}^m$ be a set of integrable functions on $[0, 1]$. There exists a $\mathbf{t} \in \bar{\Lambda}_m$ satisfying*

$$\int_0^1 u_i(x)\, h_{\mathbf{t}}(x)\, dx = 0, \qquad i = 1, \ldots, m.$$

If $\{u_i\}_{i=1}^m$ is a weak Tchebycheff (WT)-system on $[0, 1]$, then $\mathbf{t} \in \Lambda_m$.

Proof. The first statement of the theorem is based on the Borsuk Antipodality Theorem (Theorem 1.4 of Chapter II). Let

$$\Xi_{m+1} = \left\{\mathbf{x}\colon \mathbf{x} = (x_1, \ldots, x_{m+1}),\ \sum_{i=1}^{m+1} |x_i| = 1\right\}.$$

For each $\mathbf{x} \in \Xi_{m+1}$, set $t_0(\mathbf{x}) = 0$, $t_i(\mathbf{x}) = \sum_{j=1}^{i} |x_j|$, $i = 1, \ldots, m+1$. Thus $0 = t_0(\mathbf{x}) \leqq t_1(\mathbf{x}) \leqq \ldots \leqq t_m(\mathbf{x}) \leqq t_{m+1}(\mathbf{x}) = 1$. To each $\mathbf{x} \in \Xi_{m+1}$, define

$$a_i(\mathbf{x}) = \sum_{j=1}^{m+1} [\operatorname{sgn} x_j] \int_{t_{j-1}(\mathbf{x})}^{t_j(\mathbf{x})} u_i(t)\, dt,$$

and $A(\mathbf{x}) = (a_1(\mathbf{x}), \ldots, a_m(\mathbf{x}))$. Since $t_j(\mathbf{x}) = t_{j-1}(\mathbf{x})$ if and only if $x_j = 0$, it follows that $A(\mathbf{x})$ is a continuous map of Ξ_{m+1} into $\mathbb{R}^m$. Furthermore $A(\mathbf{x})$ is odd, i.e., $A(-\mathbf{x}) = -A(\mathbf{x})$. Thus by the Borsuk Antipodality Theorem there exists an $\mathbf{x}^* \in \Xi_{m+1}$ for which $a_i(\mathbf{x}^*) = 0$, $i = 1, \ldots, m$. The first part of the theorem is now proven by choosing a $\mathbf{t} \in \bar{\Lambda}_m$ and $\varepsilon \in \{-1, 1\}$ so that $h_{\mathbf{t}}(x) = \varepsilon \operatorname{sgn} x_j^*$ for $x \in (t_{j-1}(\mathbf{x}^*), t_j(\mathbf{x}^*))$, $j = 1, \ldots, m+1$. The second statement of the theorem is a consequence of Proposition 1.4 of Chapter III. □

Proof of Proposition 2.2 (continued). We are given m points $0 < s_1 < \ldots < s_m < 1$. The set $\{K(s_i, \cdot)\}_{i=1}^m$ is a WT-system on $[0,1]$. Applying Theorem 2.3 to this set of functions we obtain a $\mathbf{t} \in \Lambda_m$ for which

$$K h_{\mathbf{t}}(x) = \int_0^1 K(x, y)\, h_{\mathbf{t}}(y)\, dy$$

satisfies $K h_{\mathbf{t}}(s_i) = 0$, $i = 1, \ldots, m$. □

We attack the problem of evaluating the Kolmogorov, Gel'fand, and linear n-widths of $\mathscr{K}_\infty$ in L^q, $1 \leqq q \leqq \infty$, by first considering the extremal problem

$$\inf_{\mathbf{t} \in \Lambda_n} \| K h_{\mathbf{t}} \|_q .$$

It is this extremal problem which is the key to the determination of the n-widths of $\mathscr{K}_\infty$ in L^q. As shall be shown, since we are varying over an n-dimensional simplex, the value of the extremal problem is a lower bound for both $d_n(\mathscr{K}_\infty; L^q)$ and $d^n(\mathscr{K}_\infty; L^q)$. This lower bound does *not* depend on any particular property of the kernel K as may be seen from the proofs of Theorems 2.6 and 2.8. The NTP property of K is used in a rather crucial way to prove that this quantity is also the upper bound.

We divide the discussion of the extremal problem into two parts. In the first part we consider the case $1 \leqq q < \infty$.

Theorem 2.4. *Let K be an NTP kernel and $1 \leqq q < \infty$. There exists a $\boldsymbol{\xi}^* \in \Lambda_n$ (dependent on q) for which*

$$\inf_{\mathbf{t} \in \Lambda_n} \| K h_{\mathbf{t}} \|_q = \| K h_{\boldsymbol{\xi}^*} \|_q .$$

Furthermore $K h_{\boldsymbol{\xi}^*}$ has n zeros at $\boldsymbol{\eta}^* \in \Lambda_n$, $\operatorname{sgn} K h_{\boldsymbol{\xi}^*}(x) = h_{\boldsymbol{\eta}^*}(x)$ for $x \in (0,1)$, and

$$\int_0^1 |K h_{\boldsymbol{\xi}^*}(x)|^{q-1} [\operatorname{sgn} K h_{\boldsymbol{\xi}^*}(x)]\, K(x, \xi_i^*)\, dx = 0, \qquad i = 1, \ldots, n.$$

Proof. The fact that the infimum is attained in $\bar{\Lambda}_n$ is a standard compactness result. Thus there exists a $\boldsymbol{\xi}^* \in \Lambda_m$, $m \leqq n$, for which

$$\inf_{\mathbf{t} \in \Lambda_n} \| K h_{\mathbf{t}} \|_q = \| K h_{\boldsymbol{\xi}^*} \|_q .$$

Set $F(s_1,\ldots,s_m) = \|K h_{\mathbf{s}}\|_q$, for $\mathbf{s}\in\Lambda_m$. Since $K h_{\mathbf{s}}$ has at most m distinct zeros in $(0,1)$, F is a differentiable function of $s_i, i=1,\ldots,m$, for all $1 \leqq q < \infty$. (A problem arises only for $q=1$.) The optimality of $\boldsymbol{\xi}^*$ implies that

$$0 = \frac{\partial}{\partial s_i} F(s_1,\ldots,s_m)\Bigg|_{\mathbf{s}=\boldsymbol{\xi}^*}$$

which translates into

$$0 = \int_0^1 |K h_{\boldsymbol{\xi}^*}(x)|^{q-1} [\operatorname{sgn} K h_{\boldsymbol{\xi}^*}(x)]\, K(x,\xi_i^*)\, dx, \qquad i=1,\ldots,m.$$

By twice applying Proposition 2.2(i), we see that $K h_{\boldsymbol{\xi}^*}(x)$ has m zeros (sign changes) in $(0,1)$. Let $\boldsymbol{\eta}^* = (\eta_1^*,\ldots,\eta_m^*)\in\Lambda_m$ denote these m zeros. Then $\operatorname{sgn} K h_{\boldsymbol{\xi}^*}(x) = h_{\boldsymbol{\eta}^*}(x)$ for $x\in(0,1)$. It remains to show that $m=n$.

Assume that $m<n$. We shall construct a $\boldsymbol{\zeta}^*\in\Lambda_n$ for which $\|K h_{\boldsymbol{\zeta}^*}\|_q < \|K h_{\boldsymbol{\xi}^*}\|_q$. By Proposition 2.2(ii),

$$K\begin{pmatrix}\eta_1^*,\ldots,\eta_m^*\\ \xi_1^*,\ldots,\xi_m^*\end{pmatrix} > 0.$$

Thus for every $x_0\in[0,1]$,

$$K h_{\boldsymbol{\xi}^*}(x_0) = \int_0^1 \frac{K\begin{pmatrix}x_0,\eta_1^*,\ldots,\eta_m^*\\ y,\xi_1^*,\ldots,\xi_m^*\end{pmatrix}}{K\begin{pmatrix}\eta_1^*,\ldots,\eta_m^*\\ \xi_1^*,\ldots,\xi_m^*\end{pmatrix}} h_{\boldsymbol{\xi}^*}(y)\, dy,$$

(since $K h_{\boldsymbol{\xi}^*}(\eta_i^*) = 0$, $i=1,\ldots,m$) and

$$|K h_{\boldsymbol{\xi}^*}(x_0)| = \int_0^1 \frac{\left|K\begin{pmatrix}x_0,\eta_1^*,\ldots,\eta_m^*\\ y,\xi_1^*,\ldots,\xi_m^*\end{pmatrix}\right|}{K\begin{pmatrix}\eta_1^*,\ldots,\eta_m^*\\ \xi_1^*,\ldots,\xi_m^*\end{pmatrix}}\, dy$$

$$= \int_0^1 \left|K(x_0,y) - \sum_{i=1}^m a_i K(\eta_i^*,y)\right| dy.$$

Choose $\eta_{m+1}^*,\ldots,\eta_n^*$ so that the $\{\eta_i^*\}_{i=1}^n$ are distinct points in $(0,1)$. For ease of exposition, set

$$0 < \eta_1^* < \ldots < \eta_n^* < 1.$$

The set of functions $\{K(\eta_i^*,\cdot)\}_{i=1}^n$ is a WT-system of dimension n on $[0,1]$, and by Proposition 2.2(iii), there exists a $\boldsymbol{\zeta}^*\in\Lambda_n$ (independent of x_0) for which

$$K h_{\boldsymbol{\zeta}^*}(\eta_i^*) = \int_0^1 K(\eta_i^*,y)\, h_{\boldsymbol{\zeta}^*}(y)\, dy = 0, \qquad i=1,\ldots,n.$$

Thus

$$\begin{aligned}|K h_{\zeta^*}(x_0)| &= \left|\int_0^1 K(x_0, y)\, h_{\zeta^*}(y)\, dy\right| \\ &= \left|\int_0^1 \left[K(x_0, y) - \sum_{i=1}^m a_i K(\eta_i^*, y)\right] h_{\zeta^*}(y)\, dy\right| \\ &\leqq \int_0^1 \left|K(x_0, y) - \sum_{i=1}^m a_i K(\eta_i^*, y)\right| dy \\ &= |K h_{\xi^*}(x_0)|,\end{aligned}$$

and $|K h_{\zeta^*}(x)| \leqq |K h_{\xi^*}(x)|$ for all $x \in [0,1]$. Furthermore, $K h_{\zeta^*}(\eta_i^*) = 0$, $i = 1,\dots, n$, while $K h_{\xi^*}(\eta_i^*) = 0$ only for $i = 1,\dots, m$. Therefore

$$\| K h_{\zeta^*} \|_q < \| K h_{\xi^*} \|_q$$

and Theorem 2.4 is proved. □

The case $q = \infty$ is somewhat simpler.

Theorem 2.5. *Let K be an NTP kernel. There exists a $\xi^* \in \Lambda_n$ and $n+1$ points $0 \leqq z_1^* < \dots < z_{n+1}^* \leqq 1$ such that*

$$(K h_{\xi^*})(z_i^*)(-1)^{i+1} = \| K h_{\xi^*} \|_\infty, \qquad i = 1,\dots, n+1.$$

Furthermore,

$$\inf_{\mathbf{t} \in \Lambda_n} \| K h_{\mathbf{t}} \|_\infty = \| K h_{\xi^*} \|_\infty .$$

Proof. Various proofs of theorems of the above type exist. We choose to present what we feel is a short and elegant, albeit nonconstructive, proof of this fact.

Set $\Xi_{n+1} = \left\{\mathbf{x}\colon \mathbf{x} = (x_1,\dots, x_{n+1}),\ \sum_{i=1}^{n+1} |x_i| = 1\right\}$, and $t_0(\mathbf{x}) = 0$, $t_i(\mathbf{x}) = \sum_{j=1}^{i} |x_j|$, $i = 1,\dots, n+1$. Thus $0 = t_0(\mathbf{x}) \leqq t_1(\mathbf{x}) \leqq \dots \leqq t_n(\mathbf{x}) \leqq t_{n+1}(\mathbf{x}) = 1$. To each $\mathbf{x} \in \Xi_{n+1}$, define

$$g_{\mathbf{x}}(x) = \sum_{i=1}^{n+1} [\operatorname{sgn} x_i] \int_{t_{i-1}(\mathbf{x})}^{t_i(\mathbf{x})} K(x, y)\, dy.$$

As in the proof of Theorem 2.3, we see that $g_{\mathbf{x}}$ is a continuous function of $\mathbf{x} \in \Xi_{n+1}$, and $g_{\mathbf{x}}$ is odd, i.e., $g_{-\mathbf{x}} = -g_{\mathbf{x}}$.

Let $\{u_i\}_{i=1}^n$ be any T-system on $[0,1]$. To each $g_{\mathbf{x}}$ there exists a unique best approximant $\sum_{i=1}^n a_i(\mathbf{x}) u_i$ from span $\{u_1,\dots, u_n\}$ in the L^∞ (sup)norm. Since $\{u_1,\dots, u_n\}$ is a T-system the error function $g_{\mathbf{x}} - \sum_{i=1}^n a_i(\mathbf{x}) u_i$ must achieve its norm, alternately, at at least $n+1$ times on $[0,1]$ or vanishes identically. Set

$$A(\mathbf{x}) = (a_1(\mathbf{x}), \ldots, a_n(\mathbf{x})).$$

It follows from the properties of $g_{\mathbf{x}}$ that $A(\mathbf{x})$ is an odd continuous map of Ξ_{n+1} into $\mathbb{R}^n$. Applying the Borsuk Antipodality Theorem we obtain the existence of an $\mathbf{x}^* \in \Xi_{n+1}$ for which $A(\mathbf{x}^*) = \mathbf{0}$, i.e.,

$$\inf_{a_1, \ldots, a_n} \left\| g_{\mathbf{x}^*} - \sum_{i=1}^n a_i u_i \right\|_\infty = \| g_{\mathbf{x}^*} \|_\infty .$$

Since $g_{\mathbf{x}^*}$ cannot vanish identically, $g_{\mathbf{x}^*}$ must attain its norm alternately at $n+1$ points in $[0,1]$. Furthermore since $g_{\mathbf{x}^*} = \pm K h_{\boldsymbol{\xi}^*}$ for some $\boldsymbol{\xi}^* \in \bar{\Lambda}_n$ and $S(g_{\mathbf{x}^*}) \geqq n$, it follows from Proposition 2.2 that $\boldsymbol{\xi}^* \in \Lambda_n$. The exact orientation of the sign is also a result of Proposition 2.2.

It remains to prove that

$$\inf_{\mathbf{t} \in \Lambda_n} \| K h_{\mathbf{t}} \|_\infty = \| K h_{\boldsymbol{\xi}^*} \|_\infty .$$

Assume that this is not the case. There then exists a $\mathbf{t} \in \Lambda_n$ for which $\| K h_{\mathbf{t}} \|_\infty < \| K h_{\boldsymbol{\xi}^*} \|_\infty$. Since $K h_{\boldsymbol{\xi}^*}$ attains its norm alternately at $n+1$ points in $[0,1]$,

$$S(K h_{\boldsymbol{\xi}^*} - K h_{\mathbf{t}}) \geqq n.$$

Moreover K is totally positive and

$$(K h_{\boldsymbol{\xi}^*} - K h_{\mathbf{t}})(x) = \int_0^1 K(x, y)[h_{\boldsymbol{\xi}^*}(y) - h_{\mathbf{t}}(y)]\, dy.$$

Therefore $S(h_{\boldsymbol{\xi}^*} - h_{\mathbf{t}}) \geqq n$ by the variation diminishing property of K. However from Lemma 2.1 (ii), $S(h_{\boldsymbol{\xi}^*} - h_{\mathbf{t}}) \leqq n-1$. This contradiction proves the theorem. □

Having solved the relevant extremal problems, we are now prepared to state and prove the n-width results. Although the $\boldsymbol{\xi}^*$ of the above two theorems vary with q, we shall not explicitly exhibit this dependence.

Theorem 2.6. *Let K be an NTP kernel. Then the n-width, in the sense of Kolmogorov, of $\mathscr{K}_\infty$ in L^q is given by*

$$d_n(\mathscr{K}_\infty; L^q) = \| K h_{\boldsymbol{\xi}^*} \|_q, \quad 1 \leqq q \leqq \infty,$$

where $\boldsymbol{\xi}^* = (\xi_1^*, \ldots, \xi_n^*) \in \Lambda_n$ is as in Theorems 2.4 and 2.5. The n-dimensional subspace span $\{K(\cdot, \xi_1^*), \ldots, K(\cdot, \xi_n^*)\}$ *is optimal for $d_n(\mathscr{K}_\infty; L^q)$.*

Proof. We first prove the lower bound by a method which is now familiar to the reader. Assume for the moment that $1 < q < \infty$, and let Ξ_{n+1} be as defined in

Theorem 2.5, i.e., the unit ball in l_1^{n+1}. As previously set $t_0(\mathbf{x}) = 0$, $t_i(\mathbf{x}) = \sum_{j=1}^{i} |x_j|$, $i = 1,\ldots,n+1$ for $\mathbf{x} \in \Xi_{n+1}$, and

$$g_{\mathbf{x}}(x) = \sum_{i=1}^{n+1} [\operatorname{sgn} x_i] \int_{t_{i-1}(\mathbf{x})}^{t_i(\mathbf{x})} K(x,y)\,dy.$$

Let $X_n = \operatorname{span}\{f_1,\ldots,f_n\}$ be any n-dimensional subspace of $L^q[0,1]$. Since $1 < q < \infty$, the best L^q approximation to $g_{\mathbf{x}}$ from X_n is unique. It follows, as in Theorem 2.5, by an application of the Borsuk Antipodality Theorem, that there exists an $\mathbf{x}^* \in \Xi_{n+1}$ for which the unique best approximation to $g_{\mathbf{x}^*}$ from X_n is the zero function. Thus there exists a $\mathbf{t}^* \in \bar{\Lambda}_n$ for which

$$\begin{aligned}\sup_{\|h\|_\infty \leqq 1} \inf_{f \in X_n} \|Kh - f\|_q &\geqq \sup_{\mathbf{t} \in \Lambda_n} \inf_{f \in X_n} \|Kh_{\mathbf{t}} - f\|_q \\ &\geqq \|Kh_{\mathbf{t}^*}\|_q.\end{aligned}$$

Since $\|Kh_{\boldsymbol{\xi}^*}\|_q \leqq \|Kh_{\mathbf{t}}\|_q$ for every $\mathbf{t} \in \bar{\Lambda}_n$, we have

$$\sup_{\|h\|_\infty \leqq 1} \inf_{f \in X_n} \|Kh - f\|_q \geqq \|Kh_{\boldsymbol{\xi}^*}\|_q.$$

The right hand side is independent of the choice of X_n.Thus we obtain the lower bound $d_n(\mathscr{K}_\infty; L^q) \geqq \|Kh_{\boldsymbol{\xi}^*}\|_q$ for $1 < q < \infty$.

To prove the above result for $q = 1$ and $q = \infty$, we use continuity and the fact that the set $\{g_{\mathbf{x}}: \mathbf{x} \in \Xi_{n+1}\}$ is compact. The details are left undone. Thus the lower bound is obtained for all $q \in [1,\infty]$.

We construct a simple linear method in order to obtain the upper bound. Let $\{\eta_i^*\}_{i=1}^n$ (dependent on q) denote the n unique interior zeros of $Kh_{\boldsymbol{\xi}^*}$. Set

$$P_n h(x) = \sum_{i=1}^{n} a_i K(x, \xi_i^*)$$

where the $\{a_i\}_{i=1}^n$ are chosen so that $P_n h(\eta_i^*) = Kh(\eta_i^*)$, $i = 1,\ldots,n$. Thus

$$P_n h(x) = -\int_0^1 \frac{\begin{vmatrix} K(\eta_1^*,\xi_1^*) \ldots K(\eta_1^*,\xi_n^*) & K(\eta_1^*,y) \\ \vdots \qquad\quad \vdots & \vdots \\ K(\eta_n^*,\xi_1^*) \ldots K(\eta_n^*,\xi_n^*) & K(\eta_n^*,y) \\ K(x,\xi_1^*) \;\ldots\; K(x,\xi_n^*) & 0 \end{vmatrix}}{K\begin{pmatrix} \eta_1^*,\ldots,\eta_n^* \\ \xi_1^*,\ldots,\xi_n^* \end{pmatrix}}\, h(y)\,dy.$$

The existence of such a P_n is assured by Proposition 2.2(ii). Therefore

$$Kh(x) - P_n h(x) = \int_0^1 \frac{K\begin{pmatrix}\eta_1^*, \ldots, \eta_n^*, x \\ \xi_1^*, \ldots, \xi_n^*, y\end{pmatrix}}{K\begin{pmatrix}\eta_1^*, \ldots, \eta_n^* \\ \xi_1^*, \ldots, \xi_n^*\end{pmatrix}} h(y)\, dy.$$

The kernel

$$L(x, y) = K\begin{pmatrix}\eta_1^*, \ldots, \eta_n^*, x \\ \xi_1^*, \ldots, \xi_n^*, y\end{pmatrix} \Big/ K\begin{pmatrix}\eta_1^*, \ldots, \eta_n^* \\ \xi_1^*, \ldots, \xi_n^*\end{pmatrix}$$

satisfies $L(x, y)(-1)^{i+j} \geqq 0$ for $x \in (\eta_i^*, \eta_{i+1}^*)$ and $y \in (\xi_j^*, \xi_{j+1}^*)$, $i, j = 0, 1, \ldots, n$, where as usual $\eta_0^* = \xi_0^* = 0$ and $\eta_{n+1}^* = \xi_{n+1}^* = 1$. Thus,

$$\begin{aligned}\sup_{\|h\|_\infty \leqq 1} \|Kh - P_n h\|_q &= \sup_{\|h\|_\infty \leqq 1} \left\| \int_0^1 L(x, y)\, h(y)\, dy \right\|_q \\ &= \sup_{\|h\|_\infty \leqq 1} \left(\int_0^1 \left| \int_0^1 L(x, y)\, h(y)\, dy \right|^q dx \right)^{1/q} \\ &= \left(\int_0^1 \left(\int_0^1 |L(x, y)|\, dy \right)^q dx \right)^{1/q}\end{aligned}$$

and equality is attained for $h(y) = h_{\xi^*}(y)$. Moreover for this particular choice of h,

$$\int_0^1 L(x, y)\, h_{\xi^*}(y)\, dy = \int_0^1 K(x, y)\, h_{\xi^*}(y)\, dy,$$

since $Kh_{\xi^*}(\eta_i^*) = 0$, $i = 1, \ldots, n$. Thus

$$\sup_{\|h\|_\infty \leqq 1} \|Kh - P_n h\|_q = \|Kh_{\xi^*}\|_q. \quad \square$$

Since the upper bound was explicitly constructed by a linear method, the following result also holds.

Theorem 2.7. *Let K be an NTP kernel. The linear n-width of $\mathscr{K}_\infty$ in L^q is given by*

$$\delta_n(\mathscr{K}_\infty; L^q) = \|Kh_{\xi^*}\|_q, \quad 1 \leqq q \leqq \infty.$$

Let $\{\eta_i^*\}_{i=1}^n$ denote the n unique zeros of Kh_{ξ^*} in $(0, 1)$. The rank n linear map which assigns to each $h \in L^\infty$ the unique interpolant to Kh at the points $\{\eta_i^*\}_{i=1}^n$ from the subspace span $\{K(\cdot, \xi_1^*), \ldots, K(\cdot, \xi_n^*)\}$ *is an optimal linear map.*

The above quantity $\|Kh_{\xi^*}\|_q$ is also the value of the associated n-width in the sense of Gel'fand.

Theorem 2.8. *Let K be an NTP kernel. The n-width, in the sense of Gel'fand, of $\mathscr{K}_\infty$ in L^q is given by*

$$d^n(\mathscr{K}_\infty; L^q) = \|K h_{\xi^*}\|_q, \qquad 1 \leqq q \leqq \infty.$$

The subspace $L^n = \left\{h\colon h \in L^\infty,\ \int_0^1 K(\eta_i^*, y)\, h(y)\, dy = 0,\ i = 1, \ldots, n\right\}$ of codimension n is optimal for $d^n(\mathscr{K}_\infty; L^q)$ where the $\{\eta_i^*\}_{i=1}^n$ are the n unique zeros of $K h_{\xi^*}$ on $(0,1)$.

Proof. The upper bound is an immediate consequence of Theorem 2.7 and the inequality $d^n(\mathscr{K}_\infty; L^q) \leqq \delta_n(\mathscr{K}_\infty; L^q)$ (see Propositions 5.1 and 7.7 of Chapter II).

The proof of the lower bound presents difficulties due to the complicated nature of $(L^\infty)'$. To overcome this difficulty we note that from the proof of Theorem 8.9 of Chapter II, it suffices to consider M^n of codimension $\leqq n$ of L^∞ defined by at most n linear functionals of the predual L^1 of L^∞. Thus

$$M^n = \{h\colon h \in L^\infty,\ (g_i, h) = 0,\ i = 1, \ldots, n\}$$

for some choice of $\{g_i\}_{i=1}^n$ in L^1. For given $\{g_i\}_{i=1}^n$ in L^1, we have from the Hobby-Rice Theorem (Theorem 2.3) the existence of a $\mathbf{t} \in \bar{\Lambda}_n$ for which $(g_i, h_\mathbf{t}) = 0$, $i = 1, \ldots, n$. Thus

$$\sup_{\substack{\|h\|_\infty \leqq 1 \\ h \in M^n}} \|K h\|_q \geqq \|K h_\mathbf{t}\|_q \geqq \|K h_{\xi^*}\|_q.$$

Since $\|K h_{\xi^*}\|_q$ is independent of the particular choice of M^n, it follows that

$$d^n(\mathscr{K}_\infty; L^q) \geqq \|K h_{\xi^*}\|_q. \quad \square$$

We now consider the dual version of the above problem. The results of this next theorem are a direct consequence of Theorems 7.2, 8.9, and 8.10 of Chapter II. For given $K(x, y)$, let $K'(x, y) = K(y, x)$, the transpose of K. Thus

$$K' h(x) = \int_0^1 K(y, x)\, h(y)\, dy.$$

Theorem 2.9. *Let K be an NTP kernel. Then for all $1 \leqq p \leqq \infty$,*

$$d_n(\mathscr{K}_p; L^1) = d^n(\mathscr{K}_p; L^1) = \delta_n(\mathscr{K}_p; L^1) = \inf_{\mathbf{t} \in \Lambda_n} \|K' h_\mathbf{t}\|_{p'},$$

where $1/p + 1/p' = 1$. For each $p' \in [1, \infty]$,

$$\inf_{\mathbf{t} \in \Lambda_n} \|K' h_\mathbf{t}\|_{p'} = \|K' h_{\xi^*}\|_{p'},$$

where $\xi^* = (\xi_1^*, \ldots, \xi_n^*) \in \Lambda_n$ is dependent on p', and $\operatorname{sgn} K' h_{\xi^*}(x) = h_{\eta^*}(x)$, *for all $x \in (0,1)$, and some $\eta^* = (\eta_1^*, \ldots, \eta_n^*) \in \Lambda_n$. Moreover,*

(1) *an optimal subspace for* $d_n(\mathscr{K}_p; L^1)$ *is* $X_n = \text{span}\{K(\cdot, \eta_1^*), \ldots, K(\cdot, \eta_n^*)\}$,
(2) *an optimal subspace of codimension n for* $d^n(\mathscr{K}_p; L^1)$ *is*

$$L^n = \left\{h \colon h \in L^p, \int_0^1 K(\xi_i^*, y)\, h(y)\, dy = 0,\ i = 1, \ldots, n\right\},$$

(3) *the rank n linear operator which assigns to each* $h \in L^p$ *the unique interpolant to* Kh *at* $\{\xi_i^*\}_{i=1}^n$ *from* X_n *is optimal for* $\delta_n(\mathscr{K}_p; L^1)$.

Remark. Note that the optimal rank n linear operators for $\delta_n(\mathscr{K}_\infty; L^q)$ and $\delta_n(\mathscr{K}_p; L^1)$ are each of the form PKh where P is a rank n projection.

Let us now consider the Bernstein n-widths. These were introduced in Chapter II. We reformulate the definition for a linear operator T mapping X to Y.

Definition 2.2. Let $T \in L(X, Y)$. Then the *Bernstein n-width* is defined by

$$b_n(T(X); Y) = \sup_{X_{n+1}} \inf_{\substack{Tx \in X_{n+1} \\ Tx \neq 0}} \|Tx\|_Y / \|x\|_X$$

where X_{n+1} is any subspace of span$\{Tx \colon x \in X\}$ of dimension $\geqq n+1$.

Theorem 2.6 was first stated for $q = \infty$. The lower bound was originally proven by a different and more standard method of proof, i.e., via the Fundamental Theorem of Tichomirov (Theorem 1.5 of Chapter II). This method gives us the following result.

Theorem 2.10. *Let K be an NTP kernel. Then the Bernstein n-width of* $\mathscr{K}_\infty$ *in* L^∞ *is given by*

$$b_n(\mathscr{K}_\infty; L^\infty) = \inf_{\mathbf{t} \in \Lambda_n} \|Kh_{\mathbf{t}}\|_\infty = \|Kh_{\boldsymbol{\xi}^*}\|_\infty,$$

and $X_{n+1} = \text{span}\{u_1, \ldots, u_{n+1}\}$ *is an optimal subspace for* $b_n(\mathscr{K}_\infty; L^\infty)$, *where*

$$u_i(x) = \int_{\xi_{i-1}^*}^{\xi_i^*} K(x, y)\, dy, \qquad i = 1, \ldots, n+1.$$

Proof. Since $b_n(\mathscr{K}_\infty; L^\infty) \leqq d_n(\mathscr{K}_\infty; L^\infty)$ (Proposition 1.6 of Chapter II) it suffices to prove the lower bound. The function $\sum_{i=1}^{n+1} a_i u_i$ is in $\mathscr{K}_\infty$ if and only if $|a_i| \leqq 1$ for all $i = 1, \ldots, n+1$. Thus it is necessary to show that if $\left\|\sum_{i=1}^{n+1} a_i u_i\right\|_\infty \leqq \|Kh_{\boldsymbol{\xi}^*}\|_\infty$, then $|a_i| \leqq 1$, $i = 1, \ldots, n+1$. Assume that $\left\|\sum_{i=1}^{n+1} a_i u_i\right\|_\infty \leqq \|Kh_{\boldsymbol{\xi}^*}\|_\infty$ and set

$$|a_{i_0}| = \max_{i=1,\ldots,n+1} |a_i|.$$

If $|a_{i_0}| > 1$, then

$$\left\| \sum_{i=1}^{n+1} (a_i/a_{i_0}) u_i \right\|_\infty < \|K h_{\xi^*}\|_\infty = \left\| \sum_{i=1}^{n+1} (-1)^{i+1} u_i \right\|_\infty .$$

Since $\sum_{i=1}^{n+1} (-1)^{i+1} u_i$ attains its norm, alternately, at $n+1$ points in $[0,1]$,

$$S\left(\sum_{i=1}^{n+1} (-1)^{i+1} u_i - \sum_{i=1}^{n+1} (-1)^{i_0+1} (a_i/a_{i_0}) u_i \right) \geqq n,$$

i.e., $S\left(\sum_{i=1}^{n+1} b_i u_i \right) \geqq n$, where $b_{i_0} = 0$. Since K is TP, then by Theorem 3.3 of Chapter III,

$$S^-(b_1, \ldots, b_{n+1}) \geqq n.$$

However, since $b_{i_0} = 0$, it follows that $S^-(b_1, \ldots, b_{n+1}) \leqq n-1$. □

We wish to calculate the Bernstein n-widths of $\mathscr{K}_1$ in L^1. To this end the next proposition is introduced. This proposition is of interest in and of itself since it also provides us, via Proposition 1.8 of Chapter II, with yet another proof of the lower bound for the n-widths of $\mathscr{K}_\infty$ in L^∞.

Proposition 2.11. *Let K be an NTP kernel. Let $\{z_i^*\}_{i=1}^{n+1}$ be ordered points at which $K h_{\xi^*}$ attains its norm, alternately, as in Theorem 2.5. Let $\{\delta_i\}_{i=1}^{n+1}$ be any choice of signs, i.e., $\delta_i \in \{-1, 1\}$ for each $i = 1, \ldots, n+1$. There exists a $\mathbf{t} \in \bar{\Lambda}_n$ and an $\varepsilon \in \{-1, 1\}$ such that*

$$\operatorname{sgn} K h_{\mathbf{t}}(z_i^*) = \varepsilon \delta_i \quad \textit{and} \quad |K h_{\mathbf{t}}(z_i^*)| \geqq \|K h_{\xi^*}\|_\infty, \qquad i = 1, \ldots, n+1.$$

Proof. Consider the set of functions $\{\delta_{i+1} K(z_i^*, y) - \delta_i K(z_{i+1}^*, y)\}_{i=1}^n$. By Theorem 2.3 there exists a $\mathbf{t} \in \bar{\Lambda}_n$ for which

$$\int_0^1 (\delta_{i+1} K(z_i^*, y) - \delta_i K(z_{i+1}^*, y))\, h_{\mathbf{t}}(y)\, dy = 0, \qquad i = 1, \ldots, n,$$

i.e., $\delta_{i+1} K h_{\mathbf{t}}(z_i^*) = \delta_i K h_{\mathbf{t}}(z_{i+1}^*)$, $i = 1, \ldots, n$. If $K h_{\mathbf{t}}(z_i^*) = 0$ for some i, then $K h_{\mathbf{t}}(z_i^*) = 0$ for all $i = 1, \ldots, n+1$ and since $\mathbf{t} \in \bar{\Lambda}_n$, we obtain a contradiction to Proposition 2.2. Thus $K h_{\mathbf{t}}(z_i^*) \neq 0$ for all i. By the appropriate choice of $\varepsilon \in \{-1, 1\}$ we have therefore constructed a $\mathbf{t} \in \bar{\Lambda}_n$ for which $\operatorname{sgn} K h_{\mathbf{t}}(z_i^*) = \varepsilon \delta_i$ and $|K h_{\mathbf{t}}(z_i^*)| = c$, $i = 1, \ldots, n+1$, c a positive constant.

It remains to prove that $c \geqq \|K h_{\xi^*}\|_\infty$. Assume not. Because $K h_{\xi^*}(z_i^*)(-1)^{i+1} = \|K h_{\xi^*}\|_\infty$, $i = 1, \ldots, n+1$,

$$S(K h_{\xi^*} - K h_{\mathbf{t}}) \geqq n.$$

However K is totally positive and ξ^*, $\mathbf{t} \in \bar{\Lambda}_n$. Thus by the variation diminishing property of K and by Lemma 2.1,

$$S(K h_{\xi^*} - K h_{\mathbf{t}}) \leqq S(h_{\xi^*} - h_{\mathbf{t}}) \leqq n - 1.$$

This is a contradiction. □

On the basis of Proposition 2.11, we now prove

Theorem 2.12. *Let K be an NTP kernel. Then*

$$b_n(\mathscr{K}_1; L^1) = \inf_{\mathbf{t} \in \Lambda_n} \| K' h_{\mathbf{t}} \|_\infty = \| K' h_{\xi^*} \|_\infty,$$

and $X_{n+1} = \operatorname{span}\{K(\cdot, z_i^*)\}_{i=1}^{n+1}$ is an optimal subspace for $b_n(\mathscr{K}_1; L^1)$, where the $\{z_i^*\}_{i=1}^{n+1}$ are $n+1$ ordered points in $[0,1]$ where $K' h_{\xi^*}$ attains its norm, alternately.

Proof. It is sufficient, as in Theorem 2.10, to prove that $b_n(\mathscr{K}_1; L^1) \geqq \| K' h_{\xi^*} \|_\infty$. This will follow if we can prove that $\left\| \sum_{i=1}^{n+1} a_i K(\cdot, z_i^*) \right\|_1 \leqq \| K' h_{\xi^*} \|_\infty$ implies that $\sum_{i=1}^{n+1} |a_i| \leqq 1$.

Choose $\{\delta_i\}_{i=1}^{n+1}$ so that $\delta_i \in \{-1, 1\}$ and $\delta_i = \operatorname{sgn} a_i$ if $a_i \neq 0$, $i = 1, \ldots, n+1$. There exists, by Proposition 2.11, a $\mathbf{t} \in \bar{\Lambda}_n$ and an $\varepsilon \in \{-1, 1\}$, for which

$$\operatorname{sgn} K' h_{\mathbf{t}}(z_i^*) = \varepsilon \delta_i \quad \text{and} \quad |K' h_{\mathbf{t}}(z_i^*)| \geqq \| K' h_{\xi^*} \|_\infty, \qquad i = 1, \ldots, n+1,$$

Thus,

$$\begin{aligned} \| K' h_{\xi^*} \|_\infty &\geqq \left\| \sum_{i=1}^{n+1} a_i K(\cdot, z_i^*) \right\|_1 \\ &\geqq \left| \left(\sum_{i=1}^{n+1} a_i K(\cdot, z_i^*),\ h_{\mathbf{t}}(\cdot) \right) \right| \\ &= \left| \sum_{i=1}^{n+1} a_i\ K' h_{\mathbf{t}}(z_i^*) \right| \\ &\geqq \| K' h_{\xi^*} \|_\infty \left(\sum_{i=1}^{n+1} |a_i| \right). \end{aligned}$$

Therefore $\sum_{i=1}^{n+1} |a_i| \leqq 1$, and the theorem is proved. □

Nothing is known about the Bernstein n-widths of $\mathscr{K}_\infty$ in L^q for $1 \leqq q < \infty$, or of $\mathscr{K}_p$ in L^1 for $1 < p \leqq \infty$. It is probable that in these cases the Bernstein and Kolmogorov n-widths do not agree.

Remark. The condition $\| h \|_\infty \leqq 1$ in Theorems 2.4–2.12 may be replaced by the condition $|h(y)| \leqq w(y)$ where w is some strictly positive continuous function on $[0,1]$. This follows easily from the fact that the kernel $L(x, y) = K(x, y)\, w(y)$ is an NTP kernel if $K(x, y)$ is NTP and $w(y)$ is strictly positive. Furthermore the set

$$\{Kh\colon |h(y)| \leqq w(y),\ y \in [0,1]\}$$

is the same as the set

$$\{Lh\colon \|h\|_\infty \leqq 1\}.$$

This is one of the advantages in dealing with the class of NTP kernels rather than with a specific kernel.

The analysis of this section allows us to determine the n-widths of the following class of functions for particular K, ω, p and q. Let $V(K,\omega,p,q)$ denote those $f \in C[0,1]$ for which

$$\inf\{\|f - Kh\|_q + t\,\|h\|_p\colon h \in L^p\} \leqq \omega(t)$$

for all $t \geqq 0$.

This class is considered because of its relationship with the K-functional (not the above kernel K) used in interpolation theory, and because of the simple form of the result. We shall concern ourselves simply with the evaluation of the n-widths of this set, regarding this as an application of the methods of this section.

Our first result is the following simple upper bound which is unrelated to any particular choice of K and ω.

Proposition 2.13. *For $p, q \in [1,\infty]$ and*

$$\mathscr{K}_p = \left\{\int_0^1 K(x,y)\,h(y)\,dy\colon \|h\|_p \leqq 1\right\},$$

we have

$$d_n(V(K,\omega,p,q); L^q) \leqq \omega(d_n(\mathscr{K}_p; L^q)).$$

Proof. Let X_n be an optimal n-dimensional subspace for $d_n(\mathscr{K}_p; L^q)$. Then by definition,

$$\inf_{g \in X_n} \|Kh - g\|_q \leqq \|h\|_p\, d_n(\mathscr{K}_p; L^q).$$

Now,

$$\begin{aligned} d_n(V(K,\omega,p,q); L^q) &\leqq \sup_{f \in V(K,\omega,p,q)} \inf_{g \in X_n} \|f - g\|_q \\ &\leqq \sup_{f \in V(K,\omega,p,q)} \inf_{g \in X_n} \inf_{h \in L^p} \{\|f - Kh\|_q + \|Kh - g\|_q\} \\ &= \sup_{f \in V(K,\omega,p,q)} \inf_{h \in L^p} \{\|f - Kh\|_q + \inf_{g \in X_n} \|Kh - g\|_q\} \\ &\leqq \sup_{f \in V(K,\omega,p,q)} \inf_{h \in L^p} \{\|f - Kh\|_q + \|h\|_p\, d_n(\mathscr{K}_p; L^q)\} \\ &\leqq \omega(d_n(\mathscr{K}_p; L^q)). \quad \square \end{aligned}$$

To obtain equality in the above inequality, we must make certain assumptions. Assume that K is NTP, $p=\infty$ and that ω is a nondecreasing function on $[0,\infty)$, while $\omega(t)/t$ is a nonincreasing function. Note that if ω is a concave modulus of continuity, then it satisfies these assumptions.

Theorem 2.14. *Under the above conditions,*

$$d_n(V(K,\omega,\infty,q);L^q)=\omega(d_n(\mathscr{K}_\infty;L^q))$$

and $X_n=\operatorname{span}\{K(\cdot,\xi_i^*)\}_{i=1}^n$, *the optimal subspace for* $d_n(\mathscr{K}_\infty;L^q)$ *as given in Theorem 2.6, is also optimal here.*

Proof. By Proposition 2.13, it suffices to prove that $d_n(V(K,\omega,\infty,q);L^q)\geqq\omega(d_n(\mathscr{K}_\infty;L^q))$. Since K is NTP the following is a consequence of Theorems 2.4, 2.5 and the proof of Theorem 2.6:

Given an n-dimensional subspace X_n of L^q, there exists a $\xi\in\bar{\Lambda}_n$ for which

$$\inf_{g\in X_n}\|Kh_\xi-g\|_q=\|Kh_\xi\|_q\geqq\|Kh_{\xi^*}\|_q=d_n(\mathscr{K}_\infty;L^q).$$

For each $\xi\in\bar{\Lambda}_n$, $Kh_\xi\neq 0$. Let $C_\xi\in\mathbb{R}$ be such that $f_\xi=C_\xi Kh_\xi$ satisfies $\|f_\xi\|_q=\omega(d_n(\mathscr{K}_\infty;L^q))$. If $f_\xi\in V(K,\omega,\infty,q)$ for each $\xi\in\bar{\Lambda}_n$, then

$$\begin{aligned}\sup_{f\in V(K,\omega,\infty,q)}\inf_{g\in X_n}\|f-g\|_q&\geqq\inf_{g\in X_n}\|f_\xi-g\|_q\\&=\|f_\xi\|_q\\&=\omega(d_n(\mathscr{K}_\infty;L^q)).\end{aligned}$$

It therefore remains to prove that $f_\xi\in V(K,\omega,\infty,q)$ for all $\xi\in\bar{\Lambda}_n$. In other words we must show that

$$\inf_{h\in L^\infty}\{\|f_\xi-Kh\|_q+t\|h\|_\infty\}\leqq\omega(t)$$

for all $t\geqq 0$.

We divide the proof of this fact into two cases. We first assume that $t\leqq d_n(\mathscr{K}_\infty;L^q)$.

Now,

$$\begin{aligned}\inf_{h\in L^\infty}\{\|f_\xi-Kh\|_q+t\|h\|_\infty\}&\leqq\|f_\xi-f_\xi\|_q+tC_\xi\|h_\xi\|_\infty\\&=tC_\xi\end{aligned}$$

by the choice of $h=C_\xi h_\xi$. By definition,

$$\omega(d_n(\mathscr{K}_\infty;L^q))=\|f_\xi\|_q=C_\xi\|Kh_\xi\|_q.$$

For $\xi \in \bar{\Lambda}_n$, $\|Kh_\xi\|_q \geqq d_n(\mathscr{K}_\infty; L^q)$. Thus

$$tC_\xi \leqq t\omega(d_n(\mathscr{K}_\infty; L^q))/d_n(\mathscr{K}_\infty; L^q) \leqq \omega(t),$$

since $t \leqq d_n(\mathscr{K}_\infty; L^q)$ and $\omega(t)/t$ is a nonincreasing function.

For $t \geqq d_n(\mathscr{K}_\infty; L^q)$,

$$\begin{aligned}\inf_{h\in L^\infty} \{\|f_\xi - Kh\|_q + t\|h\|_\infty\} &\leqq \|f_\xi\|_q \\ &= \omega(d_n(\mathscr{K}_\infty; L^q)) \\ &\leqq \omega(t)\end{aligned}$$

since ω is a nondecreasing function. Therefore $f_\xi \in V(K, \omega, \infty, q)$. □

3. Exact n-Widths of $\mathscr{K}_\infty^r$ in L^q and $\mathscr{K}_p^r$ in L^1

A rich source of totally positive kernels are the Green's functions of disconjugate rth order differential equations for a large class of separated boundary conditions (see Section 4 of Chapter III). It is therefore natural to consider the class of functions

$$\mathscr{K}_p^r = \left\{\sum_{i=1}^r a_i k_i(x) + \int_0^1 K(x, y)\, h(y)\, dy : \|h\|_p \leqq 1,\ a_i \in \mathbb{R}\right\},$$

where the $\{k_i\}_{i=1}^r$ are a basis of solutions of the disconjugate equation.

We shall apply the methods of Section 2 (see also Sections 3 and 5 of Chapter IV) to the problem of determining the n-widths of this class. We therefore assume that K together with the fixed functions $\{k_i\}_{i=1}^r$ satisfy the conditions given in Section 5.2 of Chapter IV. These are

Property A

1. $\{k_i\}_{i=1}^r$ *is a* T^+*-system on* $(0, 1)$, *i.e., for any* $0 < x_1 < \ldots < x_r < 1$,

$$K\begin{pmatrix} x_1, \ldots, x_r \\ 1, \ldots, r \end{pmatrix} = \det(k_j(x_i))_{i,j=1}^r > 0.$$

2. $$K\begin{pmatrix} x_1, \ldots, x_r, & x_{r+1}, \ldots, x_{r+m} \\ 1, \ldots, r, & y_1, \ldots, y_m \end{pmatrix}$$

$$= \begin{vmatrix} k_1(x_1) & \ldots & k_r(x_1) & K(x_1, y_1) & \ldots & K(x_1, y_m) \\ \vdots & & \vdots & \vdots & & \vdots \\ k_1(x_{r+m}) & \ldots & k_r(x_{r+m}) & K(x_{r+m}, y_1) & \ldots & K(x_{r+m}, y_m) \end{vmatrix} \geqq 0$$

for every choice of $0 \leqq y_1 < \ldots < y_m \leqq 1, 0 \leqq x_1 < \ldots < x_{r+m} \leqq 1$, *and all* $m \geqq 0$. *Furthermore, for any given* $0 < y_1 < \ldots < y_m < 1$, *the above determinant is not identically zero, and for any given* $0 < x_1 < \ldots < x_{r+m} < 1$, *the above determinant is not identically zero.*

If $r = 0$, then Property A reduces to the assumption that K is a nondegenerate totally positive kernel. Note also that Property A is satisfied by the important choice $k_i(x) = x^{i-1}$, $i = 1, \ldots, r$, and $K(x, y) = (x - y)_+^{r-1}/(r-1)!$.

To determine the n-widths of $\mathscr{K}_p^r$ in L^q for $p = \infty$ or $q = 1$, we apply the ideas used in Section 2. However, the fixed r-dimensional subspace presents certain complications which must be overcome.

To ease our exposition, we introduce the following notation. Let

$$\mathscr{T}_r = \operatorname{span}\{k_1, \ldots, k_r\} \quad \text{and} \quad \mathscr{P}_m = \{k + K h_{\mathbf{t}} : k \in \mathscr{T}_r,\ \mathbf{t} \in \Lambda_m\},$$

where Λ_m, $\bar{\Lambda}_m$ and $h_{\mathbf{t}}$ for $\mathbf{t} \in \bar{\Lambda}_m$ are as defined in Section 2.

Proposition 3.1. *Assume that Property A holds and let* $f \in L^1[0, 1]$.

(i) *If* $Q = k + Kf$, *where* $k \in \mathscr{T}_r$ *and* $S(f) \leqq m$, *then* $S(Q) \leqq r + m$.

(ii) *Let* $Q = k + Kf$, *where* $k \in \mathscr{T}_r$ *and* $\operatorname{sgn} f(y) = h_{\mathbf{t}}(y)$ *for some* $\mathbf{t} \in \Lambda_m$ *a.e. in* $(0, 1)$. *If* Q *has* $r + m$ *zeros at* $s_1 < \ldots < s_{r+m}$ *in* $(0, 1)$, *then* $\operatorname{sgn} Q(x) = (-1)^{i+r}$, $x \in (s_i, s_{i+1})$, $i = 0, 1, \ldots, r + m$ (*where* $s_0 = 0$, $s_{r+m+1} = 1$).

(iii) *For* $\mathbf{t} = (t_1, \ldots, t_m) \in \Lambda_m$ *and* $\mathbf{s} = (s_1, \ldots, s_{r+m}) \in \Lambda_{r+m}$, *as above*

$$K\begin{pmatrix} s_1, \ldots, s_{r+m} \\ 1, \ldots, r, t_1, \ldots, t_m \end{pmatrix} > 0.$$

(iv) *Given* $\mathbf{s} = (s_1, \ldots, s_{r+m}) \in \Lambda_{r+m}$, *there exists a* $P \in \mathscr{P}_m$ *for which* $P(s_i) = 0$, $i = 1, \ldots, r + m$.

Proof. We first prove (i). If $r = 0$, then (i) is simply a consequence of the total positivity of K. Assume that $S(Q) = q$, and let $\mathbf{s} \in \Lambda_q$ be such that $h_{\mathbf{s}}(x)\, Q(x) \geqq 0$ for all $x \in [0, 1]$. (Either Q or $-Q$ will satisfy this condition.) If $q \leqq r$ there is nothing to prove. Thus we assume that $q > r$. Form

$$J(x, y) = K\begin{pmatrix} s_1, \ldots, s_r, x \\ 1, \ldots, r, y \end{pmatrix} \Big/ K\begin{pmatrix} s_1, \ldots, s_r \\ 1, \ldots, r \end{pmatrix}.$$

The kernel $\bar{J}(x, y) = (-1)^r h_{\mathbf{s}(r)}(x)\, J(x, y)$ is totally positive for $\mathbf{s}(r) = (s_1, \ldots, s_r) \in \Lambda_r$, since by Sylvester's determinant identity (Section 2 of Chapter III) it follows that if $0 \leqq x_1 < \ldots < x_m \leqq 1$, and $0 \leqq y_1 < \ldots < y_m \leqq 1$, then

$$\bar{J}\begin{pmatrix} x_1, \ldots, x_m \\ y_1, \ldots, y_m \end{pmatrix} = K\begin{pmatrix} w_1, \ldots, w_{r+m} \\ 1, \ldots, r, y_1, \ldots, y_m \end{pmatrix} \Big/ K\begin{pmatrix} s_1, \ldots, s_r \\ 1, \ldots, r \end{pmatrix}$$

where $0 \leqq w_1 \leqq \ldots \leqq w_{r+m} \leqq 1$, and the points w_i are an increasing rearrangement of the set of points $\{s_1, \ldots, s_r, x_1, \ldots, x_m\}$. Set

$$\bar{Q}(x) = \int_0^1 J(x, y)\, f(y)\, dy.$$

We claim that $Q = \bar{Q}$. Indeed $\bar{Q} = \bar{k} + Kf$ for some $\bar{k} \in \mathcal{T}_r$, and therefore $Q - \bar{Q} \in \mathcal{T}_r$. Furthermore, $Q(s_i) = \bar{Q}(s_i) = 0$, $i = 1, \ldots, r$ (the latter since $J(s_i, \cdot) = 0$, $i = 1, \ldots, r$) and only the zero function in $\mathcal{T}_r$ has r distinct zeros in $(0, 1)$, whence $Q = \bar{Q}$. Thus

$$(-1)^r h_{\mathbf{s}(r)}(x)\, Q(x) = \int_0^1 \bar{J}(x, y)\, f(y)\, dy.$$

Since $S(f) \leqq m$ and $\bar{J}(x, y)$ is totally positive, $S((-1)^r h_{\mathbf{s}(r)} Q) \leqq m$ and $q = S(Q) \leqq r + m$.

We now prove part (ii). Let $J(x, y)$ and $\bar{J}(x, y)$ be as defined above. Thus

$$(-1)^r h_{\mathbf{s}(r)}(x)\, Q(x) = \int_0^1 \bar{J}(x, y)\, f(y)\, dy$$

where $\bar{J}(x, y)$ is a totally positive, but *not* a nondegenerate totally positive kernel. Nonetheless, the method of proof of part (i) of Proposition 2.2 will suffice to prove the result. Let $x \in (0, 1)/\{s_1, \ldots, s_{r+m}\}$. The set of functions

$$\{\bar{J}(x, \cdot),\ \bar{J}(s_{r+1}, \cdot), \ldots, \bar{J}(s_{r+m}, \cdot)\}$$

is linearly independent and forms a weak Tchebycheff (WT)-system of order $m + 1$. The sign of the associated determinants is $(-1)^{r+i}$ for $x \in (s_i', s_{i+1}')$, $i = r, \ldots, r + m$, where $s_i' = s_i$, $i = r + 1, \ldots, r + m$, $s_r' = 0$, and $s_{r+m+1}' = 1$. The proof of (ii) now follows the proof of part (i) of Proposition 2.2.

Part (iii) is a restatement of Lemma 5.4 (see also Lemma 5.12) of Chapter IV where we note that if the determinant in the statement of the proposition is zero, then

$$J\begin{pmatrix} s_{r+1}, \ldots, s_{r+m} \\ t_1, \ldots, t_m \end{pmatrix} = 0.$$

To prove part (iv), let $\bar{J}(x, y)$ be as above. By the Hobby-Rice Theorem (Theorem 2.3) there exists a $\mathbf{t} \in \Lambda_m$ for which

$$\int_0^1 \bar{J}(s_{r+i}, y)\, h_{\mathbf{t}}(y)\, dy = 0, \quad i = 1, \ldots, m.$$

Set

$$P(x) = \int_0^1 J(x, y)\, h_{\mathbf{t}}(y)\, dy.$$

Then $P \in \mathcal{P}_m$ and $P(s_i) = 0$, $i = 1, \ldots, r + m$. $\square$

To determine the n-widths of $\mathscr{K}_\infty^r$ in L^q, we consider the following extremal problem for $n \geqq r$:

$$\inf_{P \in \mathscr{P}_{n-r}} \|P\|_q.$$

Theorem 3.2. *Assume Property A holds, $1 \leqq q < \infty$, and $n \geqq r$. There exists a $\xi^* \in \Lambda_{n-r}$ and $k^* \in \mathscr{T}_r$ (both dependent on q) for which $P_{n-r} = k^* + K h_{\xi^*}$ satisfies*

$$\inf_{P \in \mathscr{P}_{n-r}} \|P\|_q = \|P_{n-r}\|_q.$$

Furthermore P_{n-r} has n zeros at $\boldsymbol{\eta}^* \in \Lambda_n$, $\operatorname{sgn} P_{n-r}(x) = (-1)^r h_{\boldsymbol{\eta}^*}(x)$ for $x \in (0,1)$, and

$$\int_0^1 |P_{n-r}(x)|^{q-1} \operatorname{sgn}[P_{n-r}(x)]\, K(x, \xi_i^*)\, dx = 0, \qquad i = 1, \ldots, n-r,$$

$$\int_0^1 |P_{n-r}(x)|^{q-1} \operatorname{sgn}[P_{n-r}(x)]\, k_i(x)\, dx = 0, \qquad i = 1, \ldots, r.$$

Proof. The proof of Theorem 3.2 is almost totally analogous to the proof of Theorem 2.4. A standard compactness argument implies the existence of a $k^* \in \mathscr{T}_r$ and $\xi^* \in \Lambda_m$, $m \leqq n-r$, for which $P_{n-r} = k^* + K h_{\xi^*}$ satisfies

$$\inf_{P \in \mathscr{P}_{n-r}} \|P\|_q = \|P_{n-r}\|_q.$$

Since $1 \leqq q < \infty$, it follows from orthogonality considerations, as in the proof of Theorem 2.4, that

$$\int_0^1 |P_{n-r}(x)|^{q-1} \operatorname{sgn}[P_{n-r}(x)]\, K(x, \xi_i^*)\, dx = 0, \qquad i = 1, \ldots, m,$$

and

$$\int_0^1 |P_{n-r}(x)|^{q-1} \operatorname{sgn}[P_{n-r}(x)]\, k_i(x)\, dx = 0, \qquad i = 1, \ldots, r.$$

Since $\{k_1, \ldots, k_r, K(\cdot, \xi_1^*), \ldots, K(\cdot, \xi_m^*)\}$ is a WT-system of dimension $r+m$, it follows from Proposition 1.4 of Chapter III that $S(P_{n-r}) \geqq r+m$. From Proposition 3.1 (ii), P_{n-r} has at most $r+m$ zeros in $(0,1)$. Thus P_{n-r} has exactly $r+m$ zeros in $(0,1)$, each of which is a sign change. It remains to prove that $m = n-r$.

Let

$$J(x, y) = K\begin{pmatrix} \eta_1^*, \ldots, \eta_r^*, x \\ 1, \ldots, r, y \end{pmatrix} \Big/ K\begin{pmatrix} \eta_1^*, \ldots, \eta_r^* \\ 1, \ldots, r \end{pmatrix}$$

and $\bar{J}(x, y) = (-1)^r h_{\eta^*(r)}(x)\, J(x, y)$, where $\boldsymbol{\eta}^*(r) = (\eta_1^*, \ldots, \eta_r^*)$, and $\boldsymbol{\eta}^* = (\eta_1^*, \ldots, \eta_{r+m}^*)$ are the zeros of P_{n-r} in $(0, 1)$.

$$|P_{n-r}(x_0)| = \left| \int_0^1 \frac{\bar{J}\begin{pmatrix} \eta_{r+1}^*, \ldots, \eta_{r+m}^*, x_0 \\ \xi_1^*, \ldots, \xi_m^*, y \end{pmatrix}}{\bar{J}\begin{pmatrix} \eta_{r+1}^*, \ldots, \eta_{r+m}^* \\ \xi_1^*, \ldots, \xi_m^* \end{pmatrix}} h_{\xi^*}(y)\, dy \right|$$

$$= \int_0^1 \left| \bar{J}(x_0, y) - \sum_{i=r+1}^{r+m} a_i \bar{J}(\eta_i^*, y) \right| dy$$

for some choice of constants $\{a_i\}_{i=r+1}^{r+m}$. We now continue the proof as in Theorem 2.4. □

Theorem 3.3. *Assume Property A holds and $n \geqq r$. There exists a $k^* \in \mathscr{T}_r$ and a $\xi^* \in \Lambda_{n-r}$ for which $P_{n-r} = k^* + K h_{\xi^*}$ equioscillates at $n + 1$ points in $[0, 1]$, i.e., there are points $0 \leqq z_1 < \ldots < z_{n+1} \leqq 1$ satisfying*

$$P_{n-r}(z_i)(-1)^{i+r+1} = \| P_{n-r} \|_\infty, \quad i = 1, \ldots, n + 1.$$

Furthermore,

$$\inf_{P \in \mathscr{P}_{n-r}} \| P \|_\infty = \| P_{n-r} \|_\infty .$$

Proof. We parallel the proof of Theorem 2.5. Since $\{k_i\}_{i=1}^r$ is a T-system on $(0, 1)$, there exist continuous functions $\{k_i\}_{i=r+1}^n$ such that $\{k_i\}_{i=1}^n$ is a T-system on $(0, 1)$ (see Zielke [1979]). We now apply the method of proof of Theorem 2.5. We map each $\mathbf{x} \in \Xi_{n-r+1}$ into $(a_{r+1}(\mathbf{x}), \ldots, a_n(\mathbf{x}))$ where the $\{a_i(\mathbf{x})\}_{i=1}^n$ are the unique coefficients in the best L^∞-approximation $\sum_{i=1}^n a_i(\mathbf{x})\, k_i$ on $[\varepsilon, 1 - \varepsilon]$, $\varepsilon > 0$ small, to

$$g_{\mathbf{x}}(x) = \sum_{i=1}^{n-r+1} [\operatorname{sgn} x_i] \int_{t_{i-1}(\mathbf{x})}^{t_i(\mathbf{x})} K(x, y)\, dy.$$

The Borsuk Antipodality Theorem implies the existence of an $\mathbf{x}^* \in \Xi_{n-r+1}$ for which $a_i(\mathbf{x}^*) = 0$, $i = r + 1, \ldots, n$. Set $P_{n-r} = g_{\mathbf{x}^*} - \sum_{i=1}^r a_i(\mathbf{x}^*)\, k_i$. P_{n-r} equioscillates at $n + 1$ points in $[\varepsilon, 1 - \varepsilon]$, since P_{n-r} cannot be identically zero. Furthermore, from Proposition 3.1, $g_{\mathbf{x}^*} = \pm K h_{\xi^*}$ for some $\xi^* \in \Lambda_{n-r}$ depending on ε. We now let $\varepsilon \downarrow 0$ and show that the result also holds in the limit.

To prove the minimum property of P_{n-r}, let $P \in \mathscr{P}_{n-r}$ satisfy $\| P \|_\infty < \| P_{n-r} \|_\infty$. Set $P = k + K h_{\mathbf{t}}$, $k \in \mathscr{T}_r$, $\mathbf{t} \in \Lambda_{n-r}$. Thus $P_{n-r} - P = (k^* - k) + K(h_{\xi^*} - h_{\mathbf{t}})$ has at least n sign changes in $(0, 1)$. By Proposition 3.1, this implies that $S(h_{\xi^*} - h_{\mathbf{t}}) \geqq n - r$. However from Lemma 2.1 (ii), $S(h_{\xi^*} - h_{\mathbf{t}}) \leqq n - r - 1$. □

In this next theorem $\boldsymbol{\xi}^* = (\xi_1^*, \ldots, \xi_{n-r}^*) \in \Lambda_{n-r}$ and $\boldsymbol{\eta}^* = (\eta_1^*, \ldots, \eta_n^*) \in \Lambda_n$ will denote the knots and zeros of P_{n-r}, respectively. The dependence of these quantities on $q \in [1, \infty]$ is to be understood and is not explicitly stated.

Theorem 3.4. *Assume Property A holds and* $q \in [1, \infty]$. *Then*

$$d_n(\mathscr{K}_\infty^r; L^q) = d^n(\mathscr{K}_\infty^r; L^q) = \delta_n(\mathscr{K}_\infty^r; L^q) = \begin{cases} \infty, & n \leqq r-1 \\ \| P_{n-r} \|_q, & n \geqq r. \end{cases}$$

For $n \geqq r$,

(1) *the n-dimensional subspace* $X_n = \operatorname{span}\{k_1, \ldots, k_r, K(\cdot, \xi_1^*), \ldots, K(\cdot, \xi_{n-r}^*)\}$ *is optimal for* $d_n(\mathscr{K}_\infty^r; L^q)$.

(2) *The rank n linear operator* R_n *which interpolates from* X_n *to each element of* $\mathscr{K}_\infty^r$ *at the points* $\{\eta_i^*\}_{i=1}^n$ *is optimal for* $\delta_n(\mathscr{K}_\infty^r; L^q)$.

(3) *The subspace*

$$L^n = \{f\colon f \in \mathscr{K}_\infty^r,\ f(\eta_i^*) = 0,\ i = 1, \ldots, n\}$$

is optimal for $d^n(\mathscr{K}_\infty^r; L^q)$.

Remark. It is important to note that we have reverted to the original definitions of d^n and δ_n. (See the discussion in Section 7 of Chapter II.)

Proof. Since $\mathscr{T}_r$ is r-dimensional, it follows as in Section 3.1 of Chapter IV, that d_n, d^n, and δ_n are all infinite for $n \leqq r-1$. Assume that $n \geqq r$. To prove the upper bound, it suffices to show that $\delta_n(\mathscr{K}_\infty^r; L^q) \leqq \| P_{n-r} \|_q$.

Let R_n be as given above. (It is well-defined by Proposition 3.1.) Let $Q = k + Kh$, $k \in \mathscr{T}_r$. Then

$$Q(x) - R_n Q(x) = \int_0^1 \frac{K\begin{pmatrix} \eta_1^*, \ldots, \eta_n^*, x \\ 1, \ldots, r, \xi_1^*, \ldots, \xi_{n-r}^*, y \end{pmatrix}}{K\begin{pmatrix} \eta_1^*, \ldots, \eta_n^* \\ 1, \ldots, r, \xi_1^*, \ldots, \xi_{n-r}^* \end{pmatrix}} h(y)\, dy.$$

The kernel $L(x, y) = K\begin{pmatrix} \eta_1^*, \ldots, \eta_n^*, x \\ 1, \ldots, r, \xi_1^*, \ldots, \xi_{n-r}^*, y \end{pmatrix} \Big/ K\begin{pmatrix} \eta_1^*, \ldots, \eta_n^* \\ 1, \ldots, r, \xi_1^*, \ldots, \xi_{n-r}^* \end{pmatrix}$ satisfies $L(x, y)(-1)^{i+j+r} \geqq 0$ for $x \in [\eta_i^*, \eta_{i+1}^*]$, $y \in [\xi_j^*, \xi_{j+1}^*]$, $i = 0, 1, \ldots, n$; $j = 0, 1, \ldots, n-r$. Thus

$$\begin{aligned} \delta_n(\mathscr{K}_\infty^r; L^q) &\leqq \sup_{Q \in \mathscr{K}_\infty^r} \| Q - R_n Q \|_q \\ &= \sup_{\|h\|_\infty \leqq 1} \left\| \int_0^1 L(\cdot, y)\, h(y)\, dy \right\|_q \\ &= \left(\int_0^1 \left(\int_0^1 |L(x, y)|\, dy \right)^q dx \right)^{1/q}. \end{aligned}$$

The last equality is attained for $h(y) = h_{\xi^*}(y)$ and, as is easily seen, $\int_0^1 L(x, y)\, h_{\xi^*}(y)\, dy = P_{n-r}(x)$. Thus $\delta_n(\mathscr{K}_\infty^r; L^q) \leqq \| P_{n-r} \|_q$.

We must prove the lower bound for both d_n and d^n. We first consider d_n. In order that $E(\mathscr{K}_\infty^r; X_n) < \infty$, it is necessary that $\mathscr{T}_r \subseteqq X_n$. We therefore assume that

$$X_n = \text{span}\{k_1, \ldots, k_r, u_1, \ldots, u_{n-r}\}$$

for some $u_i \in L^q$, $i = 1, \ldots, n-r$. We now parallel the proof of Theorem 2.6. Use the Borsuk Antipodality Theorem to prove that for $q \in (1, \infty)$ there exists a $\mathbf{t} \in \bar{\Lambda}_{n-r}$ for which

$$\inf_{g \in X_n} \| K h_{\mathbf{t}} - g \|_q = \| K h_{\mathbf{t}} - k \|_q$$

for some $k \in \mathscr{T}_r$. This fact together with Theorem 3.2 suffices to prove that $d_n(\mathscr{K}_\infty^r; L^q) \geqq \| P_{n-r} \|_q$. To deal with the cases $q = 1$ and $q = \infty$ we perturb the norms.

Before presenting the proof of the lower bound for d^n, note that if we had considered, as in Section 2, spaces of the form

$$\{k + Kh \colon k \in \mathscr{T}_r, \ \|h\|_\infty \leqq 1, \ h \in M^n\}$$

for some M^n of codimension $\leqq n$, then we would have obtained trivial and meaningless results since the above definition totally disregards the r-dimensional subspace $\mathscr{T}_r$. We therefore consider subspaces M^n of codimension $\leqq n$ of $\mathscr{K}_\infty^r$ in L^q of the form

$$M^n = \{f \colon f \in L^q, \ (g_i, f) = 0, \ i = 1, \ldots, n\}$$

for some $g_i \in L^{q'}$, $1/q + 1/q' = 1$. (If $q = \infty$, then we work in the predual.) Difficulties now arise because the subspace L^n of the statement of the theorem is not a subspace of this form. We shall, however, not dwell on this fact since a smoothing procedure will resolve this problem (see Micchelli and Pinkus [1978]).

Let M^n be any subspace as above of codimension exactly n, and assume that $\Delta(\mathscr{K}_\infty^r; M^n) < \infty$, i.e.,

$$\sup_{f \in \mathscr{K}_\infty^r \cap M^n} \| f \|_q < \infty.$$

It necessarily follows that $M^n \cap \mathscr{T}_r = \{0\}$. Thus, if

$$M^n = \{f \colon f \in L^q, \ (g_i, f) = 0, \ i = 1, \ldots, n\},$$

where the $\{g_i\}_{i=1}^n$ are assumed to be linearly independent functions in $L^{q'}$, then the matrix

$$((g_i, k_j))_{i=1}^{n} {}_{j=1}^{r}$$

must be of full rank r. Assume without loss of generality that

$$\det((g_i, k_j))_{i,j=1}^r \neq 0.$$

Let

$$M(x,y) = \frac{\begin{vmatrix} (g_1,k_1)\dots(g_1,k_r) & (g_1,K(\cdot,y)) \\ \vdots \quad\quad \vdots & \vdots \\ (g_r,k_1)\dots(g_r,k_r) & (g_r,K(\cdot,y)) \\ k_1(x)\ \dots\ k_r(x) & K(x,y) \end{vmatrix}}{\det((g_i,k_j))_{i,j=1}^r}.$$

By the Hobby-Rice Theorem (Theorem 2.3) there exists a $\mathbf{t} \in \bar{\Lambda}_{n-r}$ for which $(g_i, M h_\mathbf{t}) = (M' g_i, h_\mathbf{t}) = 0$, $i = r+1, \dots, n$. Since $(g_i, M h) = 0$, $i = 1, \dots, r$, for any $h \in L^\infty$ it follows that $P = M h_\mathbf{t} = k + K h_\mathbf{t} \in \mathscr{K}_\infty^r \cap M^n$. Thus

$$\sup_{f \in \mathscr{K}_\infty^r \cap M^n} \|f\|_q \geqq \inf_{P \in \mathscr{P}_{n-r}} \|P\|_q = \|P_{n-r}\|_q,$$

from which we obtain $d^n(\mathscr{K}_\infty^r; L^q) \geqq \|P_{n-r}\|_q$. □

As in Section 2, these same techniques should lead us to a solution of the n-width problem for $\mathscr{K}_p^r$ in L^1. However the analysis here is again not quite as straightforward as might be expected due to the presence of the fixed r-dimensional subspace $\{k_i\}_{i=1}^r$. The extremal problem which we shall consider is not the same as the extremal problem previously considered. Rather we must look at

$$\inf\{\|K' h_\mathbf{t}\|_{p'} : \mathbf{t} \in \Lambda_n,\ (k_i, h_\mathbf{t}) = 0,\ i = 1, \dots, r\}$$

where $1/p + 1/p' = 1$ and $K'(x,y) = K(y,x)$, the transpose of K. (See Section 3 of Chapter IV for a similar situation.) Let

$$\mathscr{Q}_n = \{K' h_\mathbf{t} : \mathbf{t} \in \Lambda_n,\ (k_i, h_\mathbf{t}) = 0,\ i = 1, \dots, r\}.$$

Theorem 3.5. *Assume Property A holds, $p' \in [1, \infty)$, and $n \geqq r$. There exists a $\boldsymbol{\xi}^* \in \Lambda_n$ for which $(k_i, h_{\boldsymbol{\xi}^*}) = 0$, $i = 1, \dots, r$, and $Q_n = K' h_{\boldsymbol{\xi}^*}$ satisfies*

$$\inf_{Q \in \mathscr{Q}_n} \|Q\|_{p'} = \|Q_n\|_{p'}.$$

Furthermore, Q_n has $n - r$ distinct zeros at $\boldsymbol{\eta}^* \in \Lambda_{n-r}$, and $\operatorname{sgn} Q_n(x) = (-1)^r h_{\boldsymbol{\eta}^*}(x)$.

Proof. The above infimum problem has a solution in $\bar{\Lambda}_n$. Assume that $\boldsymbol{\xi}^* \in \Lambda_m$, $m \leqq n$, is such a solution. Because $(k_i, h_{\boldsymbol{\xi}^*}) = 0$, $i = 1, \dots, r$, we have $m \geqq r$ by Proposition 1.4 of Chapter III. The method of Lagrange multipliers implies that

$$\int_0^1 |K' h_{\boldsymbol{\xi}^*}(y)|^{p'-1} \operatorname{sgn}[K' h_{\boldsymbol{\xi}^*}(y)]\, K(\xi_j^*, y)\, dy + \sum_{i=1}^r \alpha_i k_i(\xi_j^*) = 0,$$

$j = 1, \ldots, m$, for some choice of $\{\alpha_i\}_{i=1}^r$. Thus $\sum_{i=1}^{r} \alpha_i k_i + Kf$ has m zeros at $\xi_1^*, \ldots, \xi_m^*$, where $f(y) = |K' h_{\xi^*}(y)|^{p'-1} \operatorname{sgn}[K' h_{\xi^*}(y)]$. By Proposition 3.1 (i), this implies that f, and therefore $K' h_{\xi^*}$, has at least $m - r$ sign changes in $(0, 1)$. But $K' h_{\xi^*}$ cannot have more than $m - r$ zeros in $(0, 1)$, since this possibility together with $(k_i, h_{\xi^*}) = 0$, $i = 1, \ldots, r$, and Property A, would contradict Proposition 1.4 of Chapter III. $Q_n = K' h_{\xi^*}$ therefore has exactly $m - r$ zeros (sign changes) in $(0, 1)$. The orientation of the sign is a consequence of Proposition 3.1. It remains to prove that $m = n$. The proof of this fact parallels the proof of Theorem 2.4. □

To deal with the case $p' = \infty$, we have the following analogue of Theorem 3.3.

Theorem 3.6. *Assume Property A holds and $n \geqq r$. There exists a $Q_n = K' h_{\xi^*} \in \mathcal{Q}_n$, $\xi^* \in \Lambda_n$, such that Q_n equioscillates at $n - r + 1$ points in $[0, 1]$. Furthermore,*

$$\inf_{Q \in \mathcal{Q}_n} \|Q\|_\infty = \|Q_n\|_\infty.$$

Proof. Let $\{u_i\}_{i=1}^{n-r}$ be any T-system on $[0, 1]$. To each $\mathbf{x} \in \Xi_{n+1}$ (as earlier defined) let $(a_1(\mathbf{x}), \ldots, a_{n-r}(\mathbf{x}))$ denote the coefficient vector of the unique best L^∞-approximation to

$$g_{\mathbf{x}} = \sum_{i=1}^{n+1} [\operatorname{sgn} x_i] \int_{t_{i-1}(\mathbf{x})}^{t_i(\mathbf{x})} K(x, \cdot)\, dx$$

from span $\{u_1, \ldots, u_{n-r}\}$. Map $\mathbf{x} \in \Xi_{n+1}$ into the vector

$$(a_1(\mathbf{x}), \ldots, a_{n-r}(\mathbf{x}),\ (k_1, h_{\mathbf{t}(\mathbf{x})}), \ldots, (k_r, h_{\mathbf{t}(\mathbf{x})}))$$

where $h_{\mathbf{t}(\mathbf{x})}(x) = \operatorname{sgn} x_i$, $x \in (t_{i-1}(\mathbf{x}), t_i(\mathbf{x}))$, $i = 1, \ldots, n + 1$. This map is odd and continuous. By the Borsuk Antipodality Theorem there exists an $\mathbf{x}^* \in \Xi_{n+1}$ for which $a_i(\mathbf{x}^*) = 0$, $i = 1, \ldots, n - r$, and $(k_i, h_{\mathbf{t}(\mathbf{x}^*)}) = 0$, $i = 1, \ldots, r$. The resulting $g_{\mathbf{x}^*} = Q_n = K' h_{\xi^*}$ equioscillates at $n - r + 1$ points, and $\xi^* \in \Lambda_n$ as a result of Proposition 1.4 of Chapter III. The infimum property of $K' h_{\xi^*}$ is proved in much the same way as it was proved in Theorem 3.3. □

In this next theorem, $\boldsymbol{\xi}^* = (\xi_1^*, \ldots, \xi_n^*) \in \Lambda_n$ and $\boldsymbol{\eta}^* = (\eta_1^*, \ldots, \eta_{n-r}^*) \in \Lambda_{n-r}$ will denote the knots and zeros of Q_n, respectively. (Again, we do not explicitly state the dependence on p', which is always understood.)

Theorem 3.7. *Assume Property A holds and $p \in [1, \infty]$. Then for $1/p + 1/p' = 1$,*

$$d_n(\mathcal{K}_p^r; L^1) = d^n(\mathcal{K}_p^r; L^1) = \delta_n(\mathcal{K}_p^r; L^1) = \begin{cases} \infty, & n \leqq r - 1 \\ \|Q_n\|_{p'}, & n \geqq r. \end{cases}$$

For $n \geqq r$,

(1) $X_n = \operatorname{span}\{k_1, \ldots, k_r, K(\cdot, \eta_1^*), \ldots, K(\cdot, \eta_{n-r}^*)\}$ *is optimal for* $d_n(\mathcal{K}_p^r; L^1)$.

(2) *The rank n linear operator* R_n *which maps each element of* $\mathscr{K}_p^r$ *into* X_n *by interpolation at the points* $\{\xi_i^*\}_{i=1}^n$ *is optimal* $\delta_n(\mathscr{K}_p^r; L^1)$.

(3) $L^n = \{f\colon f \in \mathscr{K}_p^r,\ f(\xi_i^*) = 0,\ i = 1, \ldots, n\}$ *is optimal for* $d^n(\mathscr{K}_p^r; L^1)$.

Proof. The n-widths are all infinite for $n \leqq r - 1$. Assume $n \geqq r$. We first prove the upper bound $\delta_n(\mathscr{K}_p^r; L^1) \leqq \|Q_n\|_{p'}$. Note that

$$K\begin{pmatrix} \xi_1^*, \ldots, \xi_n^* \\ 1, \ldots, r, \eta_1^*, \ldots, \eta_{n-r}^* \end{pmatrix} > 0$$

so that R_n, as defined above, exists. For $p' \in [1, \infty)$, this is a result of Proposition 3.1, since in the proof of Theorem 3.5 we constructed a function $k + Kf$ with zeros $\boldsymbol{\xi}^*$ for which $\operatorname{sgn} f(y) = (-1)^r h_{\boldsymbol{\eta}^*}(y)$. For $p' = \infty$, the result is also true, one method of proof being an analogue of the proof of Lemma 5.4 of Chapter IV.

Set

$$R(x, y) = K\begin{pmatrix} \xi_1^*, \ldots, \xi_n^*, x \\ 1, \ldots, r, \eta_1^*, \ldots, \eta_{n-r}^*, y \end{pmatrix} \Big/ K\begin{pmatrix} \xi_1^*, \ldots, \xi_n^* \\ 1, \ldots, r, \eta_1^*, \ldots, \eta_{n-r}^* \end{pmatrix}$$

Let $g = k + Kh$, $k \in \mathscr{T}_r$, $\|h\|_p \leqq 1$. Then

$$(g - R_n g)(x) = \int_0^1 R(x, y)\, h(y)\, dy,$$

and

$$\begin{aligned} \sup_{g \in \mathscr{K}_p^r} \|g - R_n g\|_1 &= \sup_{\|h\|_p \leqq 1} \int_0^1 \left| \int_0^1 R(x, y)\, h(y)\, dy \right| dx \\ &\leqq \left(\int_0^1 \left(\int_0^1 |R(x, y)|\, dx \right)^{p'} dy \right)^{1/p'} \\ &= \left(\int_0^1 \left| \int_0^1 R(x, y)\, h_{\boldsymbol{\xi}^*}(x)\, dx \right|^{p'} dy \right)^{1/p'}. \end{aligned}$$

Since $(k_i, h_{\boldsymbol{\xi}^*}) = 0$, $i = 1, \ldots, r$, and $K' h_{\boldsymbol{\xi}^*}(\eta_i^*) = 0$, $i = 1, \ldots, n - r$, this simplifies to

$$\sup_{g \in \mathscr{K}_p^r} \|g - R_n g\|_1 \leqq \|K' h_{\boldsymbol{\xi}^*}\|_{p'} = \|Q_n\|_{p'}.$$

Thus $\delta_n(\mathscr{K}_p^r; L^1) \leqq \|Q_n\|_{p'}$.

The following argument proves the lower bound for d_n. Let X_n be any n-dimensional subspace of L^1 for which $E(\mathscr{K}_p^r; X_n) < \infty$. Thus $\mathscr{T}_r \subseteq X_n$. By the Hobby-Rice Theorem there exists a $\mathbf{t} \in \bar{\Lambda}_n$ satisfying $(g, h_{\mathbf{t}}) = 0$ for all $g \in X_n$. Now,

$$E(\mathscr{K}_p^r; X_n) = \sup_{\|h\|_p \leqq 1} \inf_{g \in X_n} \|k + Kh - g\|_1,$$

and by Proposition 6.1 of Chapter II, this is turn equals

$$\sup_{\|h\|_p \leqq 1} \sup_{\substack{(f, X_n)=0 \\ \|f\|_\infty \leqq 1}} (k + Kh, f).$$

Since $\mathcal{T}_r \subseteqq X_n$,

$$\begin{aligned} \sup_{\|h\|_p \leqq 1} \sup_{\substack{(f, X_n)=0 \\ \|f\|_\infty \leqq 1}} (k + Kh, f) &= \sup_{\|h\|_p \leqq 1} \sup_{\substack{(f, X_n)=0 \\ \|f\|_\infty \leqq 1}} (Kh, f) \\ &= \sup_{\substack{(f, X_n)=0 \\ \|f\|_\infty \leqq 1}} \|K'f\|_{p'} \\ &\geqq \|K'h_{\mathbf{t}}\|_{p'} \\ &\geqq \|Q_n\|_{p'}. \end{aligned}$$

The right hand side is independent of X_n implying that $d_n(\mathcal{K}_p^r; L^1) \geqq \|Q_n\|_{p'}$.

It remains to prove the lower bound for d^n. Let M^n be any subspace of codimension n of L^1 for which $\Delta(\mathcal{K}_p^r; M^n) < \infty$. Thus

$$M^n = \{f\colon f \in L^1, \ (g_i, f) = 0, \ i = 1, \ldots, n\}$$

where the $\{g_i\}_{i=1}^n$ are assumed to be linearly independent functions of L^∞, and $M^n \cap \mathcal{T}_r = \{0\}$. Let $M(x, y)$ be as defined in the proof of Theorem 3.4, and set $u_i = M' g_i$, $i = r+1, \ldots, n$. (Recall that $M' g_i = 0$, $i = 1, \ldots, r$.) We claim that there exists a $\mathbf{t} \in \bar{\Lambda}_n$ for which $(k_i, h_{\mathbf{t}}) = 0$, $i = 1, \ldots, r$ and

$$\|K'h_{\mathbf{t}}\|_{p'} = \min_{\alpha_{r+1}, \ldots, \alpha_n} \left\| K'h_{\mathbf{t}} - \sum_{i=r+1}^{n} \alpha_i u_i \right\|_{p'}.$$

This is accomplished for $1 < p' < \infty$ by the method of proof of Theorem 3.6. To deal with the cases $p' = 1$ and $p' = \infty$, we perturb the norms.

Since the best $L^{p'}$-approximation to $K'h_{\mathbf{t}}$ from span $\{u_{r+1}, \ldots, u_n\}$ is zero, it follows that for $1 \leqq p' < \infty$, $(g, u_i) = 0$, $i = r+1, \ldots, n$, where $g = |K'h_{\mathbf{t}}|^{p'-1} \operatorname{sgn}[K'h_{\mathbf{t}}]$. Set $v = g/\|g\|_p$. Thus $v \in L^p$, $\|v\|_p = 1$, and $w = Mv \in M^n \cap \mathcal{K}_p^r$ by construction. Hence

$$\sup_{f \in M^n \cap \mathcal{K}_p^r} \|f\|_1 \geqq \|w\|_1 \geqq (h_{\mathbf{t}}, w)$$

and since $(k_i, h_{\mathbf{t}}) = 0$, $i = 1, \ldots, r$,

$$(h_{\mathbf{t}}, w) = (K'h_{\mathbf{t}}, v) = \|K'h_{\mathbf{t}}\|_{p'} \geqq \|Q_n\|_{p'}.$$

A similar argument proves the result for $p' = \infty$. □

Remark. We have given two distinct characterizations of $d_n(\mathcal{K}_\infty^r; L^1)$, i.e., this case is covered by both Theorem 3.4 and Theorem 3.7.

We also present analogues of Theorems 2.10 and 2.12 of Section 2. Firstly, however, we must define what we now mean by the Bernstein n-width of $\mathscr{K}_p^r$ in L^q.

Definition 3.1. For $n \geqq r$, the *Bernstein n-width* of $\mathscr{K}_p^r$ in L^q is defined as

$$b_n(\mathscr{K}_p^r; L^q) = \sup_{X_{n+1}} \inf_{\substack{k+Kh \in X_{n+1} \\ k+Kh \neq 0}} \|k + Kh\|_q / \|h\|_p$$

where X_{n+1} is any subspace of dimension $\geqq n+1$ of span$\{k + Kh \colon h \in L^p\}$ which contains $\mathscr{T}_r$.

We state, without proof, the following analogues of Theorems 2.10 and 2.12.

Theorem 3.8. *Assume Property A holds. Then for $n \geqq r$,*

$$b_n(\mathscr{K}_\infty^r; L^\infty) = \inf_{P \in \mathscr{P}_{n-r}} \|P\|_\infty = \|P_{n-r}\|_\infty$$

$$b_n(\mathscr{K}_1^r; L^1) = \inf_{Q \in \mathscr{Q}_n} \|Q\|_\infty = \|Q_n\|_\infty .$$

Furthermore,

(1) $X_{n+1} = \operatorname{span}\{k_1, \ldots, k_r, u_1, \ldots, u_{n-r+1}\}$ *is optimal for* $b_n(\mathscr{K}_\infty^r; L^\infty)$, *where*

$$u_i(x) = \int_{\xi_{i-1}^*}^{\xi_i^*} K(x, y)\, dy, \qquad i = 1, \ldots, n - r + 1.$$

(The $\{\xi_i^*\}_{i=1}^{n-r}$ are the knots of P_{n-r}.)

(2) $X_{n+1} = \operatorname{span}\{k_1, \ldots, k_r, K(\cdot, z_1^*), \ldots, K(\cdot, z_{n-r+1}^*)\}$ *is optimal for* $b_n(\mathscr{K}_1^r; L^1)$, *where the $\{z_i^*\}_{i=1}^{n-r+1}$ are $(n - r + 1)$ ordered points at which Q_n attains its norm, alternately.*

Let us now consider the main motivating example behind the results of this section, namely the Sobolev space $W_p^{(r)}$, and the set

$$B_p^{(r)} = \{f \colon f \in W_p^{(r)},\ \|f^{(r)}\|_p \leqq 1\}.$$

(This corresponds, of course, to the disconjugate rth order differential equation $D^r u = 0$.)

Definition 3.2. A *perfect spline* on $[0,1]$ of degree r with m knots $\{\xi_i\}_{i=1}^m$, $0 = \xi_0 < \xi_1 < \ldots < \xi_m < \xi_{m+1} = 1$, is any function $P(x)$ of the form

$$P(x) = \sum_{i=0}^{r-1} a_i x^i + c \sum_{j=0}^{m} (-1)^j \int_{\xi_j}^{\xi_{j+1}} (x - y)_+^{r-1}\, dy,$$

where, as usual, $x_+^k = x^k$ for $x \geqq 0$, and zero elsewhere.

Equivalently, $P(x)$ is more often written in the form

$$P(x) = \sum_{i=0}^{r-1} a_i x^i + d\left[x^r + 2\sum_{j=1}^{m}(-1)^j (x-\xi_j)_+^r\right].$$

It is a spline with m simple knots whose rth derivative, in absolute value, is a constant a.e. on $[0,1]$. Perfect splines arise in various problems of approximation theory and not only in the study of n-widths. For example, given points $0 \leqq t_1 < \ldots < t_{m+r+1} \leqq 1$, and data $\{\alpha_i\}_{i=1}^{m+r+1}$, there exists a perfect spline P of degree r with at most m knots for which $P(t_i) = \alpha_i$, $i = 1, \ldots, m+r+1$. Furthermore, if $f \in W_\infty^{(r)}$ and $f(t_i) = \alpha_i$, $i = 1, \ldots, m+r+1$, then $\| f^{(r)} \|_\infty \geqq \| P^{(r)} \|_\infty$.

Let $\mathscr{P}_m^r$ denote the subclass of perfect splines of degree r, with at most m knots, for which $|P^{(r)}(x)| = 1$ a.e. on $[0,1]$, and for $m \geqq r$, set

$$\mathscr{Q}_m^r = \{P\colon P \in \mathscr{P}_m^r,\ P^{(i)}(0) = P^{(i)}(1) = 0,\ i = 0, 1, \ldots, r-1\}.$$

Let $P_m^r \in \mathscr{P}_m^r$ be such that

$$\| P_m^r \|_p = \inf\{\| P \|_p \colon P \in \mathscr{P}_m^r\}$$

and $Q_m^r \in \mathscr{Q}_m^r$ satisfy

$$\| Q_m^r \|_p = \inf\{\| Q \|_p \colon Q \in \mathscr{Q}_m^r\},$$

$1 \leqq p \leqq \infty$. (P_m^r and Q_m^r do depend on p.) Theorems 3.2 and 3.3 imply that P_m^r has m distinct knots $\{\xi_i^*\}_{i=1}^m$ and $r+m$ distinct zeros $\{\eta_i^*\}_{i=1}^{r+m}$ in $(0,1)$. Similarly Theorems 3.5 and 3.6 imply that Q_m^r has m distinct knots $\{\tau_i^*\}_{i=1}^m$ and $m-r$ distinct zeros $\{\zeta_i^*\}_{i=1}^{m-r}$ in $(0,1)$.

From Theorems 3.4 and 3.7, we obtain the following.

Corollary 3.9. *For $q \in [1, \infty]$,*

$$d_n(B_\infty^{(r)}; L^q) = d^n(B_\infty^{(r)}; L^q) = \delta_n(B_\infty^{(r)}; L^q) = \begin{cases} \infty, & n \leqq r-1 \\ \| P_{n-r}^r \|_q, & n \geqq r \end{cases},$$

and for $n \geqq r$,

(1) $X_n = \operatorname{span}\{1, x, \ldots, x^{r-1}, (x-\xi_1^*)_+^{r-1}, \ldots, (x-\xi_{n-r}^*)_+^{r-1}\}$ *is optimal for* $d_n(B_\infty^{(r)}; L^q)$.

(2) *The linear operator which maps each element of $B_\infty^{(r)}$ to X_n by interpolation at the $\{\eta_i^*\}_{i=1}^n$ is optimal for $\delta_n(B_\infty^{(r)}; L^q)$.*

(3) $L^n = \{f\colon f \in B_\infty^{(r)},\ f(\eta_i^*) = 0,\ i = 1, \ldots, n\}$ *is optimal for $d^n(B_\infty^{(r)}; L^q)$.*

Corollary 3.10. *For $p \in [1, \infty]$, $1/p + 1/p' = 1$,*

$$d_n(B_p^{(r)}; L^1) = d^n(B_p^{(r)}; L^1) = \delta_n(B_p^{(r)}; L^1) = \begin{cases} \infty, & n \leqq r-1 \\ \| Q_n^r \|_{p'}, & n \geqq r \end{cases},$$

and for $n \geqq r$,

(1) $X_n = \text{span}\{1, x, \ldots, x^{r-1}, (x-\zeta_1^*)_+^{r-1}, \ldots, (x-\zeta_{n-r}^*)_+^{r-1}\}$ *is optimal for* $d_n(B_p^{(r)}; L^1)$.

(2) *The linear operator which maps each element of* $B_p^{(r)}$ *to* X_n *by interpolation at the* $\{\tau_i^*\}_{i=1}^n$ *is optimal for* $\delta_n(B_p^{(r)}; L^1)$.

(3) $L^n = \{f\colon f \in B_p^{(r)}, f(\tau_i^*) = 0,\ i = 1, \ldots, n\}$ *is optimal for* $d^n(B_p^{(r)}; L^1)$.

We can also specialize Theorem 3.8. We only consider $p = q = \infty$. (The case $p = q = 1$ is similar.) We state this result in a slightly altered form.

Corollary 3.11.

$$\sup_{X_n} \inf_{\substack{f^{(r)} \in X_n \\ f^{(r)} \neq 0}} \|f\|_\infty / \|f^{(r)}\|_\infty = \|P_{n-1}^r\|_\infty,$$

where X_n *is any subspace of* L^∞ *of dimension* $\geqq n$. *Equality is attained in the above for* $X_n = \text{span}\{u_1, \ldots, u_n\}$, *where*

$$u_i(x) = \begin{cases} 1, & \xi_{i-1}^* \leqq x \leqq \xi_i^* \\ 0, & \textit{otherwise} \end{cases}$$

and the $\{\xi_i^*\}_{i=1}^{n-1}$ *are the knots of* P_{n-1}^r.

In particular, for $r = 1$,

$$P_{n-1}^1(x) = -1/2n + x + 2\sum_{j=1}^{n-1} (-1)^j (x - (j/n))_+^1,$$

and $\|P_{n-1}^1\|_\infty = 1/2n$. The corollary says that given any n-dimensional subspace X_n, there necessarily exists an $f' \in X_n$ for which $\|f'\|_\infty \geqq 2n\|f\|_\infty$. However, if X_n is the space of step functions with knots (steps) at $\{j/n\}_{j=1}^{n-1}$, then for all $f' \in X_n$,

$$\|f'\|_\infty \leqq 2n\|f\|_\infty.$$

(If $X_n = \pi_{n-1}$, the space of algebraic polynomials of degree $n-1$, then from Markov's inequality (see e.g. Rivlin [1969]), there exists a $p \in \pi_n$, namely the nth Chebyshev polynomial of the first kind, for which $\|p'\|_\infty = 2n^2\|p\|_\infty$.)

The class $W_p^{(r)}$ is defined via the semi-norm $\|f^{(r)}\|_p$. For some purposes it is more convenient, when considering $B_p^{(r)}$ as a subset of L^q, to define

$$B_{p,q}^{(r)} = \{f\colon f \in W_p^{(r)},\ \|f\|_q + \|f^{(r)}\|_p \leqq 1\}.$$

Thus $B_{p,q}^{(r)}$ is a bounded subset of L^q. We easily determine the n-widths of $B_{\infty,q}^{(r)}$ in L^q and show that the optimal subspace for $d_n(B_\infty^{(r)}; L^q)$ is also optimal for $d_n(B_{\infty,q}^{(r)}; L^q)$ $(= d_n(B_\infty^{(r)}; L^q)/[1 + d_n(B_\infty^{(r)}; L^q)])$.

The upper bound is sufficiently simple to state in a general form. Let X, Y be normed linear spaces and $T \in L(X, Y)$. Set

$$B_r = \{u + Tx\colon \|x\|_X \leqq 1,\ u \in U_r\}$$

where U_r is a fixed r-dimensional subspace of Y (see Section 3.1 of Chapter IV). Define

$$B_r^* = \{u + Tx\colon \|x\|_X + \|u + Tx\|_Y \leqq 1,\ u \in U_r\}.$$

Proposition 3.12. *Let X_n be any n-dimensional subspace of Y. Then*

$$E(B_r^*; X_n) \leqq E(B_r; X_n)/[1 + E(B_r; X_n)].$$

Thus, in particular,

$$d_n(B_r^*) \leqq d_n(B_r)/[1 + d_n(B_r)].$$

Proof. For convenience, set $\lambda = E(B_r; X_n)/[1 + E(B_r; X_n)]$. Thus $\lambda \in [0, 1]$ (where $\lambda = 1$ if $E(B_r; X_n) = \infty$). By definition,

$$\inf\{\|u + Tx - y\|_Y\colon y \in X_n\} \leqq \|x\|_X\, E(B_r; X_n).$$

Thus

$$\begin{aligned}\inf_{y \in X_n} \|u + Tx - y\|_Y &= \inf_{y \in X_n} [\lambda \|u + Tx - y\|_Y + (1 - \lambda) \|u + Tx - y\|_Y] \\ &\leqq \lambda \|u + Tx\|_Y + (1 - \lambda) \inf_{y \in X_n} \|u + Tx - y\|_Y \\ &\leqq \lambda \|u + Tx\|_Y + (1 - \lambda) \|x\|_X\, E(B_r; X_n) \\ &= (\|u + Tx\|_Y + \|x\|_X)\, E(B_r; X_n)/[1 + E(B_r; X_n)]\end{aligned}$$

and therefore

$$E(B_r^*; X_n) \leqq E(B_r; X_n)/[1 + E(B_r; X_n)]. \quad \square$$

Now, let $\mathscr{K}_p^r$ be as defined at the beginning of this section (assuming Property A holds) and set

$$\mathscr{K}_{p,q}^r = \left\{\sum_1^n a_i k_i + Kh\colon \|h\|_p + \left\|\sum_1^n a_i k_i + Kh\right\|_q \leqq 1,\ a_i \in \mathbb{R}\right\}.$$

Theorem 3.13. *For $\mathscr{K}_\infty^r$ and $\mathscr{K}_{\infty,q}^r$ as above,*

$$d_n(\mathscr{K}_{\infty,q}^r; L^q) = d_n(\mathscr{K}_\infty^r; L^q)/[1 + d_n(\mathscr{K}_\infty^r; L^q)]$$

and if X_n is optimal for $d_n(\mathscr{K}_\infty^r; L^q)$, then it is also optimal for $d_n(\mathscr{K}_{\infty,q}^r; L^q)$.

Proof. We must prove the lower bound. Let X_n be any n-dimensional subspace of L^q. If X_n does not contain $\mathscr{T}_r$, then it is easily seen that $E(\mathscr{K}_{\infty,q}^r; X_n) = 1$. Thus it suffices to consider $n \geqq r$ and n-dimensional subspaces X_n containing $\mathscr{T}_r$. From the proof of Theorem 3.4 (and from Theorems 3.2 and 3.3) there exists a $\xi \in \Lambda_{n-r}$ and $k_\xi \in \mathscr{T}_r$ for which

$$\inf_{g \in X_n} \|k_\xi + Kh_\xi - g\|_q = \|k_\xi + Kh_\xi\|_q \geqq d_n(\mathscr{K}_\infty^r; L^q).$$

Thus

$$\begin{aligned} E(\mathscr{K}^r_{\infty,q}; X_n) &= \sup_{\|h\|_\infty + \|k+Kh\|_q \leqq 1} \inf_{g \in X_n} \|k + Kh - g\|_q \\ &\geqq \|k_\xi + Kh_\xi\|_q / [\|h_\xi\|_\infty + \|k_\xi + Kh_\xi\|_q] \\ &= \|k_\xi + Kh_\xi\|_q / [1 + \|k_\xi + Kh_\xi\|_q] \\ &\geqq d_n(\mathscr{K}^r_\infty; L^q)/[1 + d_n(\mathscr{K}^r_\infty; L^q)] \end{aligned}$$

since $x/(1+x)$ is an increasing function of x on $[0, \infty)$. □

4. Exact n-Widths for Periodic Functions

In this section, we consider the n-widths of two sets of related functions. Our first class is the following:

Let k be a real, 2π-periodic, continuous function. Set

$$\tilde{\mathscr{K}}_p = \left\{ (k * h)(x) = 1/2\pi \int_0^{2\pi} k(x-y)\, h(y)\, dy \colon \|h\|_p \leqq 1 \right\},$$

where

$$\|h\|_p = \begin{cases} \left(1/2\pi \int_0^{2\pi} |h(y)|^p\, dy \right)^{1/p}, & 1 \leqq p < \infty \\ \operatorname{ess\,sup} \{|h(y)| \colon 0 \leqq y \leqq 2\pi\}, & p = \infty. \end{cases}$$

We prove analogues of the results of Section 2, where $K(x, y) = k(x - y)$ is a nondegenerate cyclic variation diminishing ($NCVD$) kernel, rather than a nondegenerate totally positive kernel. Because of the nature of $NCVD$ kernels, our results, as in Section 6.1 of Chapter IV, will be mainly concerned with the calculation of the n-widths for n even, except in those cases where it can be established that trigonometric polynomials of some fixed degree (always an odd dimensional space) is also an optimal subspace. In particular, we calculate the n-widths and determine optimal subspaces for $d_{2n}(\tilde{\mathscr{K}}_p; L^q)$, $d^{2n}(\tilde{\mathscr{K}}_p; L^q)$ and $\delta_{2n}(\tilde{\mathscr{K}}_p; L^q)$ for $p = \infty$, all q; $q = 1$, all p. In addition, we obtain $d_{2n-1}(\tilde{\mathscr{K}}_p; L^q)$, $d^{2n-1}(\tilde{\mathscr{K}}_p; L^q)$, $\delta_{2n-1}(\tilde{\mathscr{K}}_p; L^q)$ and $b_{2n-1}(\tilde{\mathscr{K}}_p; L^q)$ for $p = q = \infty$, and $p = q = 1$, and $d_{2n-1}(\tilde{\mathscr{K}}_p; L^q)$ for $q = 1$, all p (and $d^{2n-1}(\tilde{\mathscr{K}}_p; L^q)$ for $p = \infty$, all q). In all these cases the values of the n-widths and the optimal subspaces are easily calculated.

The second class of functions is the following:

Set

$$\tilde{\mathscr{B}}_p = \{a + (G * h)(x) \colon \|h\|_p \leqq 1,\ h \perp 1,\ a \in \mathbb{R}\}$$

where G satisfies Property B (as defined in Section 6.2 of Chapter IV). The situation here is very much similar to that considered above for $\tilde{\mathscr{K}}_p$. We calculate the n-widths of $\tilde{\mathscr{B}}_p$ in L^q for the same p and q as above, and obtain values and optimal subspaces almost totally analogous to those obtained when considering $\tilde{\mathscr{K}}_p$. Unfortunately, the slightly different form of $\tilde{\mathscr{B}}_p$ necessitates some changes, albeit not major, in the arguments used. As such we shall develop, in parallel, our results for $\tilde{\mathscr{K}}_p$ and $\tilde{\mathscr{B}}_p$.

For convenience we first restate the relevant definitions.

Definition 4.1. Let k be a real, 2π-periodic, continuous function. Then k is *nondegenerate cyclic variation diminishing* ($NCVD$) if k is cyclic variation diminishing (CVD), and

$$\dim \operatorname{span}\{k(x_1 - \cdot), \dots, k(x_m - \cdot)\} \;(= \dim \operatorname{span}\{k(\cdot - y_1), \dots, k(\cdot - y_m)\}) = m$$

for every choice of $0 \leqq x_1 < \dots < x_m < 2\pi$ $(0 \leqq y_1 < \dots < y_m < 2\pi)$ and all m.

It should be noted that the above definition is equivalent to assuming that the rank condition holds and that

$$\operatorname{span}\{k(x_1 - \cdot), \dots, k(x_{2m-1} - \cdot)\}$$

is a WT-system for every $0 \leqq x_1 < \dots < x_{2m-1} < 2\pi$, and all m.

Definition 4.2. Let G be a real, 2π-periodic, continuous function. G satisfies *Property B* if for every choice of $0 \leqq y_1 < \dots < y_m < 2\pi$,

$$X_m = \left\{ b + \sum_{i=1}^{m} b_i G(\cdot - y_i) \colon \sum_{i=1}^{m} b_i = 0 \right\}$$

is of dimension m, and is a WT-system for m odd.

In Section 6.2 of Chapter IV, we proved various properties of G satisfying Property B, and delineated some functions which satisfy Property B, a motivating class being $\tilde{B}_p^{(r)}$.

The following notation and accompanying lemmas will be used throughout this section. Let Λ_{2m} and $\bar{\Lambda}_{2m}$ denote the open and closed, respectively, $2m$-dimensional simplex on $([0, 2\pi))^{2m}$. Due to the periodicity, we mean by Λ_{2m} the following:

$$\Lambda_{2m} = \{\boldsymbol{\xi} \colon \boldsymbol{\xi} = (\xi_1, \dots, \xi_{2m}),\ 0 \leqq \xi_1 < \dots < \xi_{2m} < 2\pi\}.$$

For each $\boldsymbol{\xi} \in \bar{\Lambda}_{2m}$, we define

$$h_{\boldsymbol{\xi}}(x) = (-1)^{i+1}, \quad x \in [\xi_{i-1}, \xi_i), \qquad i = 1, \dots, 2m+1,$$

where $\xi_0 = 0$, $\xi_{2m+1} = 2\pi$.

As an analogue of Lemma 2.1 of Section 2, we have:

Lemma 4.1. (i) *If* $\boldsymbol{\xi} \in \Lambda_{2m}$, $\boldsymbol{\eta} \in \Lambda_{2k}$, *then*

$$S_c(h_{\boldsymbol{\xi}} \pm h_{\boldsymbol{\eta}}) \leqq \min\{2m, 2k\}.$$

(ii) *If* $\boldsymbol{\xi}, \boldsymbol{\eta} \in \Lambda_{2m}$ *and* $\xi_k = \eta_{k+2r}$ *for some* k *and* r, *then*

$$S_c(h_{\boldsymbol{\xi}} - h_{\boldsymbol{\eta}}) \leqq 2(m-1).$$

One subset of functions of the above form is of particular interest and importance. Set

$$h_m(x;\alpha) = h_{\boldsymbol{\xi}(\alpha)}(x),$$

where $\boldsymbol{\xi}(\alpha) = (\xi_1(\alpha), \ldots, \xi_{2m}(\alpha))$, $\xi_i(\alpha) = \alpha + (i-1)\pi/m$, $i = 1, \ldots, 2m$. For convenience we also set

$$h_m(x) = h_m(x;0).$$

One consequence of Lemma 4.1 is the following result.

Lemma 4.2. *Let* $\boldsymbol{\xi} \in \bar{\Lambda}_{2m}$ *be such that*

$$\int_0^{2\pi} h_{\boldsymbol{\xi}}(x)\, t(x)\, dx = 0$$

for all $t \in T_{m-1}$. *Then* $h_{\boldsymbol{\xi}}(x) = h_m(x;\alpha)$ *for some choice of* $\alpha \in [0, \pi/m]$.

Proof. It is simple to show that

$$\int_0^{2\pi} h_m(x;\alpha)\, t(x)\, dx = 0$$

for all α and all $t \in T_{m-1}$, the trigonometric polynomials of degree $\leqq m-1$. Thus

$$\int_0^{2\pi} [h_m(x;\alpha) \pm h_{\boldsymbol{\xi}}(x)]\, t(x)\, dx = 0$$

for all $t \in T_{m-1}$. Since T_{m-1} is a T-system of dimension $2m-1$, it follows from Proposition 1.4 of Chapter III that either

$$S_c(h_m(\cdot;\alpha) \pm h_{\boldsymbol{\xi}}(\cdot)) \geqq 2m,$$

or $h_m(\cdot;\alpha) \pm h_{\boldsymbol{\xi}}(\cdot) = 0$. From Lemma 4.1, we see that there exists a choice of $\alpha^* \in [0, \pi/m]$ for which

$$S_c(h_m(\cdot;\alpha^*) - h_{\boldsymbol{\xi}}(\cdot)) \leqq 2(m-1).$$

Thus $h_m(\cdot;\alpha^*) = h_{\boldsymbol{\xi}}(\cdot)$. $\square$

Before entering into a discussion of the n-width problem, we consider the quantities $E(\tilde{\mathscr{K}}_\infty; T_{n-1})_\infty$ and $E(\tilde{\mathscr{B}}_\infty; T_{n-1})_\infty$. Recall that for a set A, we have

$$E(A; T_{n-1})_\infty = \sup_{f \in A} \inf_{t \in T_{n-1}} \|f - t\|_\infty.$$

We will prove that $E(\tilde{\mathscr{K}}_\infty; T_{n-1})_\infty = \|k * h_n\|_\infty$ and $E(\tilde{\mathscr{B}}_\infty; T_{n-1})_\infty = \|G * h_n\|_\infty$. This latter formula is a direct generalization of the Favard, Krein, Achieser Theorem which gives this particular result for the choice $G = 2D_r$. Preparatory to proving these formulae we have:

Proposition 4.3.

(i) *Let k be an NCVD kernel. Let t^* be the best L^1-approximation to k from T_{n-1}. There exists an $\alpha \in [0, \pi/n)$ and an $\varepsilon \in \{-1, 1\}$ such that*

$$\varepsilon(k - t^*)(x)\, h_n(x; \alpha) \geqq 0$$

for all x.

(ii) *Let G satisfy Property B. Let t^* be the best L^1-approximation to G from T_{n-1}. There exists an $\alpha \in [0, \pi/n)$ and an $\varepsilon \in \{-1, 1\}$ such that*

$$\varepsilon(G - t^*)(x)\, h_n(x; \alpha) \geqq 0$$

for all x.

Proof. One proof will suffice to prove the two statements of the proposition. Let G satisfy Property B, and assume, with no loss of generality, that $G \perp 1$. By a standard "smoothing technique" (see Chapter III) we may assume that G' exists, is 2π-periodic and continuous, and $\dim \operatorname{span}\{T_{n-1}, G, G'\} = 2n + 1$. Set

$$f_M(y) = \begin{cases} M - 1/2\pi, & y \in [0, 1/M] \\ -1/2\pi, & \text{otherwise} \end{cases}$$

and

$$g_M(y) = \begin{cases} -M^2, & y \in [0, 1/M] \\ M^2, & y \in [1/M, 2/M] \\ 0, & \text{otherwise.} \end{cases}$$

Note that $f_M \perp 1$ and $g_M \perp 1$. Let α, β be fixed constants. Given $t \in T_{n-1}$ for which $t \perp 1$, i.e., whose constant term is zero, there exists an M_0 such that for all $M \geqq M_0$,

$$S_c(\alpha f_M + \beta g_M + t) \leqq 2n.$$

It therefore follows from Proposition 6.4 of Chapter IV that

$$S_c(a + \alpha(G * f_M) + \beta(G * g_M) + (G * t)) \leqq 2n$$

for all $M \geqq M_0$, and every $a \in \mathbb{R}$. Let M tend to infinity.

$$\begin{aligned} \lim_{M\to\infty} (G * f_M)(x) &= \lim_{M\to\infty} M/2\pi \int_0^{1/M} G(x-y)\,dy - 1/2\pi \int_0^{2\pi} G(x-y)\,dy \\ &= (1/2\pi)\, G(x), \end{aligned}$$

and

$$\lim_{M\to\infty} (G * g_M)(x) = \lim_{M\to\infty} M^2/2\pi \left[-\int_0^{1/M} G(x-y)\,dy + \int_{1/M}^{2/M} G(x-y)\,dy \right]$$
$$= \lim_{M\to\infty} M^2/2\pi \int_0^{1/M} [G(x-y-(1/M)) - G(x-y)]\,dy$$
$$= (1/2\pi)\, G'(x).$$

Now $G * t \in T_{n-1}$ and since G satisfies Property B, $|a_m| \neq 0$, $m = 1, 2, \ldots$ (by the rank condition and Theorem 6.9 of Chapter IV). Thus the range of $G * t$, $t \in T_{n-1}$, $t \perp 1$, is the set of all $t \in T_{n-1}$ satisfying $t \perp 1$. However to compensate we have the free constant term a. Therefore

$$S_c((\alpha/2\pi)\, G + (\beta/2\pi)\, G' + t) \leqq 2n$$

for all $\alpha, \beta \in \mathbb{R}$, $t \in T_{n-1}$. In other words, span$\{T_{n-1}, G, G'\}$ is a WT-system of periodic functions on $[0, 2\pi)$.

Again by smoothing we assume that span$\{T_{n-1}, G, G'\}$ is, in fact, a T-system of periodic functions (see Proposition 1.9 of Chapter III).

Let t^* be the best L^1-approximant to G from T_{n-1}. Since T_{n-1} is a T-system the best approximant is unique. $G - t^*$ cannot vanish at more than $2n$ points. Thus

$$\int_0^{2\pi} \operatorname{sgn}[(G - t^*)(x)]\, t(x)\, dx = 0$$

for all $t \in T_{n-1}$, and $\operatorname{sgn}[(G - t^*)(x)] = \pm h_\xi(x)$ for some $\xi \in \bar{\Lambda}_{2n}$. By Lemma 4.2, this implies that $\varepsilon \operatorname{sgn}[(G - t^*)(x)] = h_n(x; \alpha)$ for some $\alpha \in [0, \pi/n)$ and $\varepsilon \in \{-1, 1\}$. We now "unravel" the smoothing to obtain the desired result.

Assume now that k is $NCVD$. Define

$$G_1(x) = \int_0^x (k(y) - a_0)\, dy + c$$

where $a_0 = 1/2\pi \int_0^{2\pi} k(y)\, dy$ and $c \in \mathbb{R}$. We proved in Proposition 6.7 of Chapter IV that G_1 satisfies Property B. By the above proof it therefore follows that span$\{T_{n-1}, G_1, k\}$ is a WT-system of dimension $2n+1$ of periodic functions on $[0, 2\pi)$. Replace G by k in the previous two paragraphs to obtain statement (i) of the proposition. □

Using Proposition 4.3 we can determine $E(\tilde{\mathscr{K}}_\infty; T_{n-1})_\infty$ and $E(\tilde{\mathscr{B}}_\infty; T_{n-1})_\infty$.

Theorem 4.4. (i) *Let k be an $NCVD$ kernel. Then*

$$E(\tilde{\mathscr{K}}_\infty; T_{n-1})_\infty = \| k * h_n \|_\infty .$$

(ii) *Let G satisfy Property B. Then*

$$E(\tilde{\mathscr{B}}_\infty; T_{n-1})_\infty = \| G * h_n \|_\infty .$$

Proof. We prove (ii). The proof of (i) is essentially identical. Let t^* be as given in Proposition 4.3. Then

$$\begin{aligned} E(\tilde{\mathscr{B}}_\infty; T_{n-1})_\infty &= \sup_{\substack{\|h\|_\infty \leqq 1 \\ h \perp 1}} \inf_{t \in T_{n-1}} \|G * h - t\|_\infty \\ &\leqq \sup_{\|h\|_\infty \leqq 1} \|(G - t^*) * h\|_\infty \\ &= \|G - t^*\|_1 \\ &= \left| 1/2\pi \int_0^{2\pi} (G - t^*)(x)\, h_n(x;\alpha)\, dx \right| \\ &= \left| 1/2\pi \int_0^{2\pi} G(x)\, h_n(x;\alpha)\, dx \right| \\ &\leqq \|G * h_n\|_\infty . \end{aligned}$$

Thus $E(\tilde{\mathscr{B}}_\infty; T_{n-1}) \leqq \|G * h_n\|_\infty$.

Since $h_n \perp 1$, $G * h_n \in \tilde{\mathscr{B}}_\infty$. Furthermore $(G * h_n)(x) = -(G * h_n)(x + \pi/n)$ so that $G * h_n$ attains its norm, alternately, at $2n$ points. Since T_{n-1} is a T-system of dimension $2n - 1$ it therefore follows that

$$\inf_{t \in T_{n-1}} \|G * h_n - t\|_\infty = \|G * h_n\|_\infty .$$

Thus $E(\tilde{\mathscr{B}}_\infty; T_{n-1}) = \|G * h_n\|_\infty$. □

Remark. Note that a linear method of approximation is used in the proof.

In determining the n-widths in Section 2, we considered an extremal problem. This extremal problem in our present setting translates into

$$\inf\{\|k * h_\xi\|_q : \xi \in \Lambda_{2n}\}$$

for k *NCVD*, and

$$\inf\{\|a + G * h_\xi\|_q : \xi \in \Lambda_{2n},\ h_\xi \perp 1,\ a \in \mathbb{R}\}$$

for G satisfying Property B. Unfortunately, neither k nor G is *NTP*. Thus we cannot directly appeal to the method of Section 2. We overcome this difficulty by explicitly exhibiting the solution for all q. We first consider the case where k is *NCVD*.

Theorem 4.5. *Let k be an NCVD kernel. Then for all $q \in [1, \infty]$,*

$$\inf\{\|k * h_\xi\|_q : \xi \in \Lambda_{2n}\} = \|k * h_n\|_q .$$

Proof. We separately consider the cases $q = \infty$ and $q \in [1, \infty)$.

We first prove the result when $q = \infty$. Assume $\xi \in \Lambda_{2n}$ and $\|k * h_\xi\|_\infty < \|k * h_n\|_\infty$. Since $k * h_n$ attains its norm, alternately, at $2n$ points,

$$S_c((k * h_n)(\cdot + \alpha) \pm (k * h_\xi)(\cdot)) \geqq 2n$$

for every α and $\pm$. Since k is $NCVD$, this implies that

$$S_c(h_n(\cdot + \alpha) \pm h_\xi(\cdot)) \geqq 2n$$

for every choice of α and $\pm$. A contradiction follows from Lemma 4.1 (ii). Thus Theorem 4.5 is proved for $q = \infty$.

Let $q \in [1, \infty)$. We shall assume that k is extended CVD. By this we mean that

$$k^*\begin{pmatrix} x_1, \ldots, x_{2m+1} \\ y_1, \ldots, y_{2m+1} \end{pmatrix} > 0$$

for every choice of $0 \leqq x_1 \leqq \ldots \leqq x_{2m+1} \leqq 2\pi$; $0 \leqq y_1 \leqq \ldots \leqq y_{2m+1} \leqq 2\pi$, and $m = 0, 1, \ldots$, where equality among the x_i or y_i means taking successive derivatives (see Section 3 of Chapter III). This is done by means of "smoothing". Once we have obtained our result for such functions, the theorem follows for $NCVD$ functions by a limiting procedure. The property of k being extended CVD implies that $Z_c(k * h) \leqq S_c(h)$ for all h where Z_c counts the number of zeros in $[0, 2\pi)$, counting multiplicities.

A compactness argument shows that there exists a $\boldsymbol{\xi}^* \in \bar{\Lambda}_{2n}$ for which

$$\inf\{\|k * h_{\boldsymbol{\xi}}\|_q : \boldsymbol{\xi} \in \Lambda_{2n}\} = \|k * h_{\boldsymbol{\xi}^*}\|_q .$$

Thus $\boldsymbol{\xi}^* \in \Lambda_{2m}$ for some $m \leqq n$, and $\boldsymbol{\xi}^* = (\xi_1^*, \ldots, \xi_{2m}^*)$, $0 \leqq \xi_1^* < \ldots < \xi_{2m}^* < 2\pi$. A simple perturbation argument implies the orthogonality (minimality) conditions

$$\int_0^{2\pi} |(k * h_{\boldsymbol{\xi}^*})(x)|^{q-1} [\operatorname{sgn}(k * h_{\boldsymbol{\xi}^*})(x)]\, k(x - \xi_i^*)\, dx = 0, \qquad i = 1, \ldots, 2m.$$

Now since k is extended CVD, $S_c(k * h_{\boldsymbol{\xi}^*}) \geqq 2m$. However since $\boldsymbol{\xi}^* \in \Lambda_{2m}$, $Z_c(k * h_{\boldsymbol{\xi}^*}) \leqq S_c(h_{\boldsymbol{\xi}^*}) = 2m$. Thus $k * h_{\boldsymbol{\xi}^*}$ has $2m$ distinct simple zeros. We first claim that $m = n$.

Assume that $m < n$. For each $\xi \notin \{\xi_1^*, \ldots, \xi_{2m}^*\}$ and ε sufficiently small, define $\boldsymbol{\xi}_\varepsilon^* \in \bar{\Lambda}_{2n}$ which has knots $\xi_1^*, \ldots, \xi_{2m}^*$ and the two additional knots $\xi + \varepsilon$ and $\xi - \varepsilon$. From the optimality of $\boldsymbol{\xi}^*$, it follows that

$$\int_0^{2\pi} |(k * h_{\boldsymbol{\xi}^*})(x)|^{q-1} [\operatorname{sgn}(k * h_{\boldsymbol{\xi}^*})(x)]\, k(x - \xi)\, dx = 0$$

for all $\xi \in [0, 2\pi)$. Since k is extended CVD this is impossible. Thus $\boldsymbol{\xi}^* \in \Lambda_{2n}$.

By translation we may assume that $\boldsymbol{\xi}^* = (\xi_1^*, \ldots, \xi_{2n}^*)$ satisfies $0 = \xi_1^* < \ldots < \xi_{2n}^* < 2\pi$, and

$$\delta = \xi_2^* = \xi_2^* - \xi_1^* = \min\{\xi_{i+1}^* - \xi_i^* : i = 1, \ldots, 2n\}$$

where $\xi_{2n+1}^* = 2\pi$. Assume that $h_{\boldsymbol{\xi}^*} \neq h_n$. We shall contradict the orthogonality conditions enjoyed by $k * h_{\boldsymbol{\xi}^*}$.

From Lemma 4.1 (ii),

$$S_c(h_{\xi^*}(\cdot) + h_{\xi^*}(\cdot + \delta)) \leqq 2(n-1).$$

Set $p(x) = (k * h_{\xi^*})(x)$ and $r(x) = p(x+\delta) = (k * h_{\xi^*})(x+\delta)$. Thus

$$p(x) + r(x) = 1/2\pi \int_0^{2\pi} k(x-y)\,[h_{\xi^*}(y) + h_{\xi^*}(y+\delta)]\,dy,$$

and therefore $S_c(p+r) \leqq 2(n-1)$. Since

$$\operatorname{sgn}(a+b) = \operatorname{sgn}(|a|^{q-1} \operatorname{sgn} a + |b|^{q-1} \operatorname{sgn} b)$$

for every a, b real and $q \in (1, \infty)$, it follows that

$$S_c(|p(\cdot)|^{q-1} \operatorname{sgn}(p(\cdot)) + |r(\cdot)|^{q-1} \operatorname{sgn}(r(\cdot))) \leqq 2(n-1)$$

for every $q \in [1, \infty)$.

Set $P(y) = \int_0^{2\pi} |p(x)|^{q-1} \operatorname{sgn}(p(x))\, k(x-y)\,dx$, and $R(y) = \int_0^{2\pi} |r(x)|^{q-1} \operatorname{sgn}(r(x))\, k(x-y)\,dx$. Thus $Z_c(P(\cdot) + R(\cdot)) \leqq 2(n-1)$. A simple change of variable argument shows that $R(y) = P(y+\delta)$, from which we obtain

$$Z_c(P(\cdot) + P(\cdot + \delta)) \leqq 2(n-1).$$

Now we previously proved, based on the orthogonality conditions, that $P(y)$ changes sign at ξ_i^*, $i = 1, \ldots, 2n$. By our choice of δ, $\xi_i^* < \xi_i^* + \delta \leqq \xi_{i+1}^*$, $i = 1, \ldots, 2n$, and therefore

$$S_c^+(P(\xi_1^* + \delta), \ldots, P(\xi_{2n}^* + \delta)) = 2n.$$

Thus

$$S_c^+(P(\xi_1^*) + P(\xi_1^* + \delta), \ldots, P(\xi_{2n}^*) + P(\xi_{2n}^* + \delta)) = 2n$$

which implies that

$$Z_c(P(\cdot) + P(\cdot + \delta)) \geqq 2n.$$

This is a contradiction so that $h_{\xi^*}(\cdot) = -h_{\xi^*}(\cdot + \delta)$, i.e., $h_{\xi^*} = h_n$. □

We will now consider the analogous problem for G satisfying Property B. The idea of the proof is very much the same as that of Theorem 4.5. Certain technical difficulties, however, arise. To ease our notation, let

$$\Lambda_{2m}^0 = \left\{\xi \colon \xi \in \Lambda_{2m},\ \sum_{i=0}^{2m} (-1)^i (\xi_{i+1} - \xi_i) = 0\right\}$$

where $\xi_0 = 0$, $\xi_{2m+1} = 2\pi$. Equivalently, Λ_{2m}^0 is simply the set of vectors ξ in Λ_{2m} for which $h_\xi \perp 1$.

Theorem 4.6. *Let G satisfy Property B. Then for all $q \in [1, \infty]$*

$$\inf\{\|a + G * h_\xi\|_q : \xi \in \Lambda_{2n}^0,\ a \in \mathbb{R}\} = \|G * h_n\|_q.$$

Proof. We first consider the case $q = \infty$. Assume $\xi \in \Lambda_{2n}^0$, $a \in \mathbb{R}$, and $\|a + G * h_\xi\|_\infty < \|G * h_n\|_\infty$. Since $G * h_n$ attains its L^∞-norm, alternately, at $2n$ points,

$$S_c((G * h_n)(\cdot + \alpha) \pm a \pm (G * h_\xi)(\cdot)) \geqq 2n$$

for every α and $\pm$. G satisfies Property B, and h_n, $h_\xi \perp 1$. It therefore follows from Proposition 6.4 of Chapter IV that

$$S_c(h_n(\cdot + \alpha) \pm h_\xi(\cdot)) \geqq 2n$$

for every choice of α and $\pm$. A contradiction now follows from Lemma 4.1.

We now fix $q \in [1, \infty)$. By smoothing we shall assume that for every choice of $0 \leqq x_1 < \ldots < x_{2m-1} < 2\pi$, and each m,

$$X_{2m-1} = \left\{b + \sum_{i=1}^{2m-1} b_i G(x_i - \cdot) : \sum_{i=1}^{2m-1} b_i = 0\right\}$$

is a T-system of dimension $2m - 1$. This implies, by the method of proof of Proposition 6.4 of Chapter IV that for every non-trivial h satisfying $h \perp 1$, $\tilde{Z}_c(a + G * h) \leqq S_c(h)$ for all $a \in \mathbb{R}$, where $\tilde{Z}_c$ counts (cyclically) the number of zeros of $a + G * h$, where zeros which are not sign changes are counted twice.

A compactness argument shows that there exists an $a^* \in \mathbb{R}$ and a $\xi^* \in \Lambda_{2m}^0$, $m \leqq n$, for which

$$\inf\{\|a + G * h_\xi\|_q : \xi \in \Lambda_{2n}^0,\ a \in \mathbb{R}\} = \|a^* + G * h_{\xi^*}\|_q.$$

We first prove that $\xi^* \in \Lambda_{2n}^0$. Assume that $\xi^* \in \Lambda_{2m}^0$ and $m < n$. Set

$$f(x) = |a^* + G * h_{\xi^*}(x)|^{q-1} \operatorname{sgn}[a^* + G * h_{\xi^*}(x)].$$

Since $a^* \in \mathbb{R}$ is a free variable, orthogonality (minimality) considerations immediately imply that $f \perp 1$.

For each $\xi \in (\xi_i^*, \xi_{i+1}^*)$, and $\varepsilon > 0$, sufficiently small, let $\xi_\varepsilon^* = (\xi_1^*, \ldots, \xi_{i-1}^*, \xi_i^* - 2\varepsilon, \xi - \varepsilon, \xi + \varepsilon, \xi_{i+1}^*, \ldots, \xi_{2m}^*)$. Since $m < n$, then $\xi_\varepsilon^* \in \bar{\Lambda}_{2n}^0$. From minimality considerations we now obtain

$$\int_0^{2\pi} f(x)\,[G(x - \xi_i^*) - G(x - \xi)]\,dx = 0.$$

It therefore follows that

$$a + \int_0^{2\pi} f(x)\, G(x-y)\, dx = 0,$$

for all $y \in [0, 2\pi)$, where a is some constant. Since $f \neq 0$, and $S_c(f) \leqq 2m$, this is a contradiction. Thus $m = n$ and $\boldsymbol{\xi}^* \in \Lambda_{2n}^0$.

By translation we may assume that $\boldsymbol{\xi}^* = (\xi_1^*, \ldots, \xi_{2n}^*)$ satisfies $0 = \xi_1^* < \xi_2^* < \ldots < \xi_{2n}^* < 2\pi$, and

$$\delta = \xi_2^* = \xi_2^* - \xi_1^* = \min\{\xi_{i+1}^* - \xi_i^* : i = 1, \ldots, 2n\},$$

where $\xi_{2n+1}^* = 2\pi$. Assume that $h_{\boldsymbol{\xi}^*} \neq h_n$.

From Lemma 4.1 (ii),

$$S_c(h_{\boldsymbol{\xi}^*}(\cdot) + h_{\boldsymbol{\xi}^*}(\cdot + \delta)) \leqq 2(n-1).$$

Set $p(x) = a^* + G * h_{\boldsymbol{\xi}^*}(x)$, and $r(x) = p(x+\delta) = a^* + G * h_{\boldsymbol{\xi}^*}(x+\delta)$. Since

$$p(x) + r(x) = 2a^* + 1/2\pi \int_0^{2\pi} G(x-y)\,[h_{\boldsymbol{\xi}^*}(y) + h_{\boldsymbol{\xi}^*}(y+\delta)]\, dy,$$

we obtain $S_c(p + r) \leqq 2(n-1)$. By the minimality property of the free variable a^*, we have $|p(\cdot)|^{q-1} \operatorname{sgn}(p(\cdot)) \perp 1$, and $|r(\cdot)|^{q-1} \operatorname{sgn}(r(\cdot)) \perp 1$. Set

$$P(y) = c + 1/2\pi \int_0^{2\pi} |p(x)|^{q-1} \operatorname{sgn}(p(x))\, G(x-y)\, dx$$

and

$$R(y) = c + 1/2\pi \int_0^{2\pi} |r(x)|^{q-1} \operatorname{sgn}(r(x))\, G(x-y)\, dx$$

for some $c \in \mathbb{R}$. Since $S_c(p+r) \leqq 2(n-1)$, it follows that

$$S_c(|p(\cdot)|^{q-1} \operatorname{sgn}(p(\cdot)) + |r(\cdot)|^{q-1} \operatorname{sgn}(r(\cdot))) \leqq 2(n-1)$$

for each $q \in [1, \infty)$. Thus for every $c \in \mathbb{R}$,

$$\tilde{Z}_c(P(\cdot) + R(\cdot)) \leqq 2(n-1).$$

A simple change of variable argument shows that $R(y) = P(y+\delta)$ so that

$$\tilde{Z}_c(P(\cdot) + P(\cdot + \delta)) \leqq 2(n-1).$$

In $\boldsymbol{\xi}^*$, replace ξ_i^*, ξ_{i+1}^* by $\xi_i^* + \varepsilon$, $\xi_{i+1}^* + \varepsilon$, respectively, and leave the other knots unchanged. The new vector $\boldsymbol{\xi}_\varepsilon^*$ is admissable, i.e., $\boldsymbol{\xi}_\varepsilon^* \in \Lambda_{2n}^0$, and from minimality considerations,

$$\int_0^{2\pi} |p(x)|^{q-1} \operatorname{sgn}(p(x)) [G(x-\xi_i^*) - G(x-\xi_{i+1}^*)]\, dx = 0, \qquad i = 1,\ldots,2n,$$

where $\xi_{2n+1}^* = \xi_1^*$.

We now choose $c \in \mathbb{R}$ in the definition of P so that $P(\xi_i^*) = 0$, $i = 1,\ldots,2n$. By our choice of δ, $\xi_i^* < \xi_i^* + \delta \leqq \xi_{i+1}^*$, $i = 1,\ldots,2n$, and therefore

$$S_c^+(P(\xi_1^* + \delta),\ldots,P(\xi_{2n}^* + \delta)) = 2n.$$

Thus

$$S_c^+(P(\xi_1^*) + P(\xi_1^* + \delta),\ldots,P(\xi_{2n}^*) + P(\xi_{2n}^* + \delta)) = 2n$$

which implies that

$$\tilde{Z}_c(P(\cdot) + P(\cdot + \delta)) \geqq 2n.$$

This is a contradiction, and therefore $h_{\xi^*}(\cdot) = -h_{\xi^*}(\cdot + \delta)$, i.e., $h_{\xi^*} = h_n$. It remains to prove that $a^* = 0$. Let

$$f(x;a) = |a + (G * h_n)(x)|^{q-1} \operatorname{sgn}(a + (G * h_n)(x)).$$

Since the constant term is a free variable, $f(\cdot;a^*) \perp 1$. Because $(G * h_n)(x) = -(G * h_n)(x + \pi/n)$, it follows that $f(\cdot;0) \perp 1$. Thus $a^* = 0$. □

One other preliminary result in necessary before presenting the first series of statements concerning the n-widths.

The function $k * h_n$ exhibits at least $2n$ zeros in $[0, 2\pi)$ since $(k * h_n)(x) = -(k * h_n)(x + \pi/n)$. (One can actually show that $k * h_n$ has exactly $2n$ zeros.) Let $\{\beta + \pi i/n\}_{i=0}^{2n-1}$, $\beta \in [0, \pi/n)$, denote these zeros. As an analogue of Lemma 6.2 of Chapter IV, we have:

Lemma 4.7. *Let k be an NCVD kernel. For β as above,*

$$k\begin{pmatrix} \beta, \beta + \pi/n, \ldots, \beta + (2n-1)\pi/n \\ 0, \pi/n, \ldots, (2n-1)\pi/n \end{pmatrix} \neq 0.$$

Set $X_{2n} = \operatorname{span}\{k(\cdot - \pi/n)\}_{i=0}^{2n-1}$. Let R_{2n} denote the linear map which interpolates from X_{2n} to every 2π-periodic, continuous function at $\beta + \pi i/n$, $i = 0, 1, \ldots, 2n-1$. Lemma 4.7 implies that R_{2n} is well-defined.

We can now state

Theorem 4.8. *Let k be an NCVD kernel. Then*

$$\begin{aligned} d_{2n-1}(\tilde{\mathscr{K}}_\infty; L^\infty) &= d^{2n-1}(\tilde{\mathscr{K}}_\infty; L^\infty) = \delta_{2n-1}(\tilde{\mathscr{K}}_\infty; L^\infty) \\ &= b_{2n-1}(\tilde{\mathscr{K}}_\infty; L^\infty) = \|k * h_n\|_\infty. \end{aligned}$$

Furthermore,

(1) T_{n-1} *is an optimal subspace for* $d_{2n-1}(\tilde{\mathscr{K}}_\infty; L^\infty)$.
(2) $L^{2n-1} = \{h\colon h \in L^\infty,\ h \perp T_{n-1}\}$ *is optimal for* $d^{2n-1}(\tilde{\mathscr{K}}_\infty; L^\infty)$.
(3) *The operator* R_{2n-1} *defined by* $R_{2n-1}h = t^* * h$ *is an optimal rank* $2n-1$ *operator for* $\delta_{2n-1}(\tilde{\mathscr{K}}_\infty; L^\infty)$, *where* t^* *is the best* L^1*-approximant to* k *from* T_{n-1}.
(4) $Y_{2n} = \operatorname{span}\left\{\int_{\pi(i-1)/n}^{\pi i/n} k(\cdot - y)\,dy\right\}_{i=1}^{2n}$ *is optimal for* $b_{2n-1}(\tilde{\mathscr{K}}_\infty; L^\infty)$.

Proof. From the definition of R_{2n-1} and by Theorem 4.4 (and the remark thereafter), it follows that $\delta_{2n-1}(\tilde{\mathscr{K}}_\infty; L^\infty) \leqq \|k * h_n\|_\infty$. Since $\delta_{2n-1} \geqq d_{2n-1}$, $d^{2n-1} \geqq b_{2n-1}$, it remains to prove that $b_{2n-1}(\tilde{\mathscr{K}}_\infty; L^\infty) \geqq \|k * h_n\|_\infty$. The proof of this fact parallels, almost word for word, the proof of Theorem 2.10. □

Theorem 4.9. *Let* k *be an NCVD kernel. Then for each* $q \in [1, \infty]$,

$$d_{2n}(\tilde{\mathscr{K}}_\infty; L^q) = d^{2n}(\tilde{\mathscr{K}}_\infty; L^q) = \delta_{2n}(\tilde{\mathscr{K}}_\infty; L^q) = \|k * h_n\|_q.$$

Furthermore,

(1) $X_{2n} = \operatorname{span}\{k(\cdot - \pi i/n)\}_{i=0}^{2n-1}$ *is an optimal subspace for* $d_{2n}(\tilde{\mathscr{K}}_\infty; L^q)$.
(2) $L^{2n} = \{h\colon h \in L^\infty,\ (k * h)(\pi i/n) = 0,\ i = 0, 1, \ldots, 2n-1\}$ *is optimal for* $d^{2n}(\tilde{\mathscr{K}}_\infty; L^q)$.
(3) R_{2n}, *as previoulsy defined, is an optimal rank* $2n$ *operator for* $\delta_{2n}(\tilde{\mathscr{K}}_\infty; L^q)$.

Remark. Of course the optimal subspaces and optimal map of Theorem 4.8 are also optimal in Theorem 4.9 for $q = \infty$, since the n-width values are the same. The quantity $b_{2n}(\tilde{\mathscr{K}}_\infty; L^\infty)$, however, is not known.

The proof of Theorem 4.9 totally parallels the proofs of Theorems 2.6–2.8. Let G satisfy Property B. $G * h_n$ has $2n$ zeros at $\{\gamma + \pi i/n\}_{i=0}^{2n-1}$, $\gamma \in [0, \pi/n)$. Let S_{2n} denote the interpolation operator from

$$W_{2n} = \left\{b + \sum_{i=0}^{2n-1} b_i G(\cdot - \pi i/n)\colon \sum_{i=0}^{2n-1} b_i = 0\right\}$$

which interpolates to every 2π-periodic, continuous function at $\gamma + \pi i/n$, $i = 0, 1, \ldots, 2n-1$. By Proposition 6.5 of Chapter IV S_{2n} is well-defined.

Theorem 4.10. *Let* G *satisfy Property* B. *Then*

$$d_{2n-1}(\tilde{\mathscr{B}}_\infty; L^\infty) = d^{2n-1}(\tilde{\mathscr{B}}_\infty; L^\infty) = \delta_{2n-1}(\tilde{\mathscr{B}}_\infty; L^\infty) = b_{2n-1}(\tilde{\mathscr{B}}_\infty; L^\infty) = \|G * h_n\|_\infty.$$

Furthermore,

(1) T_{n-1} *is an optimal subspace for* $d_{2n-1}(\tilde{\mathscr{B}}_\infty; L^\infty)$.
(2) $L^{2n-1} = \{f\colon f \in \tilde{\mathscr{B}}_\infty,\ f \perp T_{n-1}\}$ *is optimal for* $d^{2n-1}(\tilde{\mathscr{B}}_\infty; L^\infty)$.

(3) *For* $f = a + G * h \in \tilde{\mathscr{B}}_\infty$, *let* $S_{2n-1} f = a + t^* * h$ *where* t^* *is the best* L^1-*approximant to* G *from* T_{n-1}. *Then* S_{2n-1} *is an optimal rank* $2n-1$ *operator for* $\delta_{2n-1}(\tilde{\mathscr{B}}_\infty; L^\infty)$.

(4) $Z_{2n} = \left\{ a + \sum_{i=1}^{2n} a_i \int_{\pi(i-1)/n}^{\pi i/n} G(\cdot - y)\, dy : \sum_{i=1}^{2n} a_i = 0 \right\}$ *is an optimal subspace for* $b_{2n-1}(\tilde{\mathscr{B}}_\infty; L^\infty)$.

Remark. We have here reverted to the original definitions of d^n and δ_n (see the discussion in Section 7 of Chapter II) and to the analogue of the definition of b_n as given in Definition 3.1 of this chapter.

Proof. From the definition of S_{2n-1}, and from Theorem 4.4 (and the remark thereafter), it follows that $\delta_{2n-1}(\tilde{\mathscr{B}}_\infty; L^\infty) \leqq \| G * h_n \|_\infty$. Since $\delta_n \geqq d_n$, $d^n \geqq b_n$ it remains to prove that $b_{2n-1}(\tilde{\mathscr{B}}_\infty; L^\infty) \geqq \| G * h_n \|_\infty$. The proof of this inequality parallels the proof of Theorem 2.10. For completeness we give the proof.

We must show that

$$\| a + G * h \|_\infty \geqq \| h \|_\infty \cdot \| G * h_n \|_\infty$$

for every $a + G * h \in Z_{2n}$. For notational ease, set

$$u_i(x) = 1/2\pi \int_{\pi(i-1)/n}^{\pi i/n} G(x - y)\, dy, \qquad i = 1, \ldots, 2n.$$

The requisite condition easily translates into the following:

Prove that if $\sum_{i=1}^{2n} a_i = 0$, and

$$\left\| a + \sum_{i=1}^{2n} a_i u_i \right\|_\infty \leqq \| G^* h_n \|_\infty = \left\| \sum_{i=1}^{2n} (-1)^{i+1} u_i \right\|_\infty,$$

then $|a_i| \leqq 1$, $i = 1, \ldots, 2n$.

Assume that for some choice of $\{a_i\}_{i=1}^{2n}$ as above, we have

$$|a_{i_0}| = \max\{|a_i| : i = 1, \ldots, 2n\} > 1.$$

Then

$$\left\| a + \sum_{i=1}^{2n} a_i u_i \right\|_\infty \Big/ |a_{i_0}| < \left\| \sum_{i=1}^{2n} (-1)^{i+1} u_i \right\|_\infty.$$

Since $\sum_{i=1}^{2n} (-1)^{i+1} u_i$ attains its L^∞-norm, alternately, at $2n$ points,

$$S_c\left(\sum_{i=1}^{2n} (-1)^{i+1} u_i + (-1)^{i_0} \left[(a/a_{i_0}) + \sum_{i=1}^{2n} (a_i/a_{i_0}) u_i \right] \right) \geqq 2n,$$

i.e.,

$$S_c\left(b + \sum_{i=1}^{2n} b_i u_i\right) \geqq 2n$$

for some $\{b_i\}_{i=1}^{2n}$ satisfying $\sum_{i=1}^{2n} b_i = 0$ and $b_{i_0} = 0$. By Proposition 6.4 of Chapter IV

$$S_c\left(b + \sum_{i=1}^{2n} b_i u_i\right) \leqq S_c^-((b_1, \ldots, b_{2n})).$$

Since $b_{i_0} = 0$, it follows that

$$S_c^-((b_1, \ldots, b_{2n})) \leqq 2(n-1).$$

This contradiction proves that $b_{2n-1}(\tilde{\mathscr{B}}_\infty; L^\infty) \geqq \| G * h_n \|_\infty$. □

Analogous to Theorem 4.9, we have:

Theorem 4.11. *Let G satisfy Property B. Then for each $q \in [1, \infty]$ and $n \geqq 1$,*

$$d_{2n}(\tilde{\mathscr{B}}_\infty; L^q) = d^{2n}(\tilde{\mathscr{B}}_\infty; L^q) = \delta_{2n}(\tilde{\mathscr{B}}_\infty; L^q) = \| G * h_n \|_q.$$

Furthermore,

(1) $W_{2n} = \left\{ b + \sum_{i=0}^{2n-1} b_i G(\cdot - \pi i/n) \colon \sum_{i=0}^{2n-1} b_i = 0 \right\}$ *is an optimal subspace for* $d_{2n}(\tilde{\mathscr{B}}_\infty; L^q)$.

(2) $L^{2n} = \{f\colon f \in \tilde{\mathscr{B}}_\infty, f(\pi i/n) = 0, i = 0, 1, \ldots, 2n-1\}$ *is optimal for* $d^{2n}(\tilde{\mathscr{B}}_\infty; L^q)$.

(3) S_{2n}, *as previously defined, is an optimal rank* $2n$ *operator for* $\delta_{2n}(\tilde{\mathscr{B}}_\infty; L^q)$.

As a consequence of Theorems 7.2, 8.9 and 8.10 of Chapter II (see also Theorem 2.12 of this chapter), the following hold.

Theorem 4.12. *Let k be an NCVD kernel. Then*

$$d_{2n-1}(\tilde{\mathscr{K}}_1; L^1) = d^{2n-1}(\tilde{\mathscr{K}}_1; L^1) = \delta_{2n-1}(\tilde{\mathscr{K}}_1; L^1) = b_{2n-1}(\tilde{\mathscr{K}}_1; L^1) = \| k * h_n \|_\infty.$$

Furthermore,

(1) T_{n-1} *is optimal for* $d_{2n-1}(\tilde{\mathscr{K}}_1; L^1)$.

(2) $L^{2n-1} = \{h\colon h \in L^1,\ h \perp T_{n-1}\}$ *is optimal for* $d^{2n-1}(\tilde{\mathscr{K}}_1; L^1)$.

(3) $R_{2n-1} h = t^* * h$ *is an optimal rank* $2n-1$ *operator for* $\delta_{2n-1}(\tilde{\mathscr{K}}_1; L^1)$, *where* t^* *is the best* L^1*-approximant to* k *from* T_{n-1}.

(4) $X_{2n} = \operatorname{span}\{k(\cdot - \pi i/n)\}_{i=0}^{2n-1}$ *is optimal for* $b_{2n-1}(\tilde{\mathscr{K}}_1; L^1)$.

Theorem 4.13. *Let k be an NCVD kernel. For each $p \in [1, \infty]$,*

$$d_{2n}(\tilde{\mathscr{K}}_p; L^1) = d^{2n}(\tilde{\mathscr{K}}_p; L^1) = \delta_{2n}(\tilde{\mathscr{K}}_p; L^1) = \| k * h_n \|_{p'},$$

where $1/p + 1/p' = 1$. Furthermore,

(1) $X_{2n} = \operatorname{span}\{k(\cdot - \pi i/n)\}_{i=0}^{2n-1}$ *is optimal for* $d_{2n}(\tilde{\mathscr{K}}_p; L^1)$.
(2) $L^{2n} = \{h: h \in L^p, \quad (k * h)(\pi i/n) = 0, \quad i = 0, 1, \ldots, 2n-1\}$ *is optimal for* $d^{2n}(\tilde{\mathscr{K}}_p; L^1)$.
(3) R_{2n}, *as previously defined, is an optimal rank* $2n$ *operator for* $\delta_{2n}(\tilde{\mathscr{K}}_p; L^1)$.

Theorem 4.14. *Let G satisfy Property B. Then*

$$d_{2n-1}(\tilde{\mathscr{B}}_1; L^1) = d^{2n-1}(\tilde{\mathscr{B}}_1; L^1) = \delta_{2n-1}(\tilde{\mathscr{B}}_1; L^1) = b_{2n-1}(\tilde{\mathscr{B}}_1; L^1) = \| G * h_n \|_\infty .$$

Furthermore,

(1) T_{n-1} *is optimal for* $d_{2n-1}(\tilde{\mathscr{B}}_1; L^1)$.
(2) $L^{2n-1} = \{f: f \in \tilde{\mathscr{B}}_1, f \perp T_{n-1}\}$ *is optimal for* $d^{2n-1}(\tilde{\mathscr{B}}_1; L^1)$.
(3) S_{2n-1} *(of Theorem 4.10) is an optimal rank* $2n-1$ *operator for* $\delta_{2n-1}(\tilde{\mathscr{B}}_1; L^1)$.
(4) $W_{2n} = \left\{ b + \sum_{i=0}^{2n-1} b_i G(\cdot - \pi i/n): \sum_{i=0}^{2n-1} b_i = 0 \right\}$ *is optimal for* $b_{2n-1}(\tilde{\mathscr{B}}_1; L^1)$.

Theorem 4.15. *Let G satisfy Property B. For each* $p \in [1, \infty]$ *and* $n \geqq 1$,

$$d_{2n}(\tilde{\mathscr{B}}_p; L^1) = d^{2n}(\tilde{\mathscr{B}}_p; L^1) = \delta_{2n}(\tilde{\mathscr{B}}_p; L^1) = \| G * h_n \|_{p'},$$

where $1/p + 1/p' = 1$. *Furthermore,*

(1) $W_{2n} = \left\{ b + \sum_{i=0}^{2n-1} b_i G(\cdot - \pi i/n): \sum_{i=0}^{2n-1} b_i = 0 \right\}$ *is an optimal subspace for* $d_{2n}(\tilde{\mathscr{B}}_p; L^1)$.
(2) $L^{2n} = \{f: f \in \tilde{\mathscr{B}}_p, f(\pi i/n) = 0, i = 0, 1, \ldots, 2n-1\}$ *is optimal for* $d^{2n}(\tilde{\mathscr{B}}_p; L^1)$.
(3) S_{2n}, *as previously defined, is an optimal rank* $2n$ *operator for* $\delta_{2n}(\tilde{\mathscr{B}}_p; L^1)$.

Before continuing we list some conjectures which we believe to hold for all $1 \leqq q \leqq p \leqq \infty$.

Conjecture 1. T_{n-1} *is an optimal subspace for* $d_{2n-1}(\tilde{\mathscr{K}}_p; L^q)$ *and* $d_{2n-1}(\tilde{\mathscr{B}}_p; L^q)$.
Conjecture 2. X_{2n} *is an optimal subspace for* $d_{2n}(\tilde{\mathscr{K}}_p; L^q)$, *and* W_{2n} *is an optimal subspace for* $d_{2n}(\tilde{\mathscr{B}}_p; L^q)$.
Conjecture 3. $d_{2n-1}(\tilde{\mathscr{K}}_p; L^q) = d_{2n}(\tilde{\mathscr{K}}_p; L^q)$ *and* $d_{2n-1}(\tilde{\mathscr{B}}_p; L^q) = d_{2n}(\tilde{\mathscr{B}}_p; L^q)$.

We also expect that analogous statements hold for the Gel'fand and linear n-widths.

We have proved Conjectures 1 and 3 for $p = q \in \{1, 2, \infty\}$, and Conjecture 2 for $p = q = 2$; $p = \infty$, all q; and $q = 1$, all p. We end this section by proving that Conjectures 1 and 3 are also valid when $q = 1$, all p.

Theorem 4.16. *Let k be an NCVD kernel and G satisfy Property B. Then for* $q \in [1, \infty]$, $1/q + 1/q' = 1$,

$$d_{2n-1}(\tilde{\mathscr{K}}_{q'}; L^1) = d^{2n-1}(\tilde{\mathscr{K}}_\infty; L^q) = \| k * h_n \|_q,$$

and

$$d_{2n-1}(\tilde{\mathscr{B}}_{q'}; L^1) = d^{2n-1}(\tilde{\mathscr{B}}_\infty; L^q) = \| G * h_n \|_q .$$

Furthermore,

(1) T_{n-1} *is optimal for both* $d_{2n-1}(\tilde{\mathscr{K}}_{q'}; L^1)$ *and* $d_{2n-1}(\tilde{\mathscr{B}}_{q'}; L^1)$.
(2) $L^{2n-1} = \{h: h \in L^\infty, h \perp T_{n-1}\}$ *is optimal for* $d^{2n-1}(\tilde{\mathscr{K}}_\infty; L^q)$.
(3) $M^{2n-1} = \{f: f \in \tilde{\mathscr{B}}_\infty, f \perp T_{n-1}\}$ *is optimal for* $d^{2n-1}(\tilde{\mathscr{B}}_\infty; L^q)$.

The arguments used in proving Theorem 4.16 are essentially the same for both k *NCVD* and G satisfying Property B. As such we only prove the result in the former case.

From duality considerations, and since $d^{2n-1}(\tilde{\mathscr{K}}_\infty; L^q) \geqq d^{2n}(\tilde{\mathscr{K}}_\infty; L^q) = \|k * h_n\|_q$, it in fact suffices to prove that

$$\sup\{\|k * h\|_q : \|h\|_\infty \leqq 1, h \perp T_{n-1}\} \leqq \|k * h_n\|_q$$

for all $q \in [1, \infty)$. The case $q = \infty$ has already been dealt with in Theorem 4.8.

In what follows we assume, without loss of generality, that k is extended *CVD* and k' exists, is continuous and 2π-periodic.

We divide the proof of the theorem into a series of steps. The first set of results is concerned with delineating certain properties of $k * h_n$.

Proposition 4.17. *Assume that* $\|h\|_\infty \leqq 1$ *and* $h \perp T_{n-1}$. *If* $|(k * h)(x_0)| = |(k * h_n)(y_0)|$ *for some* $x_0, y_0 \in [0, 2\pi)$, *then*

$$|(k * h)'(x_0)| \leqq |(k * h_n)'(y_0)|.$$

Proof. By translation (and multiplying by -1 if necessary) it suffices to prove the above result if $(k * h)(x_0) = (k * h_n)(x_0)$, and $(k * h)'(x_0), (k * h_n)'(x_0) \geqq 0$. Since $\|k * h\|_\infty \leqq \|k * h_n\|_\infty$ (by Theorem 4.8) we may assume that $|(k * h)(x_0)| < \|k * h_n\|_\infty$.

To ease our exposition we shall in fact assume that $\|k * h\|_\infty < \|k * h_n\|_\infty$. This assumption is in no way restrictive since we simply consider $(1 - \varepsilon) h$ in place of h where $\varepsilon > 0$ is small.

Since $(k * h_n)(x) = -(k * h_n)(x + \pi/n)$, there exists an η such that $(k * h_n)(\eta + \pi i/n)(-1)^{i+1} = \|k * h_n\|_\infty$, $i = 0, 1, \ldots, 2n-1$. Thus

$$2n \leqq S_c(k * h_n - k * h) \leqq S_c(h_n - h) \leqq 2n$$

which implies that $S_c(k * h_n - k * h) = 2n$. Furthermore, it follows, since $\|k * h\|_\infty < \|k * h_n\|_\infty$, that $(k * h_n)(x) - (k * h)(x)$ has exactly one sign change in $(\eta + \pi i/n, \eta + \pi(i+1)/n)$ for each $i = 0, 1, \ldots, 2n-1$. However, if $(k * h)(x_0) = (k * h_n)(x_0)$ and $(k * h)'(x_0) > (k * h_n)'(x_0) \geqq 0$ for some $x_0 \in (\eta + \pi j/n, \eta + \pi(j+1)/n)$, then it is easily seen that $(k * h_n)(x) - (k * h)(x)$ has at least three sign changes in $(\eta + \pi j/n, \eta + \pi(j+1)/n)$. This is a contradiction. □

Proposition 4.18. *For* η *as above,* $k * h_n$ *is monotone in* $(\eta + \pi i/n, \eta + \pi(i+1)/n)$ *for each* i.

Proof. If $k * h_n$ is not monotone in $(\eta + \pi j/n, \eta + \pi(j+1)/n)$ there exist x_0, y_0 in this same interval for which $(k * h_n)(x_0) = (k * h_n)(y_0)$, but $|(k * h_n)'(x_0)| \neq |(k * h_n)'(y_0)|$. This contradicts Proposition 4.17. □

Let $\xi + \pi i/n$, $i = 0, 1, \ldots, 2n - 1$ denote the zeros of $k * h_n$ where $\eta < \xi < \eta + \pi/n$. The proof of this next corollary is very similar to the proof of Proposition 4.18.

Corollary 4.19. *$k * h_n$ is odd about ξ and even about η. Thus in particular $\xi = \eta + \pi/2n$.*

We now define

$$G(x) = \int_0^x (k(u) - a_0)\, du,$$

where $a_0 = 1/2\pi \int_0^{2\pi} k(u)\, du$. By Proposition 6.7 of Chapter IV, G satisfies Property B. From Theorem 4.10 and the method of proof of Propositions 4.17 and 4.18 we have:

Proposition 4.20. *For G as above, and $n \geqq 1$,*

$$\sup\{\|G * h\|_\infty : \|h\|_\infty \leqq 1,\ h \perp T_{n-1}\} = \|G * h_n\|_\infty .$$

Furthermore,

(1) *if $\|h\|_\infty \leqq 1$, $h \perp T_{n-1}$ and $|(G * h)(x_0)| = |(G * h_n)(y_0)|$, then $|(G * h)'(x_0)| \leqq |(G * h_n)'(y_0)|$.*

(2) *For η as in Proposition 4.18, $(G * h_n)(\eta + \pi i/n - \pi/2n)(-1)^i = \|G * h_n\|_\infty$, $i = 0, 1, \ldots, 2n - 1$, and $G * h_n$ is even about $\eta + \pi/2n$ and odd about η.*

For $\|h\|_\infty \leqq 1$, $h \perp T_{n-1}$, fixed, $k * h \neq k * h_n$, let $\Delta_r = [a_r, b_r]$, $r = 1, \ldots, m$ be intervals (we identify the point zero with the point 2π) satisfying

(1) $|\Delta_r| \geqq \pi/n$

(2) $(k * h)(x) \neq 0$, $x \in (a_r, b_r)$

(3) $(k * h)(x) = 0$, $x = a_r, b_r$.

Now, $[0, 2\pi] \Big/ \bigcup_{r=1}^{m} \Delta_r$ is either empty or the union of $s\,(s \leqq m)$ disjoint open intervals $\{\delta_r\}_{r=1}^s$ which satisfy the following:

If $(k * h)(x) \neq 0$ for x in δ_r, then $|\delta_r| < \pi/n$, while if $(k * h)(x)$ has zeros in δ_r, then δ_r can be subdivided into intervals of length $< \pi/n$ such that $(k * h)(x)$ vanishes at the endpoints thereof.

These sets are well-defined since $h \perp T_{n-1}$ implies $k * h \perp T_{n-1}$, which in turn implies by Proposition 1.4 of Chapter III that $k * h$ has at least $2n - 1$ zeros in $[0, 2\pi)$.

Proposition 4.21. *For δ_r as above, $1 \leqq q < \infty$,*

$$\int_{\delta_r} |(k * h)(x)|^q\, dx \leqq \int_{\delta_r} |(k * h_n)(x)|^q\, dx.$$

Proof. From the definition of δ_r, it suffices to prove that

$$\int_{x_1}^{x_2} |(k * h)(x)|^q\, dx \leqq \int_{x_1}^{x_2} |(k * h_n)(x)|^q\, dx$$

where $0 < x_2 - x_1 < \pi/n$, and $(k * h)(x_i) = 0$, $i = 1, 2$. Since

$$\min_a \int_{x_1+a}^{x_2+a} |(k * h_n)(x)|^q \, dx = \int_{x_1^*}^{x_2^*} |(k * h_n)(x)|^q \, dx$$

where $(k * h_n)((x_1^* + x_2^*)/2) = 0$, it suffices to prove that

$$\int_{\xi}^{x_2} |(k * h)(x)|^q \, dx \leqq \int_{\xi}^{x_2} |(k * h_n)(x)|^q \, dx$$

and

$$\int_{x_1}^{\xi} |(k * h)(x)|^q \, dx \leqq \int_{x_1}^{\xi} |(k * h_n)(x)|^q \, dx$$

where $\xi \in (x_1, x_2)$ satisfies $(k * h_n)(\xi) = 0$, and $k * h_n$ is monotone in (x_1, x_2). We prove the latter inequality. The proof of the former is totally analogous.

We claim that for all $x \in (x_1, \xi)$, $|(k * h)(x)| \leqq |(k * h_n)(\xi + x_1 - x)|$. Now, $(k * h)(x_1) = (k * h_n)(\xi) = 0$, and $k * h_n$ is monotone on (x_1, ξ). Assume there exists a $\zeta \in (x_1, \xi]$ for which $|(k * h)(\zeta)| > |(k * h_n)(\xi + x_1 - \zeta)|$. There then exists $x_1 \leqq \zeta_1 < \zeta_2 < \zeta$ satisfying

$$(k * h)(\zeta_1) = (k * h_n)(\xi + x_1 - \zeta_2)$$

and

$$|(k * h)'(\zeta_1)| > |(k * h_n)'(\xi + x_1 - \zeta_2)|.$$

But this contradicts Proposition 4.17. □

Proposition 4.22. *For Δ_r as above, $1 \leqq q < \infty$,*

$$\int_{\Delta_r} |(k * h)(x)|^q \, dx \leqq \int_{\Delta_r} |(k * h_n)(x)|^q \, dx.$$

Proof. For any given continuous function f and $1 < q < \infty$,

$$\int_{\Delta} |f(x)|^q \, dx = \int_0^{\|f\|_\Delta} \left(\int_{E_t} |f(x)| \, dx\right)(q-1) t^{q-2} \, dt$$

where $E_t = \Delta \cap \{x : |f(x)| \geqq t\}$, and $\|f\|_\Delta = \max\{|f(x)| : x \in \Delta\}$. (For $q = 1$, $\int_{\Delta} |f(x)| \, dx = \int_{E_0} |f(x)| \, dx$.)

Thus, for $1 < q < \infty$,

$$\int_{\Delta_r} |(k * h)(x)|^q \, dx = \int_0^{\|k*h\|_{\Delta_r}} \left(\int_{E_t} |(k * h)(x)| \, dx\right)(q-1) t^{q-2} \, dt$$

and

$$\int_{\Delta_r} |(k * h_n)(x)|^q \, dx = \int_0^{\|k*h_n\|_{\Delta_r}} \left(\int_{E_t^*} |(k * h_n)(x)| \, dx\right)(q-1) t^{q-2} \, dt,$$

where $E_t = \Delta_r \cap \{x: |(k*h)(x)| \geqq t\}$ and $E_t^* = \Delta_r \cap \{x: |(k*h_n)(x)| \geqq t\}$. Since $|\Delta_r| \geqq \pi/n$, $\|k*h\|_{\Delta_r} \leqq \|k*h\|_\infty \leqq \|k*h_n\|_\infty = \|k*h_n\|_{\Delta_r}$. Thus in order to prove the proposition for all $q \in [1, \infty)$, it suffices to prove that

$$\int_{E_t} |(k*h)(x)|\,dx \leqq \int_{E_t^*} |(k*h_n)(x)|\,dx$$

for all t.

Let $G(x) = \int_0^x (k(u) - a_0)\,du$, where $a_0 = 1/2\pi \int_0^{2\pi} k(u)\,du$. For $h \perp T_{n-1}$, $(G*h)'(x) = (k*h)(x)$. We must therefore prove that

$$\int_{E_t} |(G*h)'(x)|\,dx \leqq \int_{E_t^*} |(G*h_n)'(x)|\,dx$$

for all t, where E_t and E_t^* are as given.

We first bound the right side from below. Since $|\Delta_r| \geqq \pi/n$, $E_t^* \supseteqq e_t^*$, where $e_t^* = [0, \pi/n] \cap \{x: |(k*h_n)(x)| \geqq t\}$. It is now easily seen from Proposition 4.20 that

$$\int_{E_t^*} |(G*h_n)'(x)|\,dx \geqq \int_{e_t^*} |(G*h_n)'(x)|\,dx = 2(G*h_n)(\bar{x}),$$

where $(G*h_n)(\bar{x}) \geqq 0$, and $|(k*h_n)(\bar{x})| = t$.

Consider

$$\int_{E_t} |(G*h)'(x)|\,dx.$$

By the definition of Δ_r, $(k*h)(x)$ is of one sign on Δ_r so that $(G*h)(x)$ is monotone on E_t. Assume, with no loss of generality, that $(G*h)(x)$ is increasing thereon. Thus

$$\int_{E_t} |(G*h)'(x)|\,dx = \sum_{i=1}^{s} [(G*h)(x_i^2) - (G*h)(x_i^1)]$$

where $x_i^1 < x_i^2 < x_{i+1}^1 < x_{i+1}^2$, $i = 1, \ldots, s-1$; $(G*h)'(x_i^j) = t$, $i = 1, \ldots, s$, $j = 1, 2$; $(G*h)'(x) \geqq t$ for $x \in (x_i^1, x_i^2)$ $i = 1, \ldots, s$; and $(G*h)'(x) \geqq 0$, $x \in (x_1^1, x_s^2)$. The proposition follows if we can prove that

$$-(G*h_n)(\bar{x}) \leqq (G*h)(x_1^1) \quad \text{and} \quad (G*h)(x_s^2) \leqq (G*h_n)(\bar{x}), \qquad i = 1, \ldots, s.$$

Assume $(G*h)(x_s^2) > (G*h_n)(\bar{x})$. Now $|(G*h)'(x_s^2)| = |(G*h_n)'(\bar{x})| = t$. Since $(G*h_n)'(x) = (k*h_n)(x)$ is monotone on $(\eta + \pi i/n, \eta + \pi(i+1)/n)$, there exists a ζ for which

$$(G*h)(x_s^2) = (G*h_n)(\zeta),$$

and

$$|(G*h)'(x_s^2)| > |(G*h_n)'(\zeta)|.$$

This contradicts Proposition 4.20. Thus $(G*h)(x_s^2) \leqq (G*h_n)(\bar{x})$ and similarly $(G*h)(x_1^1) \geqq -(G*h_n)(\bar{x})$. Proposition 4.22 and Theorem 4.16 are proved. □

5. n-Widths of Rank $n + 1$ Kernels

A kernel $K(x, y) \in C([0, 1] \times [0, 1])$ is said to be of rank $n + 1$ if

$$K(x, y) = \sum_{i=1}^{n+1} u_i(x)\, v_i(y),$$

where $\{u_1, \ldots, u_{n+1}\}$ and $\{v_1, \ldots, v_{n+1}\}$ are each linearly independent subsets of $C[0, 1]$. As previously, set

$$\mathscr{K}_p = \left\{ Kh(x): Kh(x) = \int_0^1 K(x, y)\, h(y)\, dy,\ \|h\|_p \leqq 1 \right\}$$

for each $p \in [1, \infty]$.

In order to determine the n-widths of $\mathscr{K}_p$ in L^q for $p, q \in [1, \infty]$, we rely upon the following result (see Theorems 2.1, 3.6 and 4.3 of Chapter II).

Theorem 5.1. *Let A be a closed, convex, centrally symmetric subset of an $(n + 1)$-dimensional subspace X_{n+1} of X. Then,*

$$d_n(A; X) = d^n(A; X) = \delta_n(A; X) = \inf\{\|x\| : x \in \partial A\}.$$

For fixed $p, q \in [1, \infty]$, $1/p + 1/p' = 1$, set

$$\sigma = \inf_{\sum_{i=1}^{n+1} a_i b_i = 1} \left\| \sum_{i=1}^{n+1} a_i u_i \right\|_q \left\| \sum_{i=1}^{n+1} b_i v_i \right\|_{p'}.$$

It is a simple matter to prove that the infimum is actually attained for any linearly independent sets $\{u_i\}_{i=1}^{n+1}$ and $\{v_i\}_{i=1}^{n+1}$. Assume that the infimum is attained by $\bar{u} = \sum_{i=1}^{n+1} \bar{a}_i u_i$ and $\bar{v} = \sum_{i=1}^{n+1} \bar{b}_i v_i$ i.e., $\sum_{i=1}^{n+1} \bar{a}_i \bar{b}_i = 1$, and $\|\bar{u}\|_q \|\bar{v}\|_{p'} = \sigma$. Then,

Theorem 5.2. *Let K be as above. For $p, q \in [1, \infty]$,*

$$d_n(\mathscr{K}_p; L^q) = d^n(\mathscr{K}_p; L^q) = \delta_n(\mathscr{K}_p; L^q) = \sigma.$$

There exist $2n$ measures $d\bar{\alpha}_1, \ldots, d\bar{\alpha}_n, d\bar{\beta}_1, \ldots, d\bar{\beta}_n$ for which

$$\int_0^1 \bar{u}\, d\bar{\alpha}_j = \int_0^1 \bar{v}\, d\bar{\beta}_j = 0, \quad j = 1, \ldots, n,$$

and rank $\left(\int_0^1 u_i\, d\bar{\alpha}_j \right)_{i=1}^{n+1}{}_{j=1}^{n}$ = rank $\left(\int_0^1 v_i\, d\bar{\beta}_j \right)_{i=1}^{n+1}{}_{j=1}^{n}$ $= n$. *For any such $2n$ measures*

(1) span $\{\bar{\phi}_1, \ldots, \bar{\phi}_n\}$ *is optimal for* $d_n(\mathscr{K}_p; L^q)$ *where*

$$\bar{\phi}_i(x) = \int_0^1 K(x, y)\, d\bar{\beta}_i(y), \qquad i = 1, \ldots, n.$$

(2) *The linear operator* P *defined by* $Ph \in \operatorname{span}\{\bar{\phi}_1, \ldots, \bar{\phi}_n\}$, *and*

$$\int_0^1 (Kh)(x)\, d\bar{\alpha}_i(x) = \int_0^1 (Ph)(x)\, d\bar{\alpha}_i(x), \qquad i = 1, \ldots, n$$

is optimal for $\delta_n(\mathscr{K}_p; L^q)$.

(3) $L^n = \left\{ h\colon h \in L^q, \int_0^1 (Kh)(x)\, d\bar{\alpha}_i(x) = 0,\ i = 1, \ldots, n \right\}$ *is optimal for* $d^n(\mathscr{K}_p; L^q)$.

Before proving the theorem we need some lemmas.

Lemma 5.3. *Let* $u_1, \ldots, u_{n+1}$ *be linearly independent functions in* $C[0,1]$. *Given* $\mathbf{a} = (a_1, \ldots, a_{n+1}) \in \mathbb{R}^{n+1} \backslash \{\mathbf{0}\}$, *there exist measures* $d\alpha_1, \ldots, d\alpha_n$ *for which*

$$a_i = (-1)^{i+1} \det \left(\int_0^1 u_k\, d\alpha_j \right)_{\substack{k=1,\ j=1 \\ k \neq i}}^{n+1\quad n}, \qquad i = 1, \ldots, n+1.$$

Remark. The result of this lemma is equivalent to the existence of measures $d\alpha_1, \ldots, d\alpha_n$ for which

$$\int_0^1 \left(\sum_{i=1}^{n+1} a_i u_i \right) d\alpha_j = 0, \qquad j = 1, \ldots, n$$

and

$$\operatorname{rank} \left(\int_0^1 u_i\, d\alpha_j \right)_{i=1\ \ j=1}^{n+1\ \ n} = n.$$

Proof. Since the $\{u_i\}_{i=1}^{n+1}$ are linearly independent there exist distinct points $\{x_i\}_{i=1}^{n+1}$ for which the matrix $C = (c_{ij})_{i,j=1}^{n+1}$, $c_{ij} = u_j(x_i)$, is nonsingular. Let $\mathbf{c} = C\mathbf{a}$. Thus $\mathbf{c} \neq \mathbf{0}$. Let $B = (b_{ij})_{i,j=1}^{n+1}$ be any nonsingular matrix satisfying $B\mathbf{c} = \mathbf{e}^{n+1}$, where $(\mathbf{e}^{n+1})_i = \delta_{i,n+1}$, $i = 1, \ldots, n+1$. Set

$$d\alpha_i = \lambda \sum_{i=1}^{n+1} b_{ij}\, d\mu_{x_j}, \qquad i = 1, \ldots, n,$$

where $d\mu_x$ is the unit point measure at x, and λ is some nonzero constant, to be determined. Thus,

$$\begin{aligned} \int_0^1 u_j\, d\alpha_i &= \lambda \sum_{k=1}^{n+1} b_{ik} u_j(x_k) \\ &= \lambda \sum_{k=1}^{n+1} b_{ik} c_{kj} \\ &= \lambda (BC)_{ij}. \end{aligned}$$

Since $(BC)\mathbf{a} = \mathbf{e}^{n+1}$, the lemma now follows from Cramer's rule, where we use the nonsingularity of BC and choose λ appropriately. □

For $d\alpha_1, \ldots, d\alpha_m$, $d\beta_1, \ldots, d\beta_m$, any $2m$ measures on $C[0,1]$, set

$$K\begin{pmatrix} d\alpha_1, \ldots, d\alpha_m \\ d\beta_1, \ldots, d\beta_m \end{pmatrix} = \det\left(\int_0^1\int_0^1 K(x,y)\, d\alpha_i(x)\, d\beta_j(y)\right)_{i,j=1}^{m}.$$

We denote $K\begin{pmatrix} d\mu_x, d\alpha_1, \ldots, d\alpha_m \\ d\mu_y, d\beta_1, \ldots, d\beta_m \end{pmatrix}$ by $K\begin{pmatrix} x, d\alpha_1, \ldots, d\alpha_m \\ y, d\beta_1, \ldots, d\beta_m \end{pmatrix}$. Also,

$$U\begin{pmatrix} x, d\alpha_1, \ldots, d\alpha_n \\ 1, \ldots, n+1 \end{pmatrix} = \det\left(\int_0^1 u_j\, d\alpha_k\right)_{k=0}^{n}{}_{j=1}^{n+1},$$

where $d\alpha_0 = d\mu_x$, and

$$U\begin{pmatrix} d\alpha_1, \ldots, d\alpha_n \\ 1, \ldots, \hat{i}, \ldots, n+1 \end{pmatrix} = \det\left(\int_0^1 u_j\, d\alpha_k\right)_{k=1,}^{n}{}_{\substack{j=1\\ j\neq i}}^{n+1}.$$

When V replaces U, then we understand it to mean that $\{v_i\}_{i=1}^{n+1}$ has replaced $\{u_i\}_{i=1}^{n+1}$. With this notation we have

Lemma 5.4. *If* $K(x,y) = \sum_{i=1}^{n+1} u_i(x)\, v_i(y)$, *then for any* $2n$ *measures* $d\alpha_1, \ldots, d\alpha_n$, $d\beta_1, \ldots, d\beta_n$ *on* $[0,1]$,

$$K\begin{pmatrix} x, d\alpha_1, \ldots, d\alpha_n \\ y, d\beta_1, \ldots, d\beta_n \end{pmatrix} = U\begin{pmatrix} x, d\alpha_1, \ldots, d\alpha_n \\ 1, \ldots, n+1 \end{pmatrix} V\begin{pmatrix} y, d\beta_1, \ldots, d\beta_n \\ 1, \ldots, n+1 \end{pmatrix}$$

and

$$K\begin{pmatrix} d\alpha_1, \ldots, d\alpha_n \\ d\beta_1, \ldots, d\beta_n \end{pmatrix} = \sum_{i=1}^{n+1} U\begin{pmatrix} d\alpha_1, \ldots, d\alpha_n \\ 1, \ldots, \hat{i}, \ldots, n+1 \end{pmatrix} V\begin{pmatrix} d\beta_1, \ldots, d\beta_n \\ 1, \ldots, \hat{i}, \ldots, n+1 \end{pmatrix}.$$

Proof. The first equality is simply a restatement of $\det(AB) = (\det A)(\det B)$. The second equality is a consequence of the Cauchy-Binet formula (Section 2 of Chapter III). □

Proof of Theorem 5.2. We first prove the upper bound $\delta_n(\mathscr{K}_p; L^q) \leqq \sigma$. By definition,

$$\delta_n(\mathscr{K}_p; L^q) = \inf_P \sup_{\|h\|_p \leqq 1} \|Kh - Ph\|_q,$$

where P is any linear operator taking L^p into L^q of rank at most n. Let $\bar{u} = \sum_{i=1}^{n+1} \bar{a}_i u_i$, $\bar{v} = \sum_{i=1}^{n+1} \bar{b}_i v_i$ satisfy $\sum_{i=1}^{n+1} \bar{a}_i \bar{b}_i = 1$ and $\|\bar{u}\|_q\, \|\bar{v}\|_{p'} = \sigma$. By Lemma 5.3, there exist $2n$ measures $d\bar{\alpha}_1, \ldots, d\bar{\alpha}_n$, $d\bar{\beta}_1, \ldots, d\bar{\beta}_n$ for which

$$\bar{a}_i = (-1)^{i+1} \det\left(\int_0^1 u_k\, d\bar{\alpha}_j\right)_{\substack{k=1,\\ k\neq i}}^{n+1}{}_{j=1}^{n}, \qquad i = 1, \ldots, n+1$$

and

$$\bar{b}_i = (-1)^{i+1} \det \left(\int_0^1 v_k \, d\bar{\beta}_j \right)_{\substack{k=1, \; j=1 \\ k \neq i}}^{n+1 \quad n}, \qquad i = 1, \ldots, n+1.$$

By Lemma 5.4,

$$\begin{aligned} K\begin{pmatrix} d\bar{\alpha}_1, \ldots, d\bar{\alpha}_n \\ d\bar{\beta}_1, \ldots, d\bar{\beta}_n \end{pmatrix} &= \sum_{i=1}^{n+1} U\begin{pmatrix} d\bar{\alpha}_1, \ldots, d\bar{\alpha}_n \\ 1, \ldots, \hat{i}, \ldots, n+1 \end{pmatrix} V\begin{pmatrix} d\bar{\beta}_1, \ldots, d\bar{\beta}_n \\ 1, \ldots, \hat{i}, \ldots, n+1 \end{pmatrix} \\ &= \sum_{i=1}^{n+1} \bar{a}_i \bar{b}_i = 1. \end{aligned}$$

Thus for P as in the statement of the theorem (see also the remark after Lemma 5.3)

$$\begin{aligned} \delta_n(\mathcal{K}_p; L^q) &\leqq \sup_{\|h\|_p \leqq 1} \left\| \int_0^1 K\begin{pmatrix} x, d\bar{\alpha}_1, \ldots, d\bar{\alpha}_n \\ y, d\bar{\beta}_1, \ldots, d\bar{\beta}_n \end{pmatrix} h(y)\, dy \right\|_q \\ &= \sup_{\|h\|_p \leqq 1} \left\| U\begin{pmatrix} x, d\bar{\alpha}_1, \ldots, d\bar{\alpha}_n \\ 1, \ldots, n+1 \end{pmatrix} \int_0^1 V\begin{pmatrix} y, d\bar{\beta}_1, \ldots, d\bar{\beta}_n \\ 1, \ldots, n+1 \end{pmatrix} h(y)\, dy \right\|_q \\ &= \sup_{\|h\|_p \leqq 1} \left\| \bar{u}(\cdot) \int_0^1 \bar{v}(y)\, h(y)\, dy \right\|_q \\ &= \sup_{\|h\|_p \leqq 1} \|\bar{u}\|_q \left| \int_0^1 \bar{v}(y)\, h(y)\, dy \right| \\ &= \|\bar{u}\|_q \, \|\bar{v}\|_{p'} \\ &= \sigma. \end{aligned}$$

It remains to prove the lower bound. From Theorem 5.1,

$$d_n(\mathcal{K}_p; L^q) = d^n(\mathcal{K}_p; L^q) = \delta_n(\mathcal{K}_p; L^q) = \inf\{\|Kh\|_q \colon Kh \in \partial A\}$$

where $A = \{Kh\colon \|h\|_p \leq 1\}$. Now $Kh = \sum_{i=1}^{n+1} a_i u_i$, where $a_i = \int_0^1 v_i(y)\, h(y)\, dy$, $i = 1, \ldots, n+1$. Since the $\{u_i\}_{i=1}^{n+1}$ are linearly independent, the map $\mathbf{a} = (a_1, \ldots, a_{n+1}) \to \sum_{i=1}^{n+1} a_i u_i$ is a homeomorphism. Thus the boundary of A is the set of $Kh = \sum_{i=1}^{n+1} a_i u_i$ for which $\mathbf{a}$ lies on the boundary of the set

$$\mathcal{M} = \left\{ \mathbf{c} = (c_1, \ldots, c_{n+1}) \colon c_i = \int_0^1 v_i(y)\, h(y)\, dy, \; \|h\|_p \leqq 1 \right\}.$$

Since $\mathcal{M}$ is convex, closed and centrally symmetric, for every $\mathbf{a} \in \partial\mathcal{M}$ there exists a $\mathbf{b}^* \in \mathbb{R}^{n+1} \backslash \{\mathbf{0}\}$, dependent on $\mathbf{a}$, for which

$$|(\mathbf{c}, \mathbf{b}^*)| \leqq (\mathbf{a}, \mathbf{b}^*) = 1 \qquad \text{for all } \mathbf{c} \in \mathcal{M}.$$

Thus $\left|\int_0^1 \sum_{i=1}^{n+1} b_i^* v_i(y)\, h(y)\, dy\right| \leqq 1$ for all $\|h\|_p \leqq 1$ which implies that $\left\|\sum_{i=1}^{n+1} b_i^* v_i\right\|_{p'} \leqq 1$. Therefore

$$\begin{aligned}
\inf\{\|Kh\|_q : Kh \in \partial A\} &= \inf\left\{\left\|\sum_{i=1}^{n+1} a_i u_i\right\|_q : \mathbf{a} \in \partial \mathcal{M}\right\} \\
&\geqq \inf\left\{\left\|\sum_{i=1}^{n+1} a_i u_i\right\|_q \left\|\sum_{i=1}^{n+1} b_i^* v_i\right\|_{p'} : \sum_{i=1}^{n+1} a_i b_i^* = 1,\ \mathbf{a} \in \partial \mathcal{M}\right\} \\
&\geqq \inf\left\{\left\|\sum_{i=1}^{n+1} a_i u_i\right\|_q \left\|\sum_{i=1}^{n+1} b_i v_i\right\|_{p'} : \sum_{i=1}^{n+1} a_i b_i = 1\right\} \\
&= \sigma.
\end{aligned}$$

The lower and upper bounds agree. The remaining statements of the theorem follow from the remark after Lemma 5.3 and the above proof. □

When K is assumed to be NTP, as in Section 2, then it was shown that there are optimal subspaces for $d_n(\mathcal{K}_\infty; L^q)$ and $d_n(\mathcal{K}_p; L^1)$ of the form $\operatorname{span}\{K(\cdot, \xi_1), \ldots, K(\cdot, \xi_n)\}$ for some $0 < \xi_1 < \ldots < \xi_n < 1$ (dependent on p or q) and that interpolation from this subspace to Kh at some $\{\eta_i\}_{i=1}^n$ is an optimal linear method (as well as the analogous statement for d^n). These subspaces are of a particularly simple form. We will delineate criteria on K of rank $n+1$ which imply similar results for $d_n(\mathcal{K}_p; L^q)$, $d^n(\mathcal{K}_p; L^q)$ and $\delta_n(\mathcal{K}_p; L^q)$ for *all* $p, q \in [1, \infty]$.

In other words we seek conditions on K for which the $d\bar{\alpha}_1, \ldots, d\bar{\alpha}_n$ and $d\bar{\beta}_1, \ldots, d\bar{\beta}_n$, in the statement of Theorem 5.2, satisfy $d\bar{\alpha}_i = d\mu_{\eta_i}$ and $d\bar{\beta}_i = d\mu_{\xi_i}$, $i = 1, \ldots, n$ for some $0 < \eta_1 < \ldots < \eta_n < 1$ and $0 < \xi_1 < \ldots < \xi_n < 1$. Let $\bar{u} = \sum_{i=1}^{n+1} \bar{a}_i u_i$, $\bar{v} = \sum_{i=1}^{n+1} \bar{b}_i v_i$ satisfy $\|\bar{u}\|_q \|\bar{v}\|_{p'} = \sigma$ and $\sum_{i=1}^{n+1} \bar{a}_i \bar{b}_i = 1$. Then it is possible to choose $d\bar{\alpha}_i$ and $d\bar{\beta}_i$ as above if $\bar{u}(\eta_i) = 0$, $i = 1, \ldots, n$, rank $(u_j(\eta_i))_{i=1}^{n}\,{}_{j=1}^{n+1} = n$, and $\bar{v}(\xi_i) = 0$, $i = 1, \ldots, n$, rank $(v_j(\xi_i))_{i=1}^{n}\,{}_{j=1}^{n+1} = n$. We prove the following

Theorem 5.5. *Let* $\{u_i\}_{\substack{i=1\\ i \neq k}}^{n+1}$, $\{v_i\}_{\substack{i=1\\ i \neq k}}^{n+1}$, $k = 1, \ldots, n+1$, $\{u_i\}_{i=1}^{n+1}$ *and* $\{v_i\}_{i=1}^{n+1}$ *all be* T^+*-systems on* $[0,1]$. *For fixed* $p, q \in [1, \infty]$, *let* $\bar{u}$ *and* $\bar{v}$ *be as above. Then* $\bar{u}$ *and* $\bar{v}$ *each have exactly* n *distinct zeros in* $(0,1)$.

We record its consequence.

Theorem 5.6. *Let* $K(x, y) = \sum_{i=1}^{n+1} u_i(x)\, v_i(y)$ *where* $\{u_i\}_{i=1}^{n+1}$ *and* $\{v_i\}_{i=1}^{n+1}$ *satisfy the conditions of Theorem 5.5. For* $p, q \in [1, \infty]$, *let* $\{\xi_i\}_{i=1}^n$ *denote the* n *distinct zeros of* $\bar{v}$ *and* $\{\eta_i\}_{i=1}^n$ *the* n *distinct zeros of* $\bar{u}$. *Then,*

(1) $X_n = \operatorname{span}\{K(\cdot, \xi_1), \ldots, K(\cdot, \xi_n)\}$ *is optimal for* $d_n(\mathcal{K}_p; L^q)$.

(2) *The linear operator which interpolates to* Kh *from* X_n *at* $\{\eta_i\}_{i=1}^n$ *is optimal for* $\delta_n(\mathcal{K}_p; L^q)$.

(3) $L^n = \{h : h \in L^q,\ Kh(\eta_i) = 0,\ i = 1, \ldots, n\}$ *is optimal for* $d^n(\mathcal{K}_p; L^q)$.

The conditions in Theorem 5.5 on the $\{u_i\}_{i=1}^{n+1}$ and $\{v_i\}_{i=1}^{n+1}$ are used to obtain this next result.

Lemma 5.7. *Assume that* $\{u_i\}_{\substack{i=1\\i\neq k}}^{n+1}$, $k=1,\ldots,n+1$ *and* $\{u_i\}_{i=1}^{n+1}$ *are* T^+*-systems on* $[0,1]$. *If* $u=\sum_{i=1}^{n+1} a_i u_i$ *satisfies* $\tilde{Z}(u)=n$, *then* $a_i a_{i+1}<0$, $i=1,\ldots,n$. *If in addition* $u(1)>0$, *then* $a_{n+1}>0$.

Lemma 5.7 follows from the results of Section 2 of Chapter III. See in particular Corollary 2.5.

Theorem 5.5 is based upon the following proposition.

Proposition 5.8. *Let* $\{u_i\}_{i=1}^{n+1}$ *satisfy the conditions of Theorem 5.5. Let* $\{d_i\}_{i=1}^{n+1}$ *be real non-negative numbers not all zero. Then for* $q\in[1,\infty]$ *there is a unique* $\bar{u}=\sum_{i=1}^{n+1}\bar{a}_i\bar{u}_i$ *which minimizes* $\left\|\sum_{i=1}^{n+1} a_i u_i\right\|_q$ *subject to the constraint* $\sum_{i=1}^{n+1}|a_i|\,d_i=1$. *Moreover* $\bar{u}$ *has exactly n distinct zeros in* $(0,1)$.

Proof. We first consider the case $q=\infty$. Let $\bar{u}=\sum_{i=1}^{n+1}\bar{a}_i u_i$ be the polynomial which equioscillates at $n+1$ points in $[0,1]$, and is normalized such that $\bar{a}_1>0$ (i.e., $\bar{u}(0)>0$) and $\sum_{i=1}^{n+1}|\bar{a}_i|\,d_i=1$. It is well known that such a $\bar{u}$ exists and is unique. Let $u=\sum_{i=1}^{n+1}a_i u_i$ be any other polynomial $(u\neq\pm\bar{u})$ for which $\sum_{i=1}^{n+1}|a_i|\,d_i=1$. Assume $\|u\|_\infty\leqq\|\bar{u}\|_\infty$. Since $\bar{u}$ equioscillates at $n+1$ points, and $\{u_i\}_{i=1}^{n+1}$ is a T^+-system, $\bar{u}\pm u=\sum_{i=1}^{n+1}(\bar{a}_i\pm a_i)u_i$ satisfies $\tilde{Z}(\bar{u}\pm u)=n$. By Lemma 5.7, the $(\bar{a}_i\pm a_i)$ strictly alternate in sign. In fact, since $\bar{a}_1>0$, it follows that $(\bar{a}_i\pm a_i)(-1)^{i+1}>0$, $i=1,\ldots,n+1$ for each choice of $\pm$. This implies that $|\bar{a}_i|>|a_i|$, $i=1,\ldots,n+1$. Thus $1=\sum_{i=1}^{n+1}|\bar{a}_i|\,d_i>\sum_{i=1}^{n+1}|a_i|\,d_i=1$. A contradiction.

Assume $1\leqq q<\infty$. Let $u^*=\sum_{i=1}^{n+1}a_i^* u_i$ be any polynomial minimizing $\left\|\sum_{i=1}^{n+1}a_i u_i\right\|_q$ subject to $\sum_{i=1}^{n+1}|a_i|\,d_i=1$. We claim that

$$\int_0^1 |u^*(x)|^{q-1}\operatorname{sgn}[u^*(x)]\,u_k(x)\,dx=\left[\int_0^1 |u^*(x)|^q\,dx\right][\operatorname{sgn}a_k^*]\,d_k,\quad k=1,\ldots,n+1.$$

If $d_k=0$, then there is no restriction upon a_k and therefore from orthogonality considerations,

$$\int_0^1 |u^*(x)|^{q-1}\operatorname{sgn}[u^*(x)]\,u_k(x)\,dx=0.$$

If $d_k>0$ and $a_k^*\neq 0$, then the required equality is easily seen to hold by the method of Lagrange multipliers. It remains to consider the case where $d_k>0$ and

$a_k^* = 0$. However in this case we may take right and left limits to obtain

$$\pm \int_0^1 |u^*(x)|^{q-1} \operatorname{sgn}[u^*(x)]\, u_k(x)\, dx \geqq \left[\int_0^1 |u^*(x)|^q\, dx\right] d_k.$$

Since $\left[\int_0^1 |u^*(x)|^q\, dx\right] d_k > 0$, this is impossible. Thus we cannot have $d_k > 0$ and $a_k^* = 0$. This proves the desired equality for all k.

Now consider the related problem of minimizing $\left\| \sum_{i=1}^{n+1} a_i u_i \right\|_q$ subject to the constraint $\sum_{i=1}^{n+1} a_i(-1)^i d_i = 1$. Let $\bar{u} = \sum_{i=1}^{n+1} \bar{a}_i u_i$ be the unique solution to this problem, with $\sum_{i=1}^{n+1} \bar{a}_i(-1)^i d_i = 1$. (Uniqueness is easily shown.) By the method of Lagrange multipliers it follows that

$$\int_0^1 |\bar{u}(x)|^{q-1} \operatorname{sgn}[\bar{u}(x)]\, u_k(x)\, dx = \left[\int_0^1 |\bar{u}(x)|^q\, dx\right](-1)^k d_k, \qquad k = 1, \ldots, n+1.$$

Since $(-1)^k d_k$ weakly alternates in sign, and $\{u_k\}_{k=1}^{n+1}$ is a T^+-system, it is necessary that $\bar{u}$ have n sign changes in $(0,1)$. Thus, by Lemma 5.7, $\bar{a}_i \bar{a}_{i+1} < 0$, $i = 1, \ldots, n$. The constraint implies that $\bar{a}_i(-1)^i > 0$, $i = 1, \ldots, n+1$, so that $1 = \sum_{i=1}^{n+1} \bar{a}_i(-1)^i d_i = \sum_{i=1}^{n+1} |\bar{a}_i|\, d_i$. Thus the $\{\bar{a}_i\}_{i=1}^{n+1}$ are admissable in the original minimum problem, and $\|u^*\|_q \leqq \|\bar{u}\|_q$. Assume that $u^* \neq \pm \bar{u}$. From the two sets of equalities, we obtain

$$\begin{aligned} &\int_0^1 (|\bar{u}(x)|^{q-1} \operatorname{sgn}[\bar{u}(x)] \pm |u^*(x)|^{q-1} \operatorname{sgn}[u^*(x)])\, u_k(x)\, dx \\ &\quad = (\|\bar{u}\|_q^q (-1)^k \pm \|u^*\|_q^q [\operatorname{sgn} a_k^*])\, d_k, \qquad k = 1, \ldots, n+1. \end{aligned}$$

The sequence $\{(\|\bar{u}\|_q^q (-1)^k \pm \|u^*\|_q^q [\operatorname{sgn} a_k^*])\, d_k\}_{k=1}^{n+1}$ weakly alternates in sign with orientation independent of the choice $\pm$. Thus by Proposition 1.4 of Chapter III

$$|\bar{u}(x)|^{q-1} \operatorname{sgn}[\bar{u}(x)] \pm |u^*(x)|^{q-1} \operatorname{sgn}[u^*(x)]$$

exhibits at least n sign changes, and $n+1$ if

$$\{(\|\bar{u}\|_q^q (-1)^k \pm \|u^*\|_q^q [\operatorname{sgn} a_k^*])\, d_k\}_{k=1}^{n+1}$$

is an identically zero sequence.

It is easily seen that for $1 < q < \infty$,

$$\operatorname{sgn}(|\bar{u}(x)|^{q-1} \operatorname{sgn}[\bar{u}(x)] \pm |u^*(x)|^{q-1} \operatorname{sgn}[u^*(x)]) = \operatorname{sgn}(\bar{u}(x) \pm u^*(x)).$$

Since $\bar{u} \neq \pm u^*$ and $S(\bar{u} + u^*) \leqq n$, we cannot have $\|\bar{u}\|_q^q (-1)^k d_k = \pm \|u^*\|_q^q \operatorname{sgn}[a_k^*]\, d_k$ for $k = 1, \ldots, n+1$. Thus $\bar{u} \pm u^* = \sum_{i=1}^{n+1} (\bar{a}_i \pm a_i^*) u_i$ exhibits exactly n sign changes in $(0,1)$ and the orientation is fixed independent of the $\pm$. This implies by Lemma 5.7 that $|\bar{a}_i| > |a_i^*|$, $i = 1,\ldots,n+1$. If $q=1$, then the same conclusion holds. Sgn $(\operatorname{sgn}[\bar{u}(x)] \pm \operatorname{sgn}[u^*(x)])$ cannot be identically zero since $\bar{u} \neq \pm u^*$. It therefore follows that $\bar{u} \pm u^*$ has n sign changes with orientation independent of the $\pm$. This again gives $|\bar{a}_i| > |a_i^*|$, $i = 1,\ldots,n+1$.

Since $1 = \sum_{i=1}^{n+1} |\bar{a}_i|\, d_i > \sum_{i=1}^{n+1} |a_i^*|\, d_i = 1$, this contradiction implies the proposition. □

Proof of Theorem 5.5. Let $\{a_i^*\}_{i=1}^{n+1}$ and $\{b_i^*\}_{i=1}^{n+1}$ be any solution to: Minimize $\left\| \sum_{i=1}^{n+1} a_i u_i \right\|_q \left\| \sum_{i=1}^{n+1} b_i v_i \right\|_{p'}$ subject to $\sum_{i=1}^{n+1} a_i b_i = 1$. Set $u^* = \sum_{i=1}^{n+1} a_i^* u_i$ and $v^* = \sum_{i=1}^{n+1} b_i^* v_i$. Assume that u^* does not have n distinct zeros in $(0,1)$. Then

$$\|u^*\|_q \, \|v^*\|_{p'} \Big/ \left| \sum_{i=1}^{n+1} a_i^* b_i^* \right| \geqq \|u^*\|_q \, \|v^*\|_{p'} \Big/ \sum_{i=1}^{n+1} |a_i^*|\,|b_i^*|$$
$$> \|\bar{u}\|_q \, \|v^*\|_{p'} \Big/ \sum_{i=1}^{n+1} |\bar{a}_i|\,|b_i^*|,$$

where $\bar{u} = \sum_{i=1}^{n+1} \bar{a}_i u_i$ depends upon the $\{|b_i^*|\}_{i=1}^{n+1}$ (chosen as in Proposition 5.8) and $\bar{u}$ has n distinct zeros in $(0,1)$. Similarly, if v^* does not have n distinct zeros, we again apply Proposition 5.8 to obtain

$$\|u^*\|_q \, \|v^*\|_{p'} \Big/ \left| \sum_{i=1}^{n+1} a_i^* b_i^* \right| > \|\bar{u}\|_q \, \|\bar{v}\|_{p'} \Big/ \sum_{i=1}^{n+1} |\bar{a}_i|\,|\bar{b}_i|$$

where the $\{\bar{b}_i\}_{i=1}^{n+1}$ depend on the $\{|\bar{a}_i|\}_{i=1}^{n+1}$ and $\bar{v}$ has n distinct zeros. By Lemma 5.7, the $\bar{a}_i$ and $\bar{b}_i$ alternate in sign. Thus $\sum_{i=1}^{n+1} |\bar{a}_i|\,|\bar{b}_i| = \left| \sum_{i=1}^{n+1} \bar{a}_i \bar{b}_i \right|$. Therefore,

$$\|u^*\|_q \, \|v^*\|_{p'} \Big/ \left| \sum_{i=1}^{n+1} a_i^* b_i^* \right| > \|\bar{u}\|_q \, \|\bar{v}\|_{p'} \Big/ \left| \sum_{i=1}^{n+1} \bar{a}_i \bar{b}_i \right|.$$

This contradiction proves the theorem. □

Notes and References

Section 2. The results of this section are mainly due to Micchelli, Pinkus [1978]. The cases $p = q = \infty$ and $p = q = 1$ were previously considered in Micchelli, Pinkus [1977a], [1977b]. The conditions on K as stated in Micchelli, Pinkus are somewhat more restrictive than those given here. This restriction was overcome by Dyn [1982] and Theorem 2.4 in its present form is mainly due to her. Micchelli,

Pinkus [1978] also consider the problem of approximating the kernel K by finite rank kernels in various norms. This problem is intimately connected with the determination of n-widths. Brown [1982] generalizes this approach and in this way simultaneously deals with the results of Sections 2 and 3 (and generalizations thereof). Dyn [1982] also generalizes work of Micchelli, Pinkus [1978]. The proof of the Hobby-Rice Theorem (Theorem 2.3) as presented here is to be found in Pinkus [1976]. The ideas behind Theorem 2.14 are due to Sattes [1980]. He determined these n-widths with $\tilde{B}_{\infty}^{(r)}$ replacing $\mathscr{K}_{\infty}$, and $q = \infty$. Theorem 2.14 extends, in a natural way, to include the classes of functions considered in Sections 3 and 4.

Section 3. The material of this section is again based on Micchelli, Pinkus [1978], except for Theorems 3.2 and 3.5 which are due to Dyn [1983], and Proposition 3.12 and Theorem 3.13. The case $p = q = \infty$ was considered in Micchelli, Pinkus [1977a], and $p = q = 1$ in Micchelli, Pinkus [1977b]. These latter two papers used the Fundamental Theorem of Tichomirov (Theorem 1.5 of Chapter II) in obtaining the lower bounds. The motivating example behind these results is that of Tichomirov [1969] who calculated $d_n(B_{\infty}^{(r)}; L^{\infty})$ for all n. A generalization of this result is due to Chui, Smith [1975]. They determined the n-widths in L^{∞} of

$$A = \{f\colon \|Lf\|_{\infty} \leqq 1\}$$

where L is any of a restricted class of nth order disconjugate forms with constant coefficients (see Section 4 of Chapter III). Tichomirov, Babadjanov [1967] calculated $d_n(B_p^{(1)}; L^p)$ for all $p \in [1, \infty]$, and Makovoz [1972] extended this to obtain $d_n(B_p^{(1)}; L^q)$ and $d^n(B_p^{(1)}; L^q)$ for all $1 \leqq q \leqq p \leqq \infty$. (Note that $r = 1$ in both cases.)

Section 4. The results of this section were motivated by an attempt to determine $d_n(\tilde{B}_p^{(r)}; L^q)$. Tichomirov [1960a] calculated $d_n(\tilde{B}_{\infty}^{(r)}; \tilde{C})$ for all n, where $\tilde{C}$ is the class of 2π-periodic, continuous functions endowed with the L^{∞} norm. Both Makovoz [1969] and Subbotin [1970] studied $d_{2n-1}(\tilde{B}_1^{(r)}; L^1)$, and Makovoz [1972] also computed $d_{2n-1}(\tilde{B}_{\infty}^{(r)}; L^1)$, $d^{2n-1}(\tilde{B}_{\infty}^{(r)}; L^1)$ and $d^{2n-1}(\tilde{B}_{\infty}^{(r)}; L^2)$. Makovoz [1979] and Ligun [1980] also determined $d_{2n}(\tilde{B}_p^{(r)}; L^q)$ for $p = \infty$ and $q = 1$, independently of each other and of Pinkus [1979a]. In Pinkus [1979a] may be found Theorems 4.8, 4.9, 4.12 and 4.13. The idea behind Theorems 4.5 and 4.6 (for $q < \infty$) is based on a method due to Zensykbaev [1976]. The proof of the optimality of T_{n-1} for $d_{2n-1}(\tilde{\mathscr{K}}_p; L^1)$ (and for $d_{2n-1}(\tilde{\mathscr{B}}_p; L^1)$) is very much based on work of Taikov [1967a] who proved this same upper bound for $d_{2n-1}(\tilde{B}_p^{(r)}; L^1)$.

Tichomirov [1960a] also announced various other results. One of these concerned the n-widths of $\tilde{H}_{\beta}^{\infty}$ in $L^{\infty}[0, 2\pi]$ where $\tilde{H}_{\beta}^{\infty}$ is the class of functions analytic in $S_{\beta} = \{z\colon z \in \mathbb{C},\ |\mathrm{Im}\, z| < \beta\}$, real and 2π-periodic on the x-axis, and which satisfy $|\mathrm{Re}\, f(z)| \leqq 1$ for all $z \in S_{\beta}$ (see Example 6.1 of Chapter IV). The proof as outlined was incomplete (although the correct value for the n-width was stated). A correct proof was finally given by Forst [1977]. He effectively showed that the kernel K_{β} of Example 6.1 of Chapter IV is *CVD*. (A direct proof is given in Section 4 of Chapter III.)

Korneichuk [1971 a], [1974], [1976 a] and others (see e.g. Grigorian [1973], Timan [1960], Ruban [1974], Motornyi, Ruban [1975], and Ligun [1980]) have studied the n-widths of functions in $\tilde{W}_\infty^{(r)}$ for which the modulus of continuity of the $(r-1)$st derivative is bounded above by a given concave modulus of continuity. It was felt that these results were, at their present state of development, slightly beyond the scope of this work.

Section 5. The material of this section is taken from Micchelli, Pinkus [1979].

Chapter VI. Matrices and n-Widths

1. Introduction and General Remarks

Let $A = (a_{ij})_{i,j=1}^{M}$ be an $M \times M$ real matrix. (We shall deal with real square matrices for notational convenience.) For $\mathbf{x} \in \mathbb{R}^M$, set

$$\|\mathbf{x}\|_p = \begin{cases} \left(\sum_{i=1}^{M} |x_i|^p \right)^{1/p}, & 1 \leqq p < \infty, \\ \max_{i=1,\ldots,M} |x_i|, & p = \infty. \end{cases}$$

In this chapter we are concerned with the problem of evaluating the n-widths of

$$\mathscr{A}_p = \{A\mathbf{x} \colon \|\mathbf{x}\|_p \leqq 1\}$$

as subsets of l_q^M, i.e., with respect to the norm $\|\cdot\|_q$, for various choices of p, q and A. We first introduce some additional notation and recall various results from Chapter II.

For A as above set

$$\mathscr{A}_p^T = \{A^T\mathbf{x} \colon \|\mathbf{x}\|_p \leqq 1\},$$

where A^T is the transpose of A. If A is invertible define

$$\mathscr{A}_p^{-1} = \{A^{-1}\mathbf{x} \colon \|\mathbf{x}\|_p \leqq 1\}.$$

From the results of Chapter II, namely Theorems 7.2, 8.9 and 8.10, and Propositions 5.1, 5.2, 8.12 and 8.13, we have

Proposition 1.1. *For* $p, q \in [1, \infty]$, *and* $1/p + 1/p' = 1/q + 1/q' = 1$,

(a) $\delta_n(\mathscr{A}_p; l_q^M) \geqq d_n(\mathscr{A}_p; l_q^M)$, $d^n(\mathscr{A}_p; l_q^M) \geqq b_n(\mathscr{A}_p; l_q^M)$.

(b) $\delta_n(\mathscr{A}_p; l_q^M) = \delta_n(\mathscr{A}_{q'}^T; l_{p'}^M)$.

(c) $d_n(\mathscr{A}_p; l_q^M) = d^n(\mathscr{A}_{q'}^T; l_{p'}^M)$.

(d) $\delta_n(\mathscr{A}_p; l_2^M) = d_n(\mathscr{A}_p; l_2^M)$ *(and thus* $\delta_n(\mathscr{A}_2; l_q^M) = d^n(\mathscr{A}_2; l_q^M)$*).*

(e) $\delta_n(\mathscr{A}_1; l_q^M) = d_n(\mathscr{A}_1; l_q^M)$ *(and thus* $\delta_n(\mathscr{A}_p; l_\infty^M) = d^n(\mathscr{A}_p; l_\infty^M)$*).*

The following result will also prove useful.

Proposition 1.2. *If A is nonsingular, then*

$$b_n(\mathscr{A}_p; l_q^M)\, d^{M-n-1}(\mathscr{A}_q^{-1}; l_p^M) = 1$$

for $n = 0, 1, \ldots, M-1$.

Proof.
$$d^{M-n-1}(\mathscr{A}_q^{-1}; l_p^M) = \min_{X_{M-n-1}} \max_{\substack{\mathbf{x} \perp X_{M-n-1} \\ \mathbf{x} \neq \mathbf{0}}} \frac{\|A^{-1}\mathbf{x}\|_p}{\|\mathbf{x}\|_q}$$
$$= \min_{X_{M-n-1}} \max_{\substack{\mathbf{x} \perp X_{M-n-1} \\ \mathbf{x} \neq \mathbf{0}}} \left[\frac{\|\mathbf{x}\|_q}{\|A^{-1}\mathbf{x}\|_p}\right]^{-1}.$$

Setting $\mathbf{y} = A^{-1}\mathbf{x}$, and since A is invertible,

$$= \min_{X_{M-n-1}} \max_{\substack{\mathbf{y} \perp X_{M-n-1} \\ \mathbf{y} \neq \mathbf{0}}} \left[\frac{\|A\mathbf{y}\|_q}{\|\mathbf{y}\|_p}\right]^{-1}$$
$$= \left[\max_{X_{M-n-1}} \min_{\substack{\mathbf{y} \perp X_{M-n-1} \\ \mathbf{y} \neq \mathbf{0}}} \frac{\|A\mathbf{y}\|_q}{\|\mathbf{y}\|_p}\right]^{-1}$$
$$= \left[\max_{X_{n+1}} \min_{\substack{\mathbf{y} \in X_{n+1} \\ \mathbf{y} \neq \mathbf{0}}} \frac{\|A\mathbf{y}\|_q}{\|\mathbf{y}\|_p}\right]^{-1}$$
$$= [b_n(\mathscr{A}_p; l_q^M)]^{-1}. \quad \square$$

The one general case in which all the n-widths of any matrix A can be determined is when $p = q = 2$ (see Theorem 2 of Chapter I and also Theorem 2.2 of Chapter IV).

Let $\lambda_1 \geqq \ldots \geqq \lambda_M \geqq 0$ denote the M eigenvalues (listed to their multiplicity) of the positive semidefinite matrix $A^T A$, and let $\mathbf{x}^1,, \ldots, \mathbf{x}^M$ denote a corresponding set of orthonormal eigenvectors. Assume $\lambda_1 \geqq \ldots \geqq \lambda_k > \lambda_{k+1} = \ldots = \lambda_M = 0$, and for $i = 1, \ldots, k$, set $\mathbf{y}^i = \lambda_i^{-1/2} A\mathbf{x}^i$. The $\mathbf{y}^1, \ldots, \mathbf{y}^k$ so defined are orthonormal and $AA^T\mathbf{y}^i = \lambda_i \mathbf{y}^i$. Let $\mathbf{y}^{k+1}, \ldots, \mathbf{y}^M$ be solutions of $A^T\mathbf{y} = \mathbf{0}$ for which $\mathbf{y}^1, \ldots, \mathbf{y}^M$ are orthonormal. Let X denote the $M \times M$ unitary matrix with columns $\mathbf{x}^1, \ldots, \mathbf{x}^M$, and Y the corresponding matrix with columns $\mathbf{y}^1, \ldots, \mathbf{y}^M$. Then

$$A = Y\Lambda^{1/2} X^T$$

where $\Lambda^{1/2} = \operatorname{diag}\{\lambda_1^{1/2}, \ldots, \lambda_M^{1/2}\}$, i.e., the $M \times M$ diagonal matrix with diagonal entries $\lambda_1^{1/2}, \ldots, \lambda_M^{1/2}$. The $\{\lambda_i^{1/2}\}_{i=1}^M$ are the singular values (s-numbers) of A and

$$A = Y\Lambda^{1/2} X^T$$

is called the singular value decomposition of A.

Theorem 1.3. *For* $n \leqq M-1$,

$$\delta_n(\mathscr{A}_2; l_2^M) = d_n(\mathscr{A}_2; l_2^M) = d^n(\mathscr{A}_2; l_2^M) = b_n(\mathscr{A}_2; l_2^M) = \lambda_{n+1}^{1/2}.$$

Furthermore,

(1) $X_n = \operatorname{span}\{\mathbf{y}^1, \dots, \mathbf{y}^n\}$ *is an optimal subspace for* $d_n(\mathscr{A}_2; l_2^M)$.

(2) $P_n = Y\Lambda_n^{1/2} X^T$ *is an optimal linear operator for* $\delta_n(\mathscr{A}_2; l_2^M)$, *where* $\Lambda_n^{1/2} = \operatorname{diag}\{\lambda_1^{1/2}, \dots, \lambda_n^{1/2}, 0, \dots, 0\}$.

(3) $M^n = \{\mathbf{x}: (\mathbf{x}, \mathbf{x}^i) = 0,\ i = 1, \dots, n\}$ *is an optimal subspace for* $d^n(\mathscr{A}_2; l_2^M)$.

(4) X_{n+1} *is an optimal subspace for* $b_n(\mathscr{A}_2; l_2^M)$.

Little else may be said about n-widths of arbitrary matrices for all n. Of course we do have the following simple result.

Proposition 1.4.

$$\delta_0(\mathscr{A}_p; l_q^M) = d_0(\mathscr{A}_p; l_q^M) = d^0(\mathscr{A}_p; l_q^M) = b_0(\mathscr{A}_p; l_q^M) = \max_{\mathbf{x} \neq \mathbf{0}} \frac{\|A\mathbf{x}\|_q}{\|\mathbf{x}\|_p}.$$

And essentially as a result of Theorem 4.3 of Chapter II we also have

Proposition 1.5. *Let* $\operatorname{rank} A = n + 1$, *and* $X_{n+1} \subseteq \mathbb{R}^M$ *denote the* $(n+1)$-*dimensional subspace spanned by the columns of* A. *Then*

$$\delta_n(\mathscr{A}_p; l_q^M) = d_n(\mathscr{A}_p; l_q^M) = d^n(\mathscr{A}_p; l_q^M) = b_n(\mathscr{A}_p; l_q^M) = \min_{A\mathbf{x} \in \partial(\mathscr{A}_p \cap X_{n+1})} \|A\mathbf{x}\|_q.$$

Corollary 1.6.

$$\delta_{M-1}(\mathscr{A}_p; l_q^M) = d_{M-1}(\mathscr{A}_p; l_q^M) = d^{M-1}(\mathscr{A}_p; l_q^M) = b_{M-1}(\mathscr{A}_p; l_q^M) = \min_{\mathbf{x} \neq \mathbf{0}} \frac{\|A\mathbf{x}\|_q}{\|\mathbf{x}\|_p}.$$

One problem is that we generally have no way of expressing the above maximum and minimum in a more closed form. An exception to this rule is when $A = D = \operatorname{diag}\{D_1, \dots, D_M\}$, i.e., A is a diagonal matrix.

Corollary 1.7. *Let* $D = \operatorname{diag}\{D_1, \dots, D_M\}$. *Then*

$$\max_{\mathbf{x} \neq \mathbf{0}} \frac{\|D\mathbf{x}\|_q}{\|\mathbf{x}\|_p} = \begin{cases} \max\limits_{i=1,\dots,M} |D_i|, & 1 \leqq p \leqq q \leqq \infty \\ \left(\sum\limits_{i=1}^{M} |D_i|^r\right)^{1/r}, & 1 \leqq q < p \leqq \infty, \end{cases}$$

where $1/r = 1/q - 1/p$.

Corollary 1.8. *For* D *as above, and nonsingular,*

$$\min_{\mathbf{x} \neq \mathbf{0}} \frac{\|D\mathbf{x}\|_q}{\|\mathbf{x}\|_p} = \begin{cases} \left(\sum\limits_{i=1}^{M} |D_i|^r\right)^{1/r}, & 1 \leqq p < q \leqq \infty \\ \min\limits_{i=1,\dots,M} |D_i|, & 1 \leqq q \leqq p \leqq \infty, \end{cases}$$

where $1/r = 1/q - 1/p (< 0)$.

We also have

Corollary 1.9. *For any $M \times M$ real matrix A,*

$$d_0(\mathscr{A}_1; l_q^M) = \max_{j=1,\ldots,M} \left(\sum_{i=1}^{M} |a_{ij}|^q \right)^{1/q}$$

and

$$d_0(\mathscr{A}_p; l_\infty^M) = \max_{i=1,\ldots,M} \left(\sum_{j=1}^{M} |a_{ij}|^{p'} \right)^{1/p'},$$

where $1/p + 1/p' = 1$.

From the analogous result for A^{-1}, we obtain

Corollary 1.10. *If A is nonsingular then*

$$d_{M-1}(\mathscr{A}_p; l_1^M) = \min_{i=1,\ldots,M} \frac{|\det(A)|}{\left(\sum_{j=1}^{M} \left| A\begin{pmatrix} 1,\ldots,\hat{i},\ldots,M \\ 1,\ldots,\hat{j},\ldots,M \end{pmatrix} \right|^p \right)^{1/p}}$$

and

$$d_{M-1}(\mathscr{A}_\infty; l_q^M) = \min_{j=1,\ldots,M} \frac{|\det(A)|}{\left(\sum_{i=1}^{M} \left| A\begin{pmatrix} 1,\ldots,\hat{i},\ldots,M \\ 1,\ldots,\hat{j},\ldots,M \end{pmatrix} \right|^{q'} \right)^{1/q'}},$$

where $1/q + 1/q' = 1$, and $A\begin{pmatrix} 1,\ldots,\hat{i},\ldots,M \\ 1,\ldots,\hat{j},\ldots,M \end{pmatrix}$ denotes the determinant of the $(M-1) \times (M-1)$ submatrix of A obtained by deleting the ith row and jth column.

In fact it can also be shown that there exists an optimal subspace for $d_{M-1}(\mathscr{A}_\infty; l_q^M)$ which is simply the span of some $M-1$ columns of A. The choice of the $M-1$ columns does however depend on q.

In Section 2 we deal with the n-widths of diagonal matrices. In particular, in Subsection 2.1 we exactly evaluate $d_n(\mathscr{D}_p; l_q^M)$ for $q \leqq p$, and $p = 1$, $q = 2$, where

$$\mathscr{D}_p = \{D\mathbf{x} \colon \|\mathbf{x}\|_p \leqq 1\}$$

and D is an $M \times M$ diagonal matrix. In Subsection 2.2 we give estimates for $d_n(\mathscr{D}_1; l_\infty^M)$ (and $d_n(\mathscr{D}_2; l_\infty^M)$) where D is the identity matrix. These results will be used again in Chapter VII.

In Section 3 we calculate the n-widths of $\mathscr{A}_\infty$ in l_∞^M where A is a strictly totally positive matrix. It is shown that in this case there exist n columns of A which span an optimal subspace for $d_n(\mathscr{A}_\infty; l_\infty^M)$.

2. n-Widths of Diagonal Matrices

Let $D = \operatorname{diag}\{D_1, \ldots, D_M\}$ be an $M \times M$ real diagonal matrix. We can and will assume, without loss of generality, that

$$D_1 \geqq D_2 \geqq \ldots \geqq D_M > 0.$$

Set

$$\mathscr{D}_p = \{D\mathbf{x}\colon \|\mathbf{x}\|_p \leqq 1\}.$$

In this simple case it is nonetheless true that we do not know (except for $n = 0$ and $n = M - 1$) how to calculate the n-widths of $\mathscr{D}_p$ in l_q^M for all choices of $p, q \in [1, \infty]$. The known cases are presented in this first subsection.

2.1 The Exact Solution for $q \leqq p$ and $p = 1, q = 2$

It is only when $p = q$ that all the n-widths d_n, d^n, δ_n and b_n are known for all n. Let $\mathbf{e}^i$ denote the ith unit vector in $\mathbb{R}^M$. Then

Theorem 2.1. *For* $1 \leqq p \leqq \infty$,

$$\delta_n(\mathscr{D}_p; l_p^M) = d_n(\mathscr{D}_p; l_p^M) = d^n(\mathscr{D}_p; l_p^M) = b_n(\mathscr{D}_p; l_p^M) = D_{n+1}.$$

Furthermore,

(1) $X_n = \operatorname{span}\{\mathbf{e}^1, \ldots, \mathbf{e}^n\}$ *is optimal for* $d_n(\mathscr{D}_p; l_p^M)$.
(2) $P_n = \operatorname{diag}\{D_1, \ldots, D_n, 0, \ldots, 0\}$ *is optimal for* $\delta_n(\mathscr{D}_p; l_p^M)$.
(3) $L^n = \{\mathbf{x}\colon x_i = 0,\ i = 1, \ldots, n\}$ *is optimal for* $d^n(\mathscr{D}_p; l_p^M)$.
(4) X_{n+1} *is optimal for* $b_n(\mathscr{D}_p; l_p^M)$.

Proof. P_n as above is a matrix of rank n. Therefore

$$\begin{aligned}
\delta_n(\mathscr{D}_p; l_p^M) &\leqq \max_{\|\mathbf{x}\|_p \leqq 1} \|D\mathbf{x} - P_n\mathbf{x}\|_p \\
&= \max_{\mathbf{x} \neq \mathbf{0}} \frac{\left(\sum_{i=n+1}^{M} |D_i x_i|^p\right)^{1/p}}{\left(\sum_{i=1}^{M} |x_i|^p\right)^{1/p}} \\
&\leqq D_{n+1} \max_{\mathbf{x} \neq \mathbf{0}} \frac{\left(\sum_{i=n+1}^{M} |x_i|^p\right)^{1/p}}{\left(\sum_{i=1}^{M} |x_i|^p\right)^{1/p}} \\
&= D_{n+1}
\end{aligned}$$

since $D_{n+1} \geqq D_{n+2} \geqq \ldots \geqq D_M > 0$.

Now from Propositions 1.1 (a) and 1.2,

$$\begin{aligned} b_n(\mathscr{D}_p; l_p^M) &= [d^{M-n-1}(\mathscr{D}_p^{-1}; l_p^M)]^{-1} \\ &\geqq [\delta_{M-n-1}(\mathscr{D}_p^{-1}; l_p^M)]^{-1}. \end{aligned}$$

The above argument implies that

$$\delta_{M-n-1}(\mathscr{D}_p^{-1}; l_p^M) \leqq 1/D_{n+1}.$$

Thus,

$$b_n(\mathscr{D}_p; l_p^M) \geqq D_{n+1}.$$

Again by Proposition 1.1 (a) this proves that all four n-widths equal D_{n+1}. The verification of the optimality as given in (1)–(4) is left to the reader. □

Theorem 2.2. *Given* $1 \leqq q \leqq p \leqq \infty$. *Let* $1/r = 1/q - 1/p$. *Then*

$$\delta_n(\mathscr{D}_p; l_q^M) = d_n(\mathscr{D}_p; l_q^M) = d^n(\mathscr{D}_p; l_q^M) = \left(\sum_{k=n+1}^{M} D_k^r\right)^{1/r}.$$

Furthermore X_n, P_n *and* L^n *of Theorem* 2.1 *are optimal here as well.*

Note that no statement is made concerning $b_n(\mathscr{D}_p; l_q^M)$. If equality were to hold for b_n then we would, by Propositions 1.1 and 1.2, be able to calculate d_n and d^n for all p and q.

In the proof of Theorem 2.2 we utilize the following two lemmas.

Lemma 2.3. *Let* X_n *be any* n*-dimensional subspace of* $\mathbb{R}^M$, $n < M$. *There exists an* $\mathbf{x} \in \mathbb{R}^M$ *for which* $\|\mathbf{x}\|_\infty = 1$, $\mathbf{x} \perp X_n$, *and at most* n *components of* $\mathbf{x}$ *are not equal to one in absolute value.*

Proof. Set $E = \{\mathbf{x}: \|\mathbf{x}\|_\infty = 1, \mathbf{x} \perp X_n\}$. E is a closed, convex, nonempty subset of $\mathbb{R}^M$. As such it has extreme points. Let $\mathbf{x}^*$ be an extreme point of E. If $\mathbf{x}^*$ does not satisfy the hypotheses of the lemma then there exist $n+1$ distinct integers $\{i_k\}_{k=1}^{n+1}$ of $\{1, \ldots, M\}$ for which $|(\mathbf{x}^*)_{i_k}| < 1$, $k = 1, \ldots, n+1$. Set $G = \operatorname{span}\{\mathbf{e}^{i_1}, \ldots, \mathbf{e}^{i_{n+1}}\}$. There exists a $\mathbf{y} \in G$, $\mathbf{y} \neq \mathbf{0}$ for which $\mathbf{y} \perp X_n$. Thus for all δ small, $\mathbf{x}^* \pm \delta \mathbf{y} \in E$, contradicting the extreme point property of $\mathbf{x}^*$. □

Lemma 2.4. *Let* $1 \leqq s < r \leqq \infty$, $a_1, \ldots, a_{k+1} > 0$, *and* $b_j \geqq b_{k+1} > 0$, $j = 1, \ldots, k$. *Then*

$$\frac{\left(\sum_{j=1}^{k+1} b_j^s a_j\right)^{1/s}}{\left(\sum_{j=1}^{k+1} b_j^r a_j\right)^{1/r}} \geqq \frac{\left(\sum_{j=1}^{k} b_j^s a_j\right)^{1/s}}{\left(\sum_{j=1}^{k} b_j^r a_j\right)^{1/r}}.$$

Proof. If $r = \infty$, then the lemma easily follows. Assume $r < \infty$. Set $\alpha^r = \sum_{j=1}^{k} b_j^r a_j$ and $\beta^s = \sum_{j=1}^{k} b_j^s a_j$. Since $b_j \geqq b_{k+1} > 0$ for all j, and $s < r$,

$$(b_j/b_{k+1})^s \leqq (b_j/b_{k+1})^r, \quad j = 1, \ldots, k$$

and thus

$$(\beta/b_{k+1})^s \leqq (\alpha/b_{k+1})^r$$

which implies that

$$[1+(b_{k+1}^s a_{k+1}/\beta^s)]^{r/s} \geqq [1+(b_{k+1}^s a_{k+1}/\beta^s)] \geqq [1+(b_{k+1}^r a_{k+1}/\alpha^r)].$$

Therefore

$$\frac{\left(\sum_{j=1}^{k+1} b_j^s a_j\right)^{1/s}}{\left(\sum_{j=1}^{k+1} b_j^r a_j\right)^{1/r}} = \frac{(\beta^s + b_{k+1}^s a_{k+1})^{1/s}}{(\alpha^r + b_{k+1}^r a_{k+1})^{1/r}} = \frac{\beta[1+(b_{k+1}^s a_{k+1}/\beta^s)]^{1/s}}{\alpha[1+(b_{k+1}^r a_{k+1}/\alpha^r)]^{1/r}}$$

$$\geqq \frac{\beta}{\alpha}$$

$$= \frac{\left(\sum_{j=1}^{k} b_j^s a_j\right)^{1/s}}{\left(\sum_{j=1}^{k} b_j^r a_j\right)^{1/r}}. \quad \square$$

Proof of Theorem 2.2. Since $1/r = 1/q - 1/p = 1/p' - 1/q'$, it suffices by Proposition 1.1, to prove that $\delta_n(\mathscr{D}_p; l_q^M) \leqq \left(\sum_{k=n+1}^{M} D_k^r\right)^{1/r}$, and $d^n(\mathscr{D}_p; l_q^M) \geqq \left(\sum_{k=n+1}^{M} D_k^r\right)^{1/r}$.

Let P_n be as above. Then

$$\delta_n(\mathscr{D}_p; l_q^M) \leqq \max_{\mathbf{x} \neq \mathbf{0}} \frac{\|D\mathbf{x} - P_n x\|_q}{\|\mathbf{x}\|_p} = \max_{\mathbf{x} \neq \mathbf{0}} \frac{\left(\sum_{k=n+1}^{M} |D_k x_k|^q\right)^{1/q}}{\left(\sum_{k=1}^{M} |x_k|^p\right)^{1/p}}.$$

By Hölder's inequality ($1/q = 1/r + 1/p$)

$$\left(\sum_{k=n+1}^{M} |D_k x_k|^q\right)^{1/q} \leqq \left(\sum_{k=n+1}^{M} D_k^r\right)^{1/r} \left(\sum_{k=n+1}^{M} |x_k|^p\right)^{1/p}$$

and therefore $\delta_n(\mathscr{D}_p; l_q^M) \leqq \left(\sum_{k=n+1}^{M} D_k^r\right)^{1/r}$.

Now

$$d^n(\mathscr{D}_p; l_q^M) = \min_{X_n} \max_{\mathbf{x} \perp X_n,\, \mathbf{x} \neq \mathbf{0}} \frac{\|D\mathbf{x}\|_q}{\|\mathbf{x}\|_p}$$

$$= \min_{X_n} \max_{\mathbf{x} \perp X_n,\, \mathbf{x} \neq \mathbf{0}} \frac{\left(\sum_{k=1}^{M} |D_k x_k|^q\right)^{1/q}}{\left(\sum_{k=1}^{M} |x_k|^p\right)^{1/p}}.$$

Set $z_k = x_k D_k^{-q/(p-q)}$ and note that $pq/(p-q) = r$ and $D_k \neq 0, k = 1,\ldots, M$. Thus

$$d^n(\mathscr{D}_p; l_q^M) = \min_{X_n} \max_{\mathbf{z} \perp X_n,\, \mathbf{z} \neq \mathbf{0}} \frac{\left(\sum_{k=1}^{M} |z_k|^q D_k^r\right)^{1/q}}{\left(\sum_{k=1}^{M} |z_k|^p D_k^r\right)^{1/p}}.$$

Applying Lemmas 2.3 and 2.4 we obtain

$$d^n(\mathscr{D}_p; l_q^M) \geqq \frac{\left(\sum_{k=1}^{M-n} D_{i_k}^r\right)^{1/q}}{\left(\sum_{k=1}^{M-n} D_{i_k}^r\right)^{1/p}} = \left(\sum_{k=1}^{M-n} D_{i_k}^r\right)^{1/r}$$

for some $1 \leqq i_1 < \ldots < i_{M-n} \leqq M$. Since $D_1 \geqq \ldots \geqq D_M > 0$,

$$d^n(\mathscr{D}_p; l_q^M) \geqq \left(\sum_{k=n+1}^{M} D_k^r\right)^{1/r}.$$

This proves the theorem. $\square$

As a consequence of Propositions 1.1 and 1.2 and Theorem 2.2, we have

Proposition 2.5. *For* $1 \leqq p < q \leqq \infty$,

$$\delta_n(\mathscr{D}_p; l_q^M) \geqq d_n(\mathscr{D}_p; l_q^M), \quad d^n(\mathscr{D}_p; l_q^M) \geqq b_n(\mathscr{D}_p; l_q^M) = \left(\sum_{k=1}^{n+1} D_k^r\right)^{1/r}$$

where $1/r = 1/q - 1/p (< 0)$, *and* span $\{\mathbf{e}^1, \ldots, \mathbf{e}^{n+1}\}$ *is optimal for* $b_n(\mathscr{D}_p; l_q^M)$.

From Corollaries 1.7 and 1.8 equality holds for $n = 0$ and $n = M - 1$. If the D_k decrease sufficiently rapidly one can also obtain equality for the other n-widths.

Proposition 2.6. *If* $1 \leqq p < q \leqq \infty$, $1/r = 1/q - 1/p$, *and* $D_{n+2} \leqq \left(\sum_{k=1}^{n+1} D_k^r\right)^{1/r}$, *then*

$$\delta_n(\mathscr{D}_p; l_q^M) = d_n(\mathscr{D}_p; l_q^M) = d^n(\mathscr{D}_p; l_q^M) = \left(\sum_{k=1}^{n+1} D_k^r\right)^{1/r}.$$

Proof. From Propositions 1.1 and 2.5, it suffices to prove $\delta_n(\mathscr{D}_p; l_q^M) \leqq \left(\sum_{k=1}^{n+1} D_k^r\right)^{1/r}$.

We first construct the optimal P_n^*. Set $x_k^* = D_k^{r/p}$, $k = 1, \ldots, n+1$. Note that $\mathbf{x}^* = (x_1^*, \ldots, x_{n+1}^*)$ is a strictly positive vector and

$$\left(\sum_{k=1}^{n+1} (x_k^*)^p\right)^{1/p} = \left(\sum_{k=1}^{n+1} D_k^r\right)^{-1/r} \left(\sum_{k=1}^{n+1} (D_k x_k^*)^q\right)^{1/q}$$

(i.e., equality in Hölder's inequality). For $1 \leqq p < \infty$, (p cannot be infinite) choose any n linearly independent vectors $\{\mathbf{x}^i\}_{i=1}^n$ in $\mathbb{R}^{n+1}$ to satisfy

$$\sum_{k=1}^{n+1} (x_k^*)^{p-1} x_k^i = 0, \qquad i = 1, \ldots, n.$$

The choice of the $\{\mathbf{x}^i\}_{i=1}^n$ is such that

$$\|\mathbf{x}^*\|_p = \min_{a_1, \ldots, a_n} \left\| \mathbf{x}^* - \sum_{i=1}^n a_i \mathbf{x}^i \right\|_p.$$

Now $\{\mathbf{x}^1, \ldots, \mathbf{x}^n, \mathbf{x}^*\}$ forms a basis for $\mathbb{R}^{n+1}$. Let P^* be the $(n+1) \times (n+1)$ matrix defined by

$$P^* \mathbf{x}^i = D \mathbf{x}^i, \quad i = 1, \ldots, n, \quad \text{and} \quad P^* \mathbf{x}^* = \mathbf{0}.$$

Since $\mathbf{x}^* \neq \mathbf{0}$, rank $P^* \leqq n$. P_n^* is defined as the $M \times M$ matrix whose submatrix composed of the first $n+1$ rows and columns is P^*, and which is zero elsewhere. Thus rank $P_n^* \leqq n$.

We will prove that

$$\max_{\mathbf{x} \neq \mathbf{0}} \frac{\|D\mathbf{x} - P_n^* \mathbf{x}\|_q}{\|\mathbf{x}\|_p} \leqq \left(\sum_{k=1}^{n+1} D_k^r \right)^{1/r}.$$

Let $\mathbf{x}^*$ and $\{\mathbf{x}^i\}_{i=1}^n$ in $\mathbb{R}^M$ be as above with added zero entries. Every $\mathbf{x} \in \mathbb{R}^M$ may be written in the form

$$\mathbf{x} = b^* \mathbf{x}^* + \sum_{i=1}^n b_i \mathbf{x}^i + \sum_{k=n+2}^M x_k \mathbf{e}^k,$$

and $P_n^* \mathbf{x} = \sum_{i=1}^n b_i D \mathbf{x}^i$.

If $b^* = 0$, then

$$\frac{\|D\mathbf{x} - P_n^* \mathbf{x}\|_q}{\|\mathbf{x}\|_p} = \frac{\left(\sum_{k=n+2}^M |D_k x_k|^q \right)^{1/q}}{\left(\sum_{k=1}^M |x_k|^p \right)^{1/p}}$$

$$\leqq D_{n+2}$$

$$\leqq \left(\sum_{k=1}^{n+1} D_k^r \right)^{1/r}.$$

If $b^* \neq 0$, then we may assume $b^* = 1$, and

$$\frac{\|D\mathbf{x} - P_n^*\mathbf{x}\|_q}{\|\mathbf{x}\|_p} = \frac{\left(\|D\mathbf{x}^*\|_q^q + \sum_{k=n+2}^{M} |D_k x_k|^q\right)^{1/q}}{\left(\left\|\mathbf{x}^* + \sum_{i=1}^{n} b_i \mathbf{x}^i\right\|_p^p + \sum_{k=n+2}^{M} |x_k|^p\right)^{1/p}}.$$

The vector $\mathbf{x}^*$ satisfies

$$\|D\mathbf{x}^*\|_q = \|\mathbf{x}^*\|_p \left(\sum_{k=1}^{n+1} D_k^r\right)^{1/r}.$$

Therefore

$$\frac{\|D\mathbf{x} - P_n^*\mathbf{x}\|_q}{\|\mathbf{x}\|_p} \leqq \frac{\left(\|\mathbf{x}^*\|_p^q \left(\sum_{k=1}^{n+1} D_k^r\right)^{q/r} + \sum_{k=n+2}^{M} |D_k x_k|^q\right)^{1/q}}{\left(\|\mathbf{x}^*\|_p^p + \sum_{k=n+2}^{M} |x_k|^p\right)^{1/p}}.$$

Since $1 \leqq p < q \leqq \infty$, the above quantity is bounded above by

$$\max\left\{\left(\sum_{k=1}^{n+1} D_k^r\right)^{1/r}, D_{n+2}, \ldots, D_M\right\} = \left(\sum_{k=1}^{n+1} D_k^r\right)^{1/r}.$$

The proposition is proved. □

The proposition is exact in that there exist $p < q$ for which $d_n(\mathscr{D}_p; l_q^M) = \left(\sum_{k=1}^{n+1} D_k^r\right)^{1/r}$ if and only if $D_{n+2} \leqq \left(\sum_{k=1}^{n+1} D_k^r\right)^{1/r}$ (see the next theorem). If $\left(\sum_{k=1}^{n+1} D_k^r\right)^{1/r} \leqq D_{n+2}$, then we have proven that $\delta_n(\mathscr{D}_p; l_q^M) \leqq D_{n+2}$. This compares favourably with the upper bound D_{n+1} which is obtained with the choice $P_n = \operatorname{diag}\{D_1, \ldots, D_n, 0, \ldots, 0\}$.

If $p = 1$, $q = 2$ then we can calculate d_n and δ_n, but not d^n.

Theorem 2.7.

$$\delta_n(\mathscr{D}_1; l_2^M) = d_n(\mathscr{D}_1; l_2^M) = \max_{n < r \leqq M} \left(\frac{r - n}{\sum_{k=1}^{r} D_k^{-2}}\right)^{1/2}.$$

The following lemma lies at the core of the proof of the theorem.

Lemma 2.8. *For given A_k, $k = 1, \ldots, M$, satisfying $0 \leqq A_k \leqq 1$, $\sum_{k=1}^{M} A_k = n$, there exist n orthonormal vectors $\mathbf{x}^i \in \mathbb{R}^M$, $i = 1, \ldots, n$ for which $A_k = \sum_{i=1}^{n} |(\mathbf{x}^i)_k|^2$, $k = 1, \ldots, M$.*

Proof. The proof is by induction on M. For $M = n$ it suffices to take any set of orthonormal vectors in $\mathbb{R}^M$ since then $A_k = \sum_{i=1}^{M} |(\mathbf{x}^i)_k|^2 = 1$, $k = 1, \ldots, M$.

Assume that $M > n$ and that given any $\{A_k\}_{k=1}^{M-1}$ satisfying $0 \leqq A_k \leqq 1$, $\sum_{k=1}^{M-1} A_k = n$, there exist $\mathbf{x}^i \in \mathbb{R}^{M-1}$, $i = 1,\ldots,n$, orthonormal and such that $A_k = \sum_{i=1}^{n} |(\mathbf{x}^i)_k|^2$, $k = 1,\ldots, M-1$. To advance the induction let $\{B_k\}_{k=1}^{M}$ be given for which $0 \leqq B_k \leqq 1$, $\sum_{k=1}^{M} B_k = n$. Assume, without loss, that $B_M = \min_{k=1,\ldots,M} B_k$. Let $\{A_k\}_{k=1}^{M-1}$ be any sequence for which $0 \leqq B_k \leqq A_k \leqq 1$, $k = 1,\ldots, M-1$, and $\sum_{k=1}^{M-1} A_k = n$. Set $A_M = 0$. By the induction hypothesis there exist $\{\mathbf{y}^i\}_{i=1}^{n}$ in $\mathbb{R}^M$ which are orthonormal and satisfy $(\mathbf{y}^i)_M = 0$ and $A_k = \sum_{i=1}^{n} |(\mathbf{y}^i)_k|^2$, $k = 1,\ldots, M$. The idea of the proof is to go from the sequence $(A_1,\ldots, A_M)$ to the sequence $(B_1,\ldots, B_M)$ where at the jth step we decrease A_j to B_j while increasing the previous value of A_M, and where at each step we maintain an orthonormal set of n vectors.

The required transformation is the following: Given any vector $\mathbf{x} \in \mathbb{R}^M$, let $\mathbf{x}[t;j] \in \mathbb{R}^M$, where

$$(\mathbf{x}[t;j])_i = \begin{cases} x_i & i \neq j, M \\ x_j \cos t + x_M \sin t, & i = j \\ -x_j \sin t + x_M \cos t, & i = M. \end{cases}$$

If the $\{\mathbf{x}^i\}_{i=1}^{n}$ are orthonormal then so are the $\{\mathbf{x}^i[t;j]\}_{i=1}^{n}$. Furthermore if we set

$$A_k(t) = \sum_{i=1}^{n} |(\mathbf{y}^i[t;j])_k|^2, \qquad k = 1,\ldots, M,$$

then for all t, $A_k(t) = A_k(0)$, $k \neq j, M$, $\sum_{k=1}^{M} A_k(t) = n$, and $0 \leqq A_k(t) \leqq 1$, $k = 1,\ldots, M$. Thus $A_j(t) + A_M(t) = A_j(0) + A_M(0)$ for all t, and since $A_j(\pi/2) = A_M(0)$, $(A_M(\pi/2) = A_j(0))$, there exists a t_0 for which $A_j(t_0) = B_j$. Note that we are using the fact that $B_M = \min_{k=1,\ldots,M} B_k$, which implies that as we go from $(A_1,\ldots, A_M)$ to $(B_1,\ldots, B_M)$ the minimality of the last term is preserved at each step. This proves the lemma. □

We need the following two lemmas which are technical in nature.

Lemma 2.9. *If* $(r-n)\Big/\left(\sum_{k=1}^{r} D_k^{-2}\right) \geqq D_{r+1}^2$ *for some* r, $r > n$, *then for all* $j > r$,

$$(r-n)\Big/\left(\sum_{k=1}^{r} D_k^{-2}\right) \geqq (j-n)\Big/\left(\sum_{k=1}^{j} D_k^{-2}\right).$$

Proof. $(r-n)\Big/\left(\sum_{k=1}^{r} D_k^{-2}\right) \geqq D_{r+1}^2$ is equivalent to $(r-n)/D_{r+1}^2 \geqq \sum_{k=1}^{r} D_k^{-2}$. Multiplying both sides by $j-r$ and using the fact that $D_{r+1} \geqq \cdots \geqq D_M > 0$, we obtain $(r-n)\sum_{k=r+1}^{j} D_k^{-2} \geqq (j-r)\sum_{k=1}^{r} D_k^{-2}$, which is equivalent to $(r-n)\sum_{k=1}^{j} D_k^{-2} \geqq (j-n)\sum_{k=1}^{r} D_k^{-2}$. □

This same reasoning may be used to prove

Lemma 2.10. *For* $n<r$, $(r-n)\bigg/\left(\sum_{k=1}^{r} D_k^{-2}\right) \leqq D_{r+1}^2$ *if and only if* $(r-n)\bigg/\left(\sum_{k=1}^{r} D_k^{-2}\right) \leqq (r+1-n)\bigg/\left(\sum_{k=1}^{r+1} D_k^{-2}\right)$.

Proof of Theorem 2.7. From Proposition 1.1, (d) or (e), $\delta_n(\mathscr{D}_1; l_2^M) = d_n(\mathscr{D}_1; l_2^M)$. Let X_n be any n-dimensional subspace of $\mathbb{R}^M$, and let $\mathbf{x}^1, \ldots, \mathbf{x}^n$ be an orthonormal basis for X_n. Set

$$\rho_k(X_n) = \min_{\mathbf{y}\in X_n} \| D_k \mathbf{e}^k - \mathbf{y} \|_2, \qquad k=1,\ldots,M.$$

Since $\{\pm \mathbf{e}^k\}_{k=1}^M$ are the extreme points of l_1^M,

$$\begin{aligned} d_n(\mathscr{D}_1; l_2^M) &= \min_{X_n} \max_{k=1,\ldots,M} \min_{\mathbf{y}\in X_n} \| D_k \mathbf{e}^k - \mathbf{y} \|_2 \\ &= \min_{X_n} \max_{k=1,\ldots,M} \rho_k(X_n). \end{aligned}$$

Because the $\{\mathbf{x}^i\}_{i=1}^n$ are orthonormal,

$$\begin{aligned} \rho_k(X_n) &= \min_{a_1,\ldots,a_n} \left\| D_k \mathbf{e}^k - \sum_{i=1}^{n} a_i \mathbf{x}^i \right\|_2 \\ &= D_k \left[1 - \sum_{i=1}^{n} |(\mathbf{x}^i)_k|^2 \right]^{1/2}. \end{aligned}$$

Set $A_k = \sum_{i=1}^{n} |(\mathbf{x}^i)_k|^2$. Lemma 2.8 implies that varying over all n-dimensional subspaces is equivalent to varying over all $(A_1, \ldots, A_M)$ satisfying $0 \leqq A_k \leqq 1$, $\sum_{k=1}^{M} A_k = n$. Thus

$$\begin{aligned} d_n(\mathscr{D}_1; l_2^M) &= \min_{0\leqq A_k \leqq 1,\ \sum_{k=1}^{M} A_k = n} \ \max_{k=1,\ldots,M} D_k[1-A_k]^{1/2} \\ &= \min_{0\leqq C_k \leqq 1,\ \sum_{k=1}^{M} C_k = M-n} \ \max_{k=1,\ldots,M} D_k C_k^{1/2}. \end{aligned}$$

Let r be the smallest integer, $n+1 \leqq r \leqq M$ for which $(r-n)/\left(\sum_{k=1}^{r} D_k^{-2}\right) \geqq D_{r+1}^2$. (Put $D_{M+1}=0$ so that this is consistent.) Set

$$C_k^* = (r-n)\, D_k^{-2} \Big/ \left(\sum_{i=1}^{r} D_i^{-2}\right), \quad k=1,\ldots,r$$

and $C_k^*=1$, $k=r+1,\ldots,M$. Obviously $C_k^* \geqq 0$, and $\sum_{k=1}^{M} C_k^* = M-n$. To show that $\{C_k^*\}_{k=1}^M$ is admissable we must prove that $C_k^* \leqq 1$. Since $D_1 \geqq \ldots \geqq D_r$ it suffices to prove that $C_r^* \leqq 1$. Now $C_r^* \leqq 1$ if and only if $(r-n)\,D_r^{-2} \leqq \sum_{k=1}^{r} D_k^{-2}$ which is equivalent to $(r-1-n)\Big/\left(\sum_{k=1}^{r-1} D_k^{-2}\right) \leqq D_r^2$. This latter inequality is valid by our choice of r (and trivially so if $r=n+1$). The $\{C_k^*\}_{k=1}^M$ are admissable.

Since $D_k C_k^{*\,1/2} = \left((r-n)\Big/\sum_{k=1}^{r} D_k^{-2}\right)^{1/2}$, for $k=1,\ldots,r$ and $D_{r+1} \leqq \left((r-n)\Big/\sum_{k=1}^{r} D_k^{-2}\right)^{1/2}$, it follows that

$$d_n(\mathscr{D}_1; l_2^M) \leqq \left((r-n)\Big/\sum_{k=1}^{r} D_k^{-2}\right)^{1/2}.$$

To prove the lower bound assume the existence of an n-dimensional subspace X_n for which $\rho_k(X_n) < \left((r-n)\Big/\sum_{k=1}^{r} D_k^{-2}\right)^{1/2}$ for all $k=1,\ldots,M$. Since $\sum_{k=1}^{M} \rho_k^2(X_n)/D_k^2 = M-n$, and $0 \leqq \rho_k^2(X_n)/D_k^2 \leqq 1$, we have $(r-n) \leqq \sum_{k=1}^{r} \rho_k^2(X_n)/D_k^2 < \left((r-n)\Big/\sum_{i=1}^{r} D_i^{-2}\right) \sum_{k=1}^{r} D_k^{-2} = (r-n)$. This contradiction implies that

$$d_n(\mathscr{D}_1; l_2^M) = \left((r-n)\Big/\sum_{k=1}^{r} D_k^{-2}\right)^{1/2}.$$

It remains to prove that

$$\left((r-n)\Big/\sum_{k=1}^{r} D_k^{-2}\right)^{1/2} = \max_{n<j\leqq M} \left((j-n)\Big/\sum_{k=1}^{j} D_k^{-2}\right)^{1/2}.$$

This is an immediate consequence of Lemmas 2.9 and 2.10. □

We shall need the above result in Chapter VII for the case $D=I$, the identity matrix. Set

$$\mathscr{I}_p = \{\mathbf{x}\colon \|\mathbf{x}\|_p \leqq 1\}.$$

Corollary 2.11.

$$d_n(\mathcal{I}_1; l_2^M) = \delta_n(\mathcal{I}_1; l_2^M) = ((M-n)/M)^{1/2}.$$

From Proposition 1.1, (b) and (c), we see that we have also calculated $\delta_n(\mathcal{D}_2; l_\infty^M)$ and $d^n(\mathcal{D}_2; l_\infty^M)$ (and from Proposition 1.2, $b_n(\mathcal{D}_\infty; l_2^M)$). It is not true that $\delta_n(\mathcal{D}_2; l_\infty^M) = d_n(\mathcal{D}_2; l_\infty^M)$ for all n and D. To prove this we calculate $d_1(\mathcal{D}_p; l_\infty^M)$.

Proposition 2.12. *For* $1 \leqq p < \infty$,

$$d_1(\mathcal{D}_p; l_\infty^M) = D_1 D_2/(D_1^p + D_2^p)^{1/p}.$$

This implies that $d_1(\mathcal{D}_2; l_\infty^M) = \delta_1(\mathcal{D}_2; l_\infty^M)$ only when the condition of Proposition 2.6 is satisfied for $n = 1$.

Proof. The lower bound is a consequence of Proposition 2.5. It remains to find an optimal subspace. Set $\mathbf{d} = (D_1^p, \ldots, D_M^p)$. Then

$$\begin{aligned} d_1(\mathcal{D}_p; l_\infty^M) &\leqq \max_{\|\mathbf{x}\|_p \leqq 1} \min_a \| D\mathbf{x} - a\mathbf{d} \|_\infty \\ &= \max_{\|\mathbf{x}\|_p \leqq 1} \min_a \max_{i=1,\ldots,M} |D_i x_i - a D_i^p|. \end{aligned}$$

It is a classical result that since $\mathbf{d}$ is a positive vector,

$$\begin{aligned} \min_a \max_{i=1,\ldots,M} |D_i x_i - a D_i^p| &= \max_{i,j} \left| \frac{\begin{vmatrix} D_i x_i & D_i^p \\ D_j x_j & D_j^p \end{vmatrix}}{D_i^p + D_j^p} \right| \\ &= \max_{i,j} \frac{|D_i D_j^p x_i - D_j D_i^p x_j|}{D_i^p + D_j^p}. \end{aligned}$$

Thus

$$\begin{aligned} d_1(\mathcal{D}_p; l_\infty^M) &\leqq \max_{\|\mathbf{x}\|_p \leqq 1} \max_{i,j} \frac{|D_i D_j^p x_i - D_j D_i^p x_j|}{D_i^p + D_j^p} \\ &= \max_{i,j} \max_{(x_i, x_j) \neq (0,0)} \frac{D_i D_j^p |x_i| + D_j D_i^p |x_j|}{(D_i^p + D_j^p)(|x_i|^p + |x_j|^p)^{1/p}}. \end{aligned}$$

It is now easy to prove that the maximum is attained for

$$x_i = D_j, \quad x_j = D_i,$$

so that

$$\begin{aligned} d_1(\mathcal{D}_p; l_\infty^M) &\leqq \max_{i,j} D_i D_j/(D_i^p + D_j^p)^{1/p} \\ &= D_1 D_2/(D_1^p + D_2^p)^{1/p}. \quad \square \end{aligned}$$

2.2 Various Estimates for $p = 1$, $q = \infty$

From Proposition 1.1,

$$d_n(\mathscr{I}_1; l_\infty^M) = d^n(\mathscr{I}_1; l_\infty^M) = \delta_n(\mathscr{I}_1; l_\infty^M).$$

This quantity may be expressed in a variety of ways. One particular way deserves special attention because of its simple and elegant form.

$$d_n(\mathscr{I}_1; l_\infty^M) = \min_{\operatorname{rank} P \leqq n} \max_{i,j} |\delta_{ij} - p_{ij}|,$$

where $P = (p_{ij})_{i,j=1}^M$. We shall estimate this n-width both from above and below.

From previous results the following facts are known:

$$d_0(\mathscr{I}_1; l_\infty^M) = 1, \quad d_1(\mathscr{I}_1; l_\infty^M) = 1/2, \quad d_{M-1}(\mathscr{I}_1; l_\infty^M) = 1/M,$$
$$d_n(\mathscr{I}_1; l_\infty^M) \geqq 1/(n+1)$$

and since $d_n(\mathscr{I}_1; l_2^M) = [(M-n)/M]^{1/2}$ it follows that $d_n(\mathscr{I}_1; l_\infty^M) \geqq (M-n)^{1/2}/M$.

Before entering into the analysis an important digression is in order. Until now we have considered l_p^M and l_q^M as subsets of $\mathbb{R}^M$. If we had considered them as subsets of $\mathbb{C}^M$ then there would have been, in what has been so far done, no essential change in the results. This fortunate situation is no longer valid when considering $d_n(\mathscr{I}_1; l_\infty^M)$. We shall therefore write $d_n^{\mathbb{R}}(\mathscr{I}_1; l_\infty^M)$ and $d_n^{\mathbb{C}}(\mathscr{I}_1; l_\infty^M)$ (and analogously $\delta_n^{\mathbb{R}}$, $\delta_n^{\mathbb{C}}$, $d^{n,\mathbb{R}}$ and $d^{n,\mathbb{C}}$) to indicate whether we allow the approximating subspaces (rank n matrices) to have coefficients only in $\mathbb{R}$ or also in $\mathbb{C}$. Note that it is immaterial whether the set

$$\mathscr{I}_1 = \{\mathbf{x}: \|\mathbf{x}\|_\infty \leqq 1\}$$

is taken over $\mathbf{x} \in \mathbb{R}^M$ or $\mathbf{x} \in \mathbb{C}^M$ in the calculation of d_n, d^n and δ_n. The difference arises when considering the approximating subspaces or the approximating rank n matrices. As such we have

$$d_n^{\mathbb{C}}(\mathscr{I}_1; l_\infty^M) \leqq d_n^{\mathbb{R}}(\mathscr{I}_1; l_\infty^M)$$

for all n and M.

We first consider lower bounds on $d_n^{\mathbb{C}}(\mathscr{I}_1; l_\infty^M)$. The above estimates give the lower bound

$$d_n^{\mathbb{C}}(\mathscr{I}_1; l_\infty^M) \geqq \max\{1/(n+1), (M-n)^{1/2}/M\}.$$

This bound is not very good. A better lower bound is the following.

Theorem 2.13.

$$d_n^{\mathbb{C}}(\mathscr{I}_1; l_\infty^M) \geqq [1/(1 + ((M-1)\,n/(M-n))^{1/2})].$$

Proof. The proof is based on the simple inequality

$$\sum_{\substack{k=1\\k\neq i}}^{M} |x_k| \leqq \left(\sum_{\substack{k=1\\k\neq i}}^{M} |x_k|^2\right)^{1/2} (M-1)^{1/2}$$

and the result of Theorem 2.7 (Corollary 2.11), namely

$$d_n(\mathcal{I}_1; l_2^M) = d^n(\mathcal{I}_2; l_\infty^M) = \min_{X_n} \max_{\mathbf{x} \perp X_n} \frac{\|\mathbf{x}\|_\infty}{\|\mathbf{x}\|_2} = ((M-n)/M)^{1/2},$$

(which is valid whether we run over $\mathbb{R}^M$ or $\mathbb{C}^M$).

Let $\mathbf{x} \in \mathbb{C}^M$. Then for some i,

$$\begin{aligned}\|\mathbf{x}\|_1 &= \|\mathbf{x}\|_\infty + \sum_{\substack{k=1\\k\neq i}}^{M} |x_k| \leqq \|\mathbf{x}\|_\infty + \left(\sum_{\substack{k=1\\k\neq i}}^{M} |x_k|^2\right)^{1/2} (M-1)^{1/2}\\ &= \|\mathbf{x}\|_\infty + (\|\mathbf{x}\|_2^2 - \|\mathbf{x}\|_\infty^2)^{1/2} (M-1)^{1/2}.\end{aligned}$$

Thus,

$$\begin{aligned}d_n^{\mathbb{C}}(\mathcal{I}_1; l_\infty^M) = d^{n,\mathbb{C}}(\mathcal{I}_1; l_\infty^M) &= \min_{X_n} \max_{\mathbf{x} \perp X_n} \frac{\|\mathbf{x}\|_\infty}{\|\mathbf{x}\|_1}\\ &\geqq \min_{X_n} \max_{\mathbf{x} \perp X_n} \frac{\|\mathbf{x}\|_\infty}{\|\mathbf{x}\|_\infty + (\|\mathbf{x}\|_2^2 - \|\mathbf{x}\|_\infty^2)^{1/2} (M-1)^{1/2}}\\ &= \min_{X_n} \max_{\mathbf{x} \perp X_n} \frac{1}{1 + \left(\dfrac{\|\mathbf{x}\|_2^2}{\|\mathbf{x}\|_\infty^2} - 1\right)^{1/2} (M-1)^{1/2}}\\ &= \frac{1}{1 + ([d^n(\mathcal{I}_2; l_\infty^M)]^{-2} - 1)^{1/2} (M-1)^{1/2}}\\ &= \frac{1}{1 + \left(\dfrac{M}{M-n} - 1\right)^{1/2} (M-1)^{1/2}}\\ &= [1 + ((M-1)\,n/(M-n))^{1/2}]^{-1}. \quad \square\end{aligned}$$

Remark. By totally analogous reasoning

$$d_n^{\mathbb{C}}(\mathcal{I}_1; l_q^M) \geqq [1 + (M-1)^{1-q'/2}((M/n)-1)^{-q'/2}]^{-1/q'}$$

for $2 \leqq q \leqq \infty$, where $1/q + 1/q' = 1$.

This lower bound is not optimal (or even asymptotically optimal) for all M and n. From Theorem 8.11 of Chapter II, $d_n^{\mathbb{R}}(\mathcal{I}_1; l_\infty) = 1/2$ for all $n \geqq 1$, while $\lim_{M\to\infty} [1 + ((M-1)\,n/(M-n))^{1/2}]^{-1} = [1 + n^{1/2}]^{-1}$ for fixed n.

We wish to determine possible n and M for which the above lower bound is attainable. The next few results are concerned with this problem.

Proposition 2.14. *The following are equivalent:*

(1) $d_n^{\mathbb{C}}(\mathcal{I}_1; l_\infty^M) = [1 + (n(M-1)/(M-n))^{1/2}]^{-1}$.

(2) $d_{M-n}^{\mathbb{C}}(\mathcal{I}_1; l_\infty^M) = [1 + ((M-n)(M-1)/n)^{1/2}]^{-1}$.

(3) *There exists a rank n matrix $P = (p_{ij})_{i,j=1}^M$ satisfying $|p_{ij}| = ((M-n)/n(M-1))^{1/2}$ for all $i \neq j$, and $p_{ii} = 1$.*

(4) *There exists a rank $M-n$ matrix $Q = (q_{ij})_{i,j=1}^M$ satisfying $|q_{ij}| = (n/(M-n)(M-1))^{1/2}$ for all $i \neq j$, and $q_{ii} = 1$.*

(5) *There exists an $M \times n$ matrix X such that $\bar{X}^T X = I$, and $P = (M/n) X \bar{X}^T$ is as in* (3).

(6) *There exists an $M \times (M-n)$ matrix Y such that $\bar{Y}^T Y = I$, and $Q = (M/(M-n)) Y \bar{Y}^T$ is as in* (4).

Similarly we have

Proposition 2.15. *The following are equivalent:*

(1) $d_n^{\mathbb{R}}(\mathcal{I}_1; l_\infty^M) = [1 + (n(M-1)/(M-n))^{1/2}]^{-1}$.

(2) $d_{M-n}^{\mathbb{R}}(\mathcal{I}_1; l_\infty^M) = [1 + ((M-n)(M-1)/n)^{1/2}]^{-1}$.

(3) *There exists a real rank n matrix $P = (p_{ij})_{i,j=1}^M$ satisfying $|p_{ij}| = ((M-n)/n(M-1))^{1/2}$, for all $i \neq j$, and $p_{ii} = 1$.*

(4) *There exists a real rank $M-n$ matrix $Q = (q_{ij})_{i,j=1}^M$, satisfying $|q_{ij}| = (n/(M-n)(M-1))^{1/2}$ for all $i \neq j$, and $q_{ii} = 1$.*

(5) *There exists a real $M \times n$ matrix X such that $X^T X = I$, and $P = (M/n) X X^T$ is as in* (3).

(6) *There exists a real $M \times (M-n)$ matrix Y such that $Y^T Y = I$, and $Q = (M/(M-n)) Y Y^T$ is as in* (4).

Propositions 2.14 and 2.15 are identical except that the word real has been inserted in the appropriate places in the statement of Proposition 2.15. This is also true of the proofs. We therefore prove only Proposition 2.14.

Proof of Proposition 2.14. It is easily seen that (3) implies (1) and (4) implies (2). For example, if P satisfies (3), then

$$d_n^{\mathbb{C}}(\mathcal{I}_1; l_\infty^M) \leqq \max_{i,j=1,\ldots,M} |\delta_{ij} - \gamma p_{ij}|$$

for any choice of γ. The optimal choice is $\gamma = [1 + ((M-n)/n(M-1))^{1/2}]^{-1}$. With this choice it follows that $d_n^{\mathbb{C}}(\mathcal{I}_1; l_\infty^M) \leqq [1 + ((M-1)n/(M-n))^{1/2}]^{-1}$, which together with Theorem 2.13 proves (1). Similarly one proves that (4) implies (2). We shall prove that (1) implies (4). Exchanging n and $M-n$, it will follow that (2) implies (3). This will prove the equivalence of (1), (2), (3) and (4).

From the proof of Theorem 2.13 it may be seen that in order for (1) to hold it is necessary that two conditions obtain. The first condition is that there exists an X_n optimal for $d_n(\mathcal{I}_1; l_2^M)$ which is also optimal for $d_n^{\mathbb{C}}(\mathcal{I}_1; l_\infty^M)$. The second condition is that if $\min\{\|\mathbf{e}^k - \mathbf{x}\|_2 : \mathbf{x} \in X_n\} = ((M-n)/M)^{1/2} = d_n(\mathcal{I}_1; l_2^M)$, then there exists a $\mathbf{q}^k \perp X_n$ for which $|(\mathbf{q}^k)_i|$ is a constant for $i \neq k$, and

$\|\mathbf{q}^k\|_\infty/\|\mathbf{q}^k\|_2 = ((M-n)/M)^{1/2}$. (This second condition comes from equality in the simple form of Hölder's inequality as used in the proof of Theorem 2.13.)

If X_n is optimal for $d_n(\mathscr{I}_1; l_2^M)$ then it is of exact dimension n and

$$\max_{k=1,\dots,M} \min_{\mathbf{x}\in X_n} \|\mathbf{e}^k - \mathbf{x}\|_2 = ((M-n)/M)^{1/2}.$$

In the proof of Theorem 2.7 we proved that $\sum_{k=1}^{M} (\min\{\|\mathbf{e}^k - \mathbf{x}\|_2^2 : \mathbf{x}\in X_n\}) = M-n$, which implies that $\min\{\|\mathbf{e}^k - \mathbf{x}\|_2 : \mathbf{x}\in X_n\} = ((M-n)/M)^{1/2}$ for all $k = 1,\dots, M$. For each $k = 1,\dots, M$, let $\mathbf{q}^k$ be as above and normalized so that $(\mathbf{q}^k)_k = 1$. (This is possible.) Since $\|\mathbf{q}^k\|_\infty/\|\mathbf{q}^k\|_2 = ((M-n)/M)^{1/2}$, it follows that $|(\mathbf{q}^k)_i| = (n/(M-n)(M-1))^{1/2}$ for all $i \neq k$. Let Q denote the $M \times M$ matrix with kth column equal to $\mathbf{q}^k$. Since X_n is n-dimensional and $\mathbf{q}^k \perp X_n$, $k = 1,\dots, M$, it follows that Q has rank at most $M-n$. Since the span of the columns of Q is optimal for $d_{M-n}^{\mathbb{C}}(\mathscr{I}_1; l_\infty^M)$ and thus for $d_{M-n}(\mathscr{I}_1; l_2^M)$ (from the proof of (4) implies (2) and the above) it follows that Q is of exact rank $M-n$. This proves (4).

Obviously (5) implies (3) and (6) implies (4). To complete the proof we will show that (3) implies (5). Let P be as in (3). Let $\mathbf{p}^k$ denote the kth column of P, and X_n the n-dimensional subspace spanned by the columns of P. Since X_n is optimal for $d_n^{\mathbb{C}}(\mathscr{I}_1; l_\infty^M)$, it is also optimal for $d_n(\mathscr{I}_1; l_2^M)$. In fact a simple calculation shows that $\|\mathbf{e}^k - (n/M)\,\mathbf{p}^k\|_2 = ((M-n)/M)^{1/2}$ for $k = 1,\dots, M$. Thus $(n/M)\,\mathbf{p}^k$ is the best approximation to $\mathbf{e}^k$ from X_n in l_2^M. Let $\{\mathbf{x}^1,\dots,\mathbf{x}^n\}$ be an orthonormal basis for X_n, and let X be the $M \times n$ matrix with kth column equal to $\mathbf{x}^k$. Thus $\bar{X}^T X = I$. The best l_2^M-approximation to $\mathbf{e}^k$ from X_n is given by $\sum_{i=1}^{n} (\mathbf{e}^k, \mathbf{x}^i)\,\mathbf{x}^i$. Thus

$$(n/M)\,\mathbf{p}^k = \sum_{i=1}^{n} (\mathbf{e}^k, \mathbf{x}^i)\,\mathbf{x}^i = \sum_{i=1}^{n} (\bar{\mathbf{x}}^i)_k\,\mathbf{x}^i.$$

From here it is easily seen that $(n/M)\,P = X\bar{X}^T$. This proves the proposition. □

Let X be as above in (5). Let $\mathbf{z}^1,\dots,\mathbf{z}^M$ denote the row vectors of X (vectors in $\mathbb{R}^n$ or $\mathbb{C}^n$) and set $\mathbf{y}^i = (M/n)^{1/2}\,\mathbf{z}^i$, $i = 1,\dots, M$. Thus $(\mathbf{y}^i, \mathbf{y}^i) = 1$ and $|(\mathbf{y}^i, \mathbf{y}^j)| = ((M-n)/n(M-1))^{1/2}$ for $i \neq j$. We have M vectors on the unit sphere of $\mathbb{R}^n$ or $\mathbb{C}^n$ with inner products constant in absolute value. The existence of sets of vectors of the above form is a topic (called equiangular lines) which has been investigated in graph theory and combinatorics. (Theorem 2.7 or equivalently Theorem 2.13 implies that there cannot exist M vectors $\mathbf{y}^1,\dots,\mathbf{y}^M$ on the unit sphere of $\mathbb{R}^n$ or $\mathbb{C}^n$ for which $|(\mathbf{y}^i, \mathbf{y}^j)| < ((M-n)/n(M-1))^{1/2}$ for all $i \neq j$. This is a very indirect proof of this result.) As a consequence of the investigations of this latter topic we have the following result:

Theorem 2.16.

(1) *If* $d_n^{\mathbb{C}}(\mathscr{I}_1; l_\infty^M) = [1 + ((M-1)\,n/(M-n))^{1/2}]^{-1}$, *and* $1 < n < M-1$, *then*

$$M \leq \min\{n^2, (M-n)^2\}.$$

(2) *If* $d_n^{\mathbb{R}}(\mathscr{I}_1; l_\infty^M) = [1 + ((M-1)\,n/(M-n))^{1/2}]^{-1}$, *and* $1 < n < M-1$, *then*

$$M \leqq \min\{n(n+1)/2,\ (M-n)(M-n+1)/2\}.$$

Proof. From the equivalence of (1) and (2) in Propositions 2.14 and 2.15, it suffices to prove that $M \leqq n^2$ in (1) and $M \leqq n(n+1)/2$ in (2).

Let $\mathbf{y}^1, \ldots, \mathbf{y}^M$ be vectors in $\mathbb{R}^n$ or $\mathbb{C}^n$ for which $\|\mathbf{y}^i\|_2 = 1$, $|(\mathbf{y}^i, \mathbf{y}^j)| = ((M-n)/n(M-1))^{1/2}$, for $i \neq j$. Such vectors exist by Propositions 2.14 and 2.15 as normalized row vectors of X. Recall that $P = (p_{ij})_{i,j=1}^M$, $p_{ij} = (\mathbf{y}^i, \mathbf{y}^j)$ is as in (3). Set $R_i = (\mathbf{y}^i)(\bar{\mathbf{y}}^i)^T$, $i = 1, \ldots, M$. Thus R_i is an $n \times n$ Hermitian matrix of rank 1. We will prove that the $\{R_i\}_{i=1}^M$ are linearly independent. Since there exist at most n^2 linearly independent Hermitian matrices in $\mathbb{C}^{n \times n}$, and at most $n(n+1)/2$ linearly independent Hermitian (symmetric) matrices in $\mathbb{R}^{n \times n}$, the result will follow.

It remains to prove that the $\{R_i\}_{i=1}^M$ are linearly independent. Assume not. There then exists an $\boldsymbol{\alpha} = (\alpha_1, \ldots, \alpha_M) \neq \mathbf{0}$ for which $\sum_{i=1}^M \alpha_i R_i = 0$, i.e.,

$$\sum_{i=1}^M \alpha_i \mathbf{y}_k^i \bar{\mathbf{y}}_r^i = 0, \qquad k, r = 1, \ldots, n.$$

Multiply the above equality by $\bar{\mathbf{y}}_k^j \mathbf{y}_r^j$ and sum on k, r. Thus

$$0 = \sum_{i=1}^M \alpha_i \left(\sum_{k=1}^n \mathbf{y}_k^i \bar{\mathbf{y}}_k^j \right) \left(\sum_{r=1}^n \mathbf{y}_r^j \bar{\mathbf{y}}_r^i \right).$$

Since $\sum_{r=1}^n \mathbf{y}_r^j \bar{\mathbf{y}}_r^i = p_{ji} = \bar{p}_{ij}$, we obtain

$$0 = \sum_{i=1}^M \alpha_i p_{ij} \bar{p}_{ij} = \sum_{i=1}^M \alpha_i |p_{ij}|^2, \qquad j = 1, \ldots, M.$$

The vector $\boldsymbol{\alpha}$ is an eigenvector, with eigenvalue zero, of the matrix

$$R = (|p_{ij}|^2)_{i,j=1}^M.$$

Since $|p_{ii}|^2 = 1$, $|p_{ij}|^2 = (M-n)/n(M-1)$, $i \neq j$, it follows that

$$R = (M(n-1)/n(M-1))\, I + ((M-n)/n(M-1))\, J,$$

where J is the matrix all of whose entries are equal to one. Since $\boldsymbol{\alpha}$ is an eigenvector of R with eigenvalue zero, it is an eigenvector of J with eigenvalue $-M(n-1)/(M-n)$. However J has no such eigenvalue. (J has $M-1$ eigenvalues equal to zero and one eigenvalue equal to M.) This contradiction implies that the $\{R_i\}_{i=1}^M$ are linearly independent. □

Theorem 2.16 gives necessary conditions on M and n for the lower bound of Theorem 2.13 to be attainable. These conditions are often not sufficient. It is therefore of interest to determine when this lower bound is attained. As an example set $M = 2n$.

Proposition 2.17. *If there exists a* $2n \times 2n$ *skew Hadamard matrix* S, *then* $d_n^{\mathbb{C}}(\mathscr{I}_1; l_\infty^{2n}) = [1 + (2n-1)^{1/2}]^{-1}$.

S is an $m \times m$ Hadamard matrix if S has entries ± 1 and $SS^T = mI$. (If S is skew Hadamard, then $S + S^T = 2I$.)

Proof. Set $A = ((2n-1)^{1/2} - i)\, I + iS$. Then

$$\begin{aligned} AA^T &= (((2n-1)^{1/2} - i)\, I + iS)\, (((2n-1)^{1/2} - i)\, I + iS^T) \\ &= (2n - 2 - 2i(2n-1)^{1/2})\, I + (i(2n-1)^{1/2} + 1)\,(S + S^T) - SS^T \\ &= (2n - 2 - 2i(2n-1)^{1/2})\, I + 2(i(2n-1)^{1/2} + 1)\, I - 2nI \\ &= 0. \end{aligned}$$

This implies that rank $A \leqq n$. (The columns of A^T are eigenvectors of A with eigenvalue zero. If rank A = rank $A^T > n$, then there are at least $n+1$ linearly independent eigenvectors of A with eigenvalue zero. A contradiction.) In fact rank $A = n$ since $A + \bar{A} = 2(2n-1)^{1/2} I$ which implies that rank A + rank $\bar{A} \geqq 2n$. Now $(A)_{ii} = (2n-1)^{1/2}$ and $|(A)_{ij}| = 1$ for all $i \neq j$. The matrix $P = (2n-1)^{-1/2} A$ satisfies condition (3) of Proposition 2.14. □

It is conjectured that skew Hadamard matrices exist for all k, $k(\bmod 4) = 0$ (and $k = 1, 2$). They are known to exist for many values of k, e.g. $k = 2^r$.

It is conjectured by Melkman [1981] that $d_n^{\mathbb{C}}(\mathscr{I}_1; l_\infty^{2n}) = [1 + (2n-1)^{1/2}]^{-1}$ for all n. This conjecture is known to be true for all n up to 50 except for two values.

When considering $d_n^{\mathbb{R}}(\mathscr{I}_1; l_\infty^{2n})$ the above conjecture is false. For example, by Theorem 2.16 and Proposition 2.17 $d_2^{\mathbb{R}}(\mathscr{I}_1; l_\infty^4) > d_2^{\mathbb{C}}(\mathscr{I}_1; l_\infty^4)$ and this is not the only case with strict inequality. On the other hand there are an infinite number of n for which $d_n^{\mathbb{R}}(\mathscr{I}_1; l_\infty^{2n}) = [1 + (2n-1)^{1/2}]^{-1}$. Moreover this lower bound does not seem to be far off due to this next result.

Proposition 2.18. *Given n, assume that there exists a* $k \times k$ *Hadamard matrix where* $n \leqq k \leqq n + n^{1/2}$. *Then*

$$d_n^{\mathbb{R}}(\mathscr{I}_1; l_\infty^{n+k}) \leqq [1 + n^{1/2}]^{-1}.$$

For $k \geqq n$, $d_n^{\mathbb{R}}(\mathscr{I}_1; l_\infty^{2n}) \leqq d_n^{\mathbb{R}}(\mathscr{I}_1; l_\infty^{n+k})$.

Proof. Let H denote the $k \times k$ Hadamard matrix. Let A denote the $k \times n$ matrix obtained from H by deleting any $k - n$ columns of H. Set

$$B = \begin{pmatrix} n^{-1/2} A \\ I \end{pmatrix}$$

where I is the $n \times n$ identity matrix. Thus B is $(n+k) \times n$. Define $P = BB^T$. P is of rank n, and

$$P = \begin{pmatrix} n^{-1} AA^T & n^{-1/2} A \\ n^{-1/2} A^T & I \end{pmatrix}$$

Since $HH^T = kI$, $n^{-1}(AA^T)_{ii} = 1$, and $|n^{-1}(AA^T)|_{ij} \leqq (k-n)/n \leqq n^{-1/2}$ for $i \neq j$ and $n \leqq k \leqq n + n^{1/2}$. Thus $(P)_{ii} = 1$ and $|(P)_{ij}| \leqq n^{-1/2}$ for all $i \neq j$, from which it easily follows that $d_n^{\mathbb{R}}(\mathscr{I}_1; l_\infty^{n+k}) \leqq [1 + n^{1/2}]^{-1}$. □

If there are in fact Hadamard matrices of order k for $k = 1, 2$ and all $k \pmod 4 = 0$ (these are known to exist to $k = 264$), then we would obtain the upper bound $[1 + n^{1/2}]^{-1}$ for $d_n^{\mathbb{R}}(\mathscr{I}_1; l_\infty^{2n})$ for all n except $n = 5$. However for $n = 5$ let A be the 5×5 matrix with diagonal entries equal to one and off-diagonal entries equal to minus one, and apply the reasoning of the above proof. (Actually $d_5^{\mathbb{R}}(\mathscr{I}_1; l_\infty^{10}) = 1/4$. i.e., attains the lower bound of Theorem 2.13, see Melkman [1981].)

Numerous other instances exist where the n-widths are known to equal the lower bound of Theorem 2.13 (see Melkman [1980], [1981]). Obviously the determination of $d_n^{\mathbb{C}}(\mathscr{I}_1; l_\infty^M)$ and $d_n^{\mathbb{R}}(\mathscr{I}_1; l_\infty^M)$ for all M and n is an exceedingly difficult if not impossible task.

We now obtain more general upper bounds on $d_n(\mathscr{I}_1; l_\infty^M)$. Various upper bounds have been proven. We list these below.

(1) Ismagilov [1974] and Glushkin [1974]. $d_n^{\mathbb{C}}(\mathscr{I}_1; l_\infty^M) \leqq CM^{1/2}/n$, where C is a constant independent of M and n.
(2) Kashin [1974]. $d_n^{\mathbb{R}}(\mathscr{I}_1; l_\infty^M) \leqq 2((\ln M)/n)^{1/2}$.
(3) Kashin [1975]. For $n \leqq M \leqq n^\lambda$, $d_n^{\mathbb{R}}(\mathscr{I}_1; l_\infty^M) \leqq C_\lambda/n^{1/2}$, where C_λ depends only on λ.
(4) Kashin [1977b]. $d_n^{\mathbb{R}}(\mathscr{I}_1; l_\infty^M) \leqq (C/n^{1/2})(1 + \ln(M/n))^{1/2}$, where C is a constant independent of M and n.
(5) Hollig [1979a]. $d_n^{\mathbb{R}}(\mathscr{I}_1; l_\infty^M) \leqq (8/n^{1/2})((\ln M)/\ln n))$.

We shall prove the estimates (2) and (5). The proof of the estimate (4) is more complicated and will not be presented here. (Recall that we always have $d_n^{\mathbb{R}}(\mathscr{I}_1; l_\infty^M) \leqq 1/2$ for $n \geqq 1$.)

The idea in the proof of both the estimates (2) and (5) is to construct an $M \times n$ real matrix A such that $P = AA^T$, $P = (p_{ij})$, satisfies $p_{ii} = 1$, $i = 1, \ldots, M$, and $|p_{ij}| \leqq \gamma$ for all $i \neq j$ (with the appropriate γ). Since A is $M \times n$, the rank of P is at most n and it follows by the method of proof of Proposition 2.14 that $d_n^{\mathbb{R}}(\mathscr{I}_1; l_\infty^M) \leqq [1 + \gamma^{-1}]^{-1} \leqq \gamma$.

Theorem 2.19. *For $n < M$,*

$$d_n^{\mathbb{R}}(\mathscr{I}_1; l_\infty^M) \leqq 2((\ln M)/n)^{1/2}.$$

Proof. We claim that there exist M vectors $\mathbf{y}^i \in \mathbb{R}^n$, $i = 1, \ldots, M$ for which

(i) $\|\mathbf{y}^i\|_2 = n^{1/2}$, $i = 1, \ldots, M$
(ii) $|(\mathbf{y}^i, \mathbf{y}^j)| \leqq 2(n \ln M)^{1/2}$, for $i \neq j$.

If we can construct such vectors, then define A as the $M \times n$ matrix whose ith row is $n^{-1/2}\mathbf{y}^i$ and set $P = AA^T$. It therefore remains to prove the claim.

Let $\mathbf{x}^1, \ldots, \mathbf{x}^{2^n}$ be all possible distinct vectors in $\mathbb{R}^n$ such that $(\mathbf{x}^i)_j \in \{-1, 1\}$ for $i = 1, \ldots, 2^n; j = 1, \ldots, n$. Let $C = (c_{ij})$ be the $2^n \times 2^n$ matrix given by $c_{ij} = (\mathbf{x}^i, \mathbf{x}^j)$. (Thus $c_{ii} = n, i = 1, \ldots, 2^n$.) Each row of C consists of the same numbers to within a permutation. Let a be any fixed positive number, and let $\alpha_i(a)$ be the number of c_{ij} ($j = 1, \ldots, 2^n$) for which $|c_{ij}| > a$. Thus $\alpha_i(a) = \alpha(a)$, i.e., it is independent of i. If $\alpha(a) < 2^n/M$ then there exist M vectors $\mathbf{x}^{i_1}, \ldots, \mathbf{x}^{i_M}$ for which $|(\mathbf{x}^{i_j}, \mathbf{x}^{i_k})| \leqq a$ for all $j \neq k$. To see this select $\mathbf{x}^{i_1}$ arbitrarily and choose $\mathbf{x}^{i_2}$ so that $|(\mathbf{x}^{i_1}, \mathbf{x}^{i_2})| \leqq a$. If $M \geqq 3$, then we claim that there exists an $\mathbf{x}^{i_3}$ for which $|(\mathbf{x}^{i_j}, \mathbf{x}^{i_3})| \leqq a, j = 1, 2$. The number of vectors $\mathbf{x}^k$ for which $|(\mathbf{x}^i, \mathbf{x}^k)| > a$ is less that $2^n/M$. Thus the number of vectors $\mathbf{x}^k$ for which $|(\mathbf{x}^{i_1}, \mathbf{x}^k)| > a$ or $|(\mathbf{x}^{i_2}, \mathbf{x}^k)| > a$ is less than $2^n/M + 2^n/M < 2^n$. In this way we continue the process until we have selected $\mathbf{x}^{i_1}, \ldots, \mathbf{x}^{i_M}$.

It therefore remains to prove that $\alpha(2(n \ln M)^{1/2}) < 2^n/M$. Let $r_k(x)$ denote the kth Rademacher function on $[0, 1]$, i.e., $r_k(x) = \operatorname{sgn}(\sin 2^k \pi x)$, $k = 0, 1, \ldots$. Since $\alpha(a) = \alpha_i(a)$ where $\mathbf{x}^i$ is the vector all of whose coefficients are equal to one, it is easy to see that

$$\alpha(a)/2^n = \mu\left\{x \colon x \in [0, 1], \left|\sum_{k=1}^{n} r_k(x)\right| > a\right\},$$

where μ is the usual Lebesgue measure. It is known (see e.g. Davie [1973]) that

$$\mu\left\{x \colon x \in [0, 1], \left|\sum_{k=1}^{n} r_k(x)\right| > b\right\} < \exp\{-b^2/4n\}.$$

Thus $\alpha(2(n \ln M)^{1/2})/2^n < M^{-1}$ and the theorem is proved. □

Theorem 2.20.

$$d_n^{\mathbb{R}}(\mathscr{I}_1; l_\infty^M) \leqq (8/n^{1/2})((\ln M)/(\ln n)).$$

This upper bound is based on an ingenious use of algebra to produce a lower bound in the following combinatorial problem:

Given positive integers m, p, n ($p \leqq n$), let $N = N(n, p, m)$ be the maximum number of subsets $S_1, \ldots, S_N$ of $\{1, \ldots, n\}$ such that

(i) $|S_i| = p, i = 1, \ldots, N$,
(ii) $|S_i \cap S_j| < m, i \neq j; i, j = 1, \ldots, N$,

where $|S_i|$ indicates the number of elements in S_i.

We cannot determine $N(n, p, m)$. However the following estimate holds.

Proposition 2.21. *If p is a power of a prime, then $N(p^2, p, m) \geqq p^m$.*

Proof. Since p is a power of a prime there exists a field of order p, denoted $GF(p)$. Identify the numbers $1, \ldots, p^2$ with the points of the finite Euclidean plane

$$E = \{(x, y) \colon x, y \in GF(p)\}.$$

Over $GF(p)$ there exist exactly p^m distinct polynomials $Q_1, \ldots, Q_{p^m}$ of degree $< m$, i.e., $Q_i(x) = \sum_{j=0}^{m-1} a_j^i x^j$, $a_j^i \in GF(p)$. Set

$$S_i = \{(x, y)\colon Q_i(x) = y\}.$$

Thus $|S_i| = p$. To show that $|S_i \cap S_j| < m$ for $i \neq j$, assume that $(x, y) \in S_i \cap S_j$. Then $Q_i(x) - Q_j(x) = 0$. Since $Q_i - Q_j$ is a polynomial of degree $< m$ over $GF(p)$ it has at most $m - 1$ zeros. Thus $|S_i \cap S_j| < m$, and $N(p^2, p, m) \geqq p^m$. □

Proof of Theorem 2.20. Let $n < M$ and let p be the largest power of a prime for which $p^2 \leqq n$. (It actually suffices to take p as a power of two.) Thus $n \leqq 4p^2$. Let m be the smallest integer for which $p^m \geqq M$, and let $S_1, \ldots, S_{p^m}$ be as defined in Proposition 2.21. We construct an $M \times p^2$ matrix $A = (a_{ij})$ as follows

$$a_{ij} = \begin{cases} p^{-1/2}, & j \in S_i \\ 0, & j \notin S_i, \end{cases}$$

$i = 1, \ldots, M;\ j = 1, \ldots, p^2$. Set $P = AA^T$. Then $P = (p_{ij})$ is an $M \times M$ matrix of rank $\leqq p^2 \leqq n$, and $p_{ii} = 1$, $0 \leqq p_{ij} \leqq (m-1)/p$, for $i \neq j$. This implies, as indicated earlier that $d_n^{\mathbb{R}}(\mathscr{I}_1; l_\infty^M) \leqq (m-1)/p$. Since $p^{m-1} \leqq M$ and $p^4 \geqq n$, it follows that $m - 1 \leqq 4(\ln M)/(\ln n)$. Since $n \leqq 4p^2$, $1/p \leqq 2/n^{1/2}$ and thus $d_n^{\mathbb{R}}(\mathscr{I}_1; l_\infty^M) \leqq (8/n^{1/2})((\ln M)/(\ln n))$. □

For $M = n^\lambda$ this provides an upper bound of $8\lambda/n^{1/2}$ (proving (3)). The lower bound from Theorem 2.13 is also of the order of $n^{-1/2}$. Thus in this case we have asymptotically optimal upper and lower bounds.

In the next chapter we need two additional results which we shall not prove. The first result is a non-trivial extension of the above theorem.

Theorem 2.22. *There exists an $M \times M$ real matrix $P = (p_{ij})$ of rank n and a constant C, independent of M and n, such that*

$$\max_{\|\mathbf{x}\|_1 \leqq 1} \|(I - P)\mathbf{x}\|_\infty = \max_{i,j} |\delta_{ij} - p_{ij}| \leqq C n^{-1/2}((\ln M)/(\ln n)),$$

and

$$\max_{\|\mathbf{x}\|_2 \leqq 1} \|(I - P)\mathbf{x}\|_2 \leqq CM/n.$$

The point here is that it is the same matrix P used in both estimates. Thus from the Riesz-Thorin Interpolation Theorem (see e.g. Bergh, Löfstrom [1976]), we obtain

Corollary 2.23. *Let $1 \leqq p \leqq 2$, $1/p + 1/p' = 1$. Then*

$$\delta_n(\mathscr{I}_p; l_{p'}^M) \leqq C n^{1/2 - 1/p}(M/n)^{2/p'}((\ln M)/(\ln n))^{(2/p)-1},$$

where C is a constant independent of M and n.

It is interesting the number of different techniques which have been brought to bear on the problem of evaluating and estimating $d_n(\mathscr{I}_1; l_\infty^M)$. Melkman [1980] uses another idea to estimate $d_n(\mathscr{I}_1; l_\infty^M)$ from above. To illustrate his idea we first consider the case $n = 2$.

Proposition 2.24. *For* $M > 2$

$$d_2^{\mathbb{R}}(\mathscr{I}_1; l_\infty^M) \leqq (1 + \sec \pi/M)^{-1}.$$

Proof. Set $\mathbf{x}^k = (\cos(k-1)\pi/M, \sin(k-1)\pi/M) \in \mathbb{R}^2$, $k = 1, \ldots, M$. The vectors $\mathbf{x}^k$ satisfy $\|\mathbf{x}^k\|_2 = 1$ and $|(\mathbf{x}^i, \mathbf{x}^j)| \leqq \cos \pi/M$ for $i \neq j$. Let X denote the $M \times 2$ matrix with $\mathbf{x}^k$ as the kth row. Set $P = (1 + \cos \pi/M)^{-1} XX^T$. Since rank $P = 2$, it follows that

$$d_2^{\mathbb{R}}(\mathscr{I}_1; l_\infty^M) \leqq \max_{i,j} |\delta_{ij} - p_{ij}| = (1 + \sec \pi/M)^{-1}. \quad \square$$

It is conjectured that this upper bound is exact.

The idea used above was that of packing M spherical caps of largest possible spherical radius $\theta = \pi/2M$ on the unit hemisphere in $\mathbb{R}^2$. This same idea is used by Melkman [1980] in $\mathbb{R}^n$. He shows that if $P(\theta)$ is the maximum number of spherical caps of spherical radius θ which can be packed (i.e., they are disjoint) on the unit sphere in $\mathbb{R}^n$ so that the centers of the caps subtend angles between 2θ and $180° - 2\theta$, then

$$P(\theta) \geqq \frac{\sqrt{\pi}}{2} \frac{\Gamma((n-1)/2)}{\Gamma(n/2)} \left(\int_0^{2\theta} \sin^{n-2}\psi \, d\psi \right)^{-1}.$$

If $P(\theta) \geqq M$, then it follows that $d_n^{\mathbb{R}}(\mathscr{I}_1; l_\infty^M) \leqq (1 + \sec 2\theta)^{-1}$. Melkman uses these inequalities to obtain estimates for $d_n^{\mathbb{R}}(\mathscr{I}_1; l_\infty^M)$.

The second important result, alluded to previously, is concerned with an estimate for $d_n^{\mathbb{R}}(\mathscr{I}_2; l_\infty^M)$. From previous results of this chapter $d_n^{\mathbb{R}}(\mathscr{I}_2; l_\infty^M) \leqq \delta_n^{\mathbb{R}}(\mathscr{I}_2; l_\infty^M) = \delta_n^{\mathbb{R}}(\mathscr{I}_1; l_2^M) = ((M-n)/M)^{1/2}$. Unfortunately this upper bound is far from optimal. Another method of obtaining upper bounds for $d_n^{\mathbb{R}}(\mathscr{I}_2; l_\infty^M)$ is the following.

Proposition 2.25. *For* $1 \leqq p \leqq q \leqq \infty$,

$$d_n(\mathscr{I}_p; l_q^M) \leqq [d_n(\mathscr{I}_1; l_q^M)]^{1-(q'/p')},$$

where $1/p + 1/p' = 1/q + 1/q' = 1$. *In particular,*

$$d_n(\mathscr{I}_2; l_\infty^M) \leqq [d_n(\mathscr{I}_1; l_\infty^M)]^{1/2}.$$

Proof. From Proposition 1.1,

$$d_n(\mathscr{I}_p; l_q^M) = d^n(\mathscr{I}_{q'}; l_{p'}^M) = \inf_{X_n} \sup_{\substack{\|\mathbf{x}\|_{q'} \leqq 1 \\ \mathbf{x} \perp X_n}} \|\mathbf{x}\|_{p'}.$$

Since $\|\mathbf{x}\|_{p'} \leqq \|\mathbf{x}\|_{q'}^{q'/p'} \|\mathbf{x}\|_{\infty}^{1-(q'/p')}$,

$$\begin{aligned} d^n(\mathscr{I}_{q'}; l_{p'}^M) &\leqq \inf_{X_n} \sup_{\|\mathbf{x}\|_{q'} \leqq 1,\ \mathbf{x} \perp X_n} \|\mathbf{x}\|_{\infty}^{1-(q'/p')} \\ &= [d^n(\mathscr{I}_{q'}; l_{\infty}^M)]^{1-(q'/p')} \\ &= [d_n(\mathscr{I}_1; l_q^M)]^{1-(q'/p')}. \quad \square \end{aligned}$$

This simple proposition provides us with upper bounds for $d_n^{\mathbb{R}}(\mathscr{I}_2; l_{\infty}^M)$ which are substantially better that $((M-n)/M)^{1/2}$. However these upper bounds are still insufficient for the applications to be found in the next chapter. The following result will be used.

Theorem 2.26. *For $n < M$ there exists a constant C, independent of n and M, for which*

$$d_n^{\mathbb{R}}(\mathscr{I}_2; l_{\infty}^M) \leqq C n^{-1/2}(1 + \ln(M/n))^{3/2}.$$

The known proof of Theorem 2.26 is beyond the scope of this monograph. However we briefly outline one of the main ideas.

The main content of the proof of Theorem 2.26 is the construction of an $M \times n$ matrix $A = (a_{ij})_{i=1}^{M}{}_{j=1}^{n}$ with the following properties:

Let $\mathbf{a}^i$ denote the ith row of A. Then,

(i) every n rows are linearly independent.

(ii) For any distinct $i_1, \ldots, i_{n+1} \in \{1, \ldots, M\}$ and

$$\mathbf{a}^{i_{n+1}} = \sum_{k=1}^{n} \lambda_k \mathbf{a}^{i_k},$$

the vector $\boldsymbol{\lambda} = (\lambda_1, \ldots, \lambda_n)$ satisfies

$$\frac{\|\boldsymbol{\lambda}\|_2 + 1}{\|\boldsymbol{\lambda}\|_1} \leqq C n^{-1/2}(1 + \ln(M/n))^{3/2}.$$

Assuming that such a matrix has been constructed, let us prove the theorem.

Let $\{\mathbf{y}^i\}_{i=1}^n$ denote the columns of A (in $\mathbb{R}^M$) and set $X_n = \operatorname{span}\{\mathbf{y}^1, \ldots, \mathbf{y}^n\}$. We must show that

$$\sup_{\|\mathbf{x}\|_2 \leqq 1} \inf_{\mathbf{y} \in X_n} \|\mathbf{x} - \mathbf{y}\|_{\infty} \leqq C n^{-1/2}(1 + \ln(M/n))^{3/2}.$$

Given $\mathbf{x}$, it is well-known that $\inf_{\mathbf{y} \in X_n} \|\mathbf{x} - \mathbf{y}\|_{\infty} \leqq \rho$ if and only if for each choice of distinct $i_1, \ldots, i_{n+1}$ in $\{1, \ldots, M\}$

$$\inf_{\mathbf{y} \in X_n} \max_{i = i_1, \ldots, i_{n+1}} |(\mathbf{x} - \mathbf{y})_i| \leqq \rho.$$

Fix $i_1, \ldots, i_{n+1}$ in $\{1, \ldots, M\}$. The $n \times n$ submatrix of A composed of rows $i_1, \ldots, i_n$ is, by (i), nonsingular. Let B denote its inverse. Thus $B = (b_{st})_{s,t=1}^n$, where

$$b_{st} = (-1)^{s+t} \frac{A\begin{pmatrix} i_1, \ldots, \hat{i}_t, \ldots, i_n \\ 1, \ldots, \hat{s}, \ldots, n \end{pmatrix}}{A\begin{pmatrix} i_1, \ldots, i_n \\ 1, \ldots, n \end{pmatrix}}.$$

Set $\mathbf{z}^t = \sum_{s=1}^n b_{st} \mathbf{y}^s$. Thus $X_n = \operatorname{span}\{\mathbf{z}^1, \ldots, \mathbf{z}^n\}$. It follows from the definition of B that $(\mathbf{z}^t)_{i_m} = \delta_{mt}$ for $m, t = 1, \ldots, n$. Since $\mathbf{a}^{i_{n+1}} = \sum_{k=1}^n \lambda_k \mathbf{a}^{i_k}$, it is easily seen that $(\mathbf{z}^t)_{i_{n+1}} = \lambda_t$, $t = 1, \ldots, n$. (We have simply defined a new and simpler basis for X_n on $i_1, \ldots, i_{n+1}$.) It is well-known that

$$\inf_{\alpha_1, \ldots, \alpha_n} \sup_{i = i_1, \ldots, i_{n+1}} \left| \left(\mathbf{x} - \sum_{j=1}^n \alpha_j \mathbf{z}^j \right)_i \right| = \frac{\left| \det \begin{pmatrix} x_{i_1} & 1 \ldots 0 \\ \vdots & \vdots \quad \vdots \\ x_{i_n} & 0 \ldots 1 \\ x_{i_{n+1}} & \lambda_1 \ldots \lambda_n \end{pmatrix} \right|}{\sum_{j=1}^n |\lambda_j|}$$

$$= \left| \sum_{j=1}^n \lambda_j x_{i_j} - x_{i_{n+1}} \right| \Big/ \sum_{j=1}^n |\lambda_j|$$

$$\leqq \left(\sum_{j=1}^n |\lambda_j|^2 + 1 \right)^{1/2} \left(\sum_{j=1}^{n+1} |x_{i_j}|^2 \right)^{1/2} \Big/ \sum_{j=1}^n |\lambda_j|$$

$$\leqq (\|\boldsymbol{\lambda}\|_2 + 1) \, \|\mathbf{x}\|_2 / \|\boldsymbol{\lambda}\|_1.$$

Thus

$$\sup_{\|\mathbf{x}\|_2 \leqq 1} \inf_{\mathbf{y} \in X_n} \|\mathbf{x} - \mathbf{y}\|_\infty \leqq \frac{\|\boldsymbol{\lambda}\|_2 + 1}{\|\boldsymbol{\lambda}\|_1} \leqq C n^{-1/2} (1 + \ln(M/n))^{3/2}.$$

The proof of the existence of a matrix A, as specified above, is probabilistic in nature and is too complicated to present here. A simpler and constructive proof of the existence of such a matrix would be of interest.

3. n-Widths of Strictly Totally Positive Matrices

If A is a strictly totally positive (STP) matrix, then analogous to the results obtained in Section 2 of Chapter V, it is possible to determine $d_n(\mathscr{A}_p; l_q^M)$, $d^n(\mathscr{A}_p; l_q^M)$ and $\delta_n(\mathscr{A}_p; l_q^M)$ for $p = \infty$, or $q = 1$. When $p = \infty$, then the result is especially pleasing. In this case there exists an optimal n-dimensional subspace for d_n which is simply the span of n columns of A. To ease our exposition we consider only the case $p = q = \infty$.

Let $A = (a_{ij})_{i,j=1}^{M}$ and let $\mathbf{a}^j$ denote the jth column vector of A. For $0 < n < M$, set

$$J = \{\mathbf{j}: \mathbf{j} = (j_1, \ldots, j_n),\ 1 \leqq j_1 < \ldots < j_n \leqq M\}$$

and

$$I = \{\mathbf{i}: \mathbf{i} = (i_1, \ldots, i_{n+1}),\ 1 \leqq i_1 < \ldots < i_{n+1} \leqq M\}.$$

Definition 3.1. Given $\mathbf{x} \in \mathbb{R}^M$, we say that $\mathbf{x}$ *alternates between* $\mathbf{j} = (j_1, \ldots, j_n) \in J$ provided that there exists a sign $\varepsilon \in \{-1, 1\}$ such that $x_k = (-1)^{i+1}\varepsilon$, $j_{i-1} < k < j_i$, $i = 1, \ldots, n+1$ (where $j_0 = 0$, $j_{n+1} = M+1$). (Note that no restriction is placed on the components $x_{j_1}, \ldots, x_{j_n}$.) We say that $\mathbf{x}$ alternates *positively* between $\mathbf{j} \in J$ if in addition $\varepsilon = 1$ in the above.

If $\mathbf{x}$ alternates between some $\mathbf{j} \in J$, then $S^-(\mathbf{x}) \leqq n$.

Definition 3.2. A vector $\mathbf{y} \in \mathbb{R}^M \backslash \{\mathbf{0}\}$ is said to *equioscillate on* $\mathbf{i} = (i_1, \ldots, i_{n+1}) \in I$ if there exists a sign $\varepsilon \in \{-1, 1\}$ for which

$$y_{i_k} = (-1)^{k+1} \varepsilon \, \|\mathbf{y}\|_\infty, \qquad k = 1, \ldots, n+1.$$

It is said to equioscillate *positively* on $\mathbf{i} \in I$ if $\varepsilon = 1$ in the above.

If $\mathbf{y}$ equioscillates on $\mathbf{i} \in I$, then $S^-(\mathbf{y}) \geqq n$.
For each $\mathbf{i} \in I$, $\mathbf{j} \in J$, we define

$$G(\mathbf{i}, \mathbf{j}) = \frac{\sum_{j=1}^{M} \left| A \begin{pmatrix} i_1, \ldots, i_{n+1} \\ j_1, \ldots, j_n, j \end{pmatrix} \right|}{\sum_{s=1}^{n+1} A \begin{pmatrix} i_1, \ldots, \hat{i}_s, \ldots, i_{n+1} \\ j_1, \ldots, j_n \end{pmatrix}}.$$

(A is STP so that the denominator is positive.)
With these definitions we can now proceed to our first result.

Proposition 3.1. *For* $A \in \mathbb{R}^{M \times M}$ STP, $0 < n < M$ *and* $\mathbf{j} \in J$,

$$\max_{\|\mathbf{x}\|_\infty \leqq 1} \min_{\alpha_j} \|A\mathbf{x} - \sum_{j \in \mathbf{j}} \alpha_j \mathbf{a}^j\|_\infty = \max_{\mathbf{i} \in I} G(\mathbf{i}, \mathbf{j}).$$

Proof. For fixed $\mathbf{z} \in \mathbb{R}^M$ and $\mathbf{j} \in J$, set

$$E(\mathbf{z}; \mathbf{j}) = \min_{\alpha_j} \|\mathbf{z} - \sum_{j \in \mathbf{j}} \alpha_j \mathbf{a}^j\|_\infty.$$

It is well-known that since the $\{\mathbf{a}^j\}_{j \in \mathbf{j}}$ form a T-system (i.e., the associated $M \times n$ matrix is SSC_n)

$$E(\mathbf{z}; \mathbf{j}) = \max_{\mathbf{i} \in I} \frac{\left| \det \begin{pmatrix} a_{i_1 j_1} & \ldots a_{i_1 j_n} & z_{i_1} \\ \vdots & \vdots & \vdots \\ a_{i_{n+1} j_1} & \ldots a_{i_{n+1} j_n} & z_{i_{n+1}} \end{pmatrix} \right|}{\sum_{s=1}^{n+1} A \begin{pmatrix} i_1, \ldots, \hat{i}_s, \ldots, i_{n+1} \\ j_1, \ldots, j_n \end{pmatrix}}$$

(this fact was also used at the end of the previous section) and that the minimum error equioscillates on $\mathbf{i} \in I$ at which the above maximum is attained.

By a multilinear expansion of the last column of the numerator,

$$E(A\mathbf{x};\mathbf{j}) = \max_{\mathbf{i}\in I} \frac{\left|\sum_{j=1}^{M} x_j A\begin{pmatrix} i_1,\ldots,i_{n+1} \\ j_1,\ldots,j_n,j \end{pmatrix}\right|}{\sum_{s=1}^{n+1} A\begin{pmatrix} i_1,\ldots,\hat{i}_s,\ldots,i_{n+1} \\ j_1,\ldots,j_n \end{pmatrix}}.$$

For each $\mathbf{i} \in I$ the maximum of the above quantity over $\|\mathbf{x}\|_\infty \leqq 1$ is obviously attained by choosing

$$x_j = \varepsilon \operatorname{sgn} A\begin{pmatrix} i_1,\ldots,i_{n+1} \\ j_1,\ldots,j_n,j \end{pmatrix}, \quad j \notin \mathbf{j},\ \varepsilon \in \{-1,1\}.$$

Since A is STP this choice is independent of $\mathbf{i} \in I$, i.e., we choose $\mathbf{x}$ to alternate between $\mathbf{j} \in J$. Thus

$$\begin{aligned}\max_{\|\mathbf{x}\|_\infty \leqq 1} \min_{\alpha_j} \|A\mathbf{x} - \sum_{j\in\mathbf{j}} \alpha_j \mathbf{a}^j\|_\infty &= \max_{\mathbf{i}\in I} \frac{\sum_{j=1}^{M} \left|A\begin{pmatrix} i_1,\ldots,i_{n+1} \\ j_1,\ldots,j_n,j \end{pmatrix}\right|}{\sum_{s=1}^{n+1} A\begin{pmatrix} i_1,\ldots,\hat{i}_s,\ldots,i_{n+1} \\ j_1,\ldots,j_n \end{pmatrix}} \\ &= \max_{\mathbf{i}\in I} G(\mathbf{i},\mathbf{j}).\end{aligned}$$

This proves the proposition. □

Theorem 3.2. *Let $A \in \mathbb{R}^{M\times M}$ be STP and $0 < n < M$. There exists a vector $\mathbf{x}^0 \in \mathbb{R}^M$ for which*

(i) $\|\mathbf{x}^0\|_\infty = 1$
(ii) $\mathbf{x}^0$ *alternates between* $\mathbf{j}^0 \in J$
(iii) $A\mathbf{x}^0$ *equioscillates on* $\mathbf{i}^0 \in I$.

Theorem 3.2 is the finite dimensional analogue of Theorem 2.5 of Chapter V. One method of proving it is by paralleling the proof of Theorem 2.5. We prefer, however, to present a different somewhat constructive proof.

Proof. For each $\mathbf{j} \in J$ let

$$\max_{\|\mathbf{x}\|_\infty \leqq 1} \min_{\alpha_j} \|A\mathbf{x} - \sum_{j\in\mathbf{j}} \alpha_j \mathbf{a}^j\|_\infty = \|A\mathbf{x}_\mathbf{j}^* - \sum_{j\in\mathbf{j}} \alpha_j^* \mathbf{a}^j\|_\infty.$$

Set

$$(\mathbf{x_j})_k = \begin{cases} (\mathbf{x}_\mathbf{j}^*)_k, & k \notin \mathbf{j} \\ (\mathbf{x}_\mathbf{j}^*)_k - \alpha_k^*, & k \in \mathbf{j}. \end{cases}$$

Then $\mathbf{x}_\mathbf{j}$ alternates between $\mathbf{j} \in J$ and $A\mathbf{x}_j$ equioscillates on some $\mathbf{i}(\mathbf{j}) \in I$. Let $\mathbf{i}^0 = (i_1^0, \ldots, i_{n+1}^0) \in I$ and $\mathbf{j}^0 = (j_1^0, \ldots, j_n^0) \in J$ satisfy

$$\min_{\mathbf{j} \in J} \max_{\mathbf{i} \in I} G(\mathbf{i}, \mathbf{j}) = \max_{\mathbf{i} \in I} G(\mathbf{i}, \mathbf{j}^0) = G(\mathbf{i}^0, \mathbf{j}^0).$$

Set $\mathbf{x}^0 = \mathbf{x}_\mathbf{j}^0$. Then

(1) $\mathbf{x}^0$ alternates between $\mathbf{j}^0 \in J$

(2) $A\mathbf{x}^0$ equioscillates on $\mathbf{i}^0 \in I$

(3) $\|A\mathbf{x}^0\|_\infty \leqq \|A\mathbf{x}_\mathbf{j}\|_\infty$ for every $\mathbf{j} \in J$.

To prove the theorem we show that (3) implies that $\|\mathbf{x}^0\|_\infty = 1$.

For every $\mathbf{j} \in J$, the definition of $\mathbf{x}_\mathbf{j}$ implies that $\|\mathbf{x}_\mathbf{j}\|_\infty \geqq 1$, and $S^-(\mathbf{x}_\mathbf{j}) \leqq n$. Since $A\mathbf{x}_\mathbf{j}$ equioscillates on some $\mathbf{i} \in I$, $S^-(A\mathbf{x}_\mathbf{j}) \geqq n$. Therefore by Theorem 2.4 of Chapter III,

$$S^-(A\mathbf{x}_\mathbf{j}) = S^+(A\mathbf{x}_\mathbf{j}) = S^-(\mathbf{x}_\mathbf{j}) = n,$$

and the sign patterns of $\mathbf{x}_\mathbf{j}$ and $A\mathbf{x}_\mathbf{j}$ agree, i.e., the sign of the first component of $A\mathbf{x}_\mathbf{j}$ (necessarily nonzero here) agrees with the sign of the first nonzero component of $\mathbf{x}_\mathbf{j}$. Multiplying $\mathbf{x}_\mathbf{j}$ by -1, if necessary, we assume that $\mathbf{x}_\mathbf{j}$ alternates positively between $\mathbf{j}$ (and $A\mathbf{x}_\mathbf{j}$ equioscillates positively on some $\mathbf{i} \in I$).

Our object is to prove that $|(\mathbf{x}^0)_{j_k^0}| \leqq 1$, $k = 1, \ldots, n$. First suppose that $j_k^0 > k$ and let r be the largest integer less than j_k^0 for which $r \neq j_i^0$, $i = 1, \ldots, k-1$. Define $\mathbf{j} = (j_1, \ldots, j_n) \in J$ where $\{j_1, \ldots, j_n\}$ is the set of indices $\{j_1^0, \ldots, j_{k-1}^0, j_{k+1}^0, \ldots, j_n^0, r\}$ arranged in increasing order. Since $\mathbf{x}_\mathbf{j}$ alternates positively on $\mathbf{j}$,

$$(\mathbf{x}_\mathbf{j})_{j_k^0} = (-1)^k,$$

and

$$(\mathbf{x}_\mathbf{j} - \mathbf{x}^0)_m = 0, \quad m \notin \{j_1^0, \ldots, j_n^0, r\}.$$

If $\mathbf{x}_\mathbf{j} = \mathbf{x}^0$, then $|(\mathbf{x}^0)_{j_k^0}| = 1$. Assume $\mathbf{x}_\mathbf{j} \neq \mathbf{x}^0$. Thus $S^-(\mathbf{x}_\mathbf{j} - \mathbf{x}^0) \leqq n$. Since $A\mathbf{x}_\mathbf{j}$ equioscillates on some $\mathbf{i} \in I$ and $\|A\mathbf{x}_\mathbf{j}\|_\infty \geqq \|A\mathbf{x}^0\|_\infty$,

$$S^+(A\mathbf{x}_\mathbf{j} - A\mathbf{x}^0) \geqq n.$$

Thus, by Theorem 2.4 of Chapter III,

$$S^+(A\mathbf{x}_\mathbf{j} - A\mathbf{x}^0) = S^-(\mathbf{x}_\mathbf{j} - \mathbf{x}^0) = n.$$

If $\|A\mathbf{x}_\mathbf{j}\|_\infty = \|A\mathbf{x}^0\|_\infty$, then it is easily shown that $S^+(A\mathbf{x}_\mathbf{j} - A\mathbf{x}^0) \geqq n + 1$, a contradiction. Thus $\|A\mathbf{x}_\mathbf{j}\|_\infty > \|A\mathbf{x}^0\|_\infty$, and the sign pattern of $A\mathbf{x}_\mathbf{j} - A\mathbf{x}^0$ starts positively. Hence

$$\operatorname{sgn}((\mathbf{x}_\mathbf{j})_{j_k^0} - (\mathbf{x}^0)_{j_k^0}) = (-1)^k,$$

and since $(\mathbf{x}_\mathbf{j})_{j_k^0} = (-1)^k$, this implies that $1 > (\mathbf{x}^0)_{j_k^0}(-1)^k$. Similarly, if $j_k^0 < M - n + k$ let r be the smallest integer greater than j_k^0 for which $r \neq j_i^0$, $i > k$, and as above it follows that $1 \geqq (\mathbf{x}^0)_{j_k^0}(-1)^{k+1}$. Thus if $k < j_k^0 < M - n + k$, the desired result holds.

If $j_k^0 = k$, then $j_i^0 = i$, $i = 1,\ldots,k-1$, and $\operatorname{sgn}(\mathbf{x}^0)_{j_k^0} = (-1)^{k+1}$ since $S^-(\mathbf{x}^0) = n$ and $\mathbf{x}^0$ alternates positively between $\mathbf{j}^0$. However $n < M$ which implies that $j_k^0 = k < M - n + k$ so that $|(\mathbf{x}^0)_{j_k^0}| = (-1)^{k+1}(\mathbf{x}^0)_{j_k^0} \leqq 1$. Similarly if $j_k^0 = M - n + k$, then $j_k^0 > k$, and $|(\mathbf{x}^0)_{j_k^0}| = (-1)^k(\mathbf{x}^0)_{j_k^0} \leqq 1$. In all cases $\|\mathbf{x}^0\|_\infty \leqq 1$. $\square$

Theorem 3.3. *Let $A \in \mathbb{R}^{M \times M}$ be STP, $0 < n < M$. Let $\mathbf{x}^0$ and $\mathbf{j}^0 \in J$ be as above. Then*

$$d_n(\mathscr{A}_\infty; l_\infty^M) = \|A\mathbf{x}^0\|_\infty$$

and $X_n = \operatorname{span}\{\mathbf{a}^j : j \in \mathbf{j}^0\}$ is an optimal subspace.

Proof. By the definition of $\mathbf{x}^0$, $d_n(\mathscr{A}_\infty; l_\infty^M) \leqq \|A\mathbf{x}^0\|_\infty$. Since $S^-(\mathbf{x}^0) = n$, and $\mathbf{x}^0$ alternates positively on $\mathbf{j}^0$, there exists $\{r_1,\ldots,r_n\}$, $0 = r_0 < r_1 < \ldots < r_{n+1} = M$, such that $\operatorname{sgn}(\mathbf{x}^0)_i = (-1)^{k+1}$ or zero, for $r_{k-1} < i \leqq r_k$, $k = 1,\ldots,n+1$, and for each $k \in \{1,\ldots,n+1\}$ there exists an i, $r_{k-1} < i \leqq r_k$, for which $(\mathbf{x}^0)_i \neq 0$.

Set $\mathbf{b}^k = \sum_{i=r_{k-1}+1}^{r_k} |(\mathbf{x}^0)_i|\,\mathbf{a}^i$, $k = 1,\ldots,n+1$. Let B be the $M \times (n+1)$ matrix composed of the columns $\mathbf{b}^1,\ldots,\mathbf{b}^{n+1}$. The matrix B is STP. To prove the lower bound it suffices (by Theorem 1.6 of Chapter II) to prove that if $\left\|\sum_{k=1}^{n+1} \alpha_k \mathbf{b}^k\right\|_\infty \leqq \|A\mathbf{x}^0\|_\infty$, then $|\alpha_k| \leqq 1$, $k = 1,\ldots,n+1$.

Assume that $\left\|\sum_{k=1}^{n+1} \alpha_k \mathbf{b}^k\right\|_\infty \leqq \|A\mathbf{x}^0\|_\infty$ and $|\alpha_{k_0}| = \max_{k=1,\ldots,n+1} |\alpha_k| > 1$. Now $A\mathbf{x}^0 = \sum_{k=1}^{n+1} (-1)^{k+1}\mathbf{b}^k$ and $A\mathbf{x}^0$ equioscillates on $\mathbf{i}^0$. Therefore, since $A\mathbf{x}^0 \neq \pm \sum_{k=1}^{n+1} \frac{\alpha_k}{\alpha_{k_0}} \mathbf{b}^k$,

$$S^+\left(\sum_{k=1}^{n+1} (-1)^{k+1}\mathbf{b}^k - (-1)^{k_0+1}\sum_{k=1}^{n+1} \frac{\alpha_k}{\alpha_{k_0}} \mathbf{b}^k\right) \geqq n.$$

The vector with coefficients $\{(-1)^{k+1} - (-1)^{k_0+1}\,\alpha_k/\alpha_{k_0}\}_{k=1}^{n+1}$ has at most $n-1$ weak sign changes since its k_0 coefficient is zero. This is a contradiction to Theorem 2.4 of Chapter III. Thus $|\alpha_k| \leqq 1$, $k = 1,\ldots,n+1$. $\square$

The above proof parallels the proof of Theorem 2.10 of Chapter V. In other words we have also proven

Theorem 3.4. *For A, $\mathbf{x}^0$ and $\{\mathbf{b}^1,\ldots,\mathbf{b}^{n+1}\}$ as above,*

$$b_n(\mathscr{A}_\infty; l_\infty^M) = \|A\mathbf{x}^0\|_\infty$$

and $X_{n+1} = \operatorname{span}\{\mathbf{b}^1,\ldots,\mathbf{b}^{n+1}\}$ is an optimal subspace.

Theorems 3.2 and 3.3 imply that $d_n(\mathscr{A}_\infty; l_\infty^M) = \min_{\mathbf{j} \in J} \max_{\mathbf{i} \in I} G(\mathbf{i},\mathbf{j})$. In fact we also have the following.

Proposition 3.5. *For A as above*

$$\min_{\mathbf{j}\in J}\max_{\mathbf{i}\in I} G(\mathbf{i},\mathbf{j}) = \max_{\mathbf{i}\in I}\min_{\mathbf{j}\in J} G(\mathbf{i},\mathbf{j}).$$

Proof. Since $\max_{\mathbf{i}\in I}\min_{\mathbf{j}\in J} G(\mathbf{i},\mathbf{j}) \leqq \min_{\mathbf{j}\in J}\max_{\mathbf{i}\in I} G(\mathbf{i},\mathbf{j})$, it suffices to prove that $\min_{\mathbf{j}\in J} G(\mathbf{i}^0,\mathbf{j}) \geqq \|A\mathbf{x}^0\|_\infty$. This inequality is equivalent to the statement that if $\mathbf{x}$ alternates on some $\mathbf{j}\in J$, then

$$\|A\mathbf{x}^0\|_\infty \leqq \max_{i\in\mathbf{i}^0} |(A\mathbf{x})_i|.$$

Assume to the contrary that there exists an $\mathbf{x}$ which alternates on some $\mathbf{j}\in J$ for which $|(A\mathbf{x})_i| < \|A\mathbf{x}^0\|_\infty$, all $i\in\mathbf{i}^0$. $A\mathbf{x}^0$ equioscillates on $\mathbf{i}^0$ and thus by Theorem 2.4 of Chapter III it follows that for every sufficiently small $\delta > 0$,

$$n \leqq S^-(A\mathbf{x}^0 \pm (1+\delta)A\mathbf{x}) \leqq S^-(\mathbf{x}^0 \pm (1+\delta)\mathbf{x}) \leqq n.$$

The latter inequality holds since $\mathbf{x}$ alternates on some $\mathbf{j}\in J$. The sign pattern of $A\mathbf{x}^0 \pm (1+\delta)A\mathbf{x}$ is determined by $A\mathbf{x}^0$, while that of $\mathbf{x}^0 \pm (1+\delta)\mathbf{x}$ is determined by $\pm(1+\delta)\mathbf{x}$. This contradicts Theorem 2.4 of Chapter III which states that the sign patterns must agree. □

From Theorem 3.4 we see that $\|A\mathbf{x}^0\|_\infty$ is a lower bound for $d^n(\mathscr{A}_\infty; l_\infty^M)$. Therefore to prove that $d^n(\mathscr{A}_\infty; l_\infty^M) = \delta_n(A_\infty; l_\infty^M) = \|A\mathbf{x}^0\|_\infty$, we must construct a rank n matrix $B\in\mathbb{R}^{M\times M}$ for which

$$\sup_{\|\mathbf{x}\|_\infty\leqq 1} \|A\mathbf{x} - B\mathbf{x}\|_\infty \leqq \|A\mathbf{x}^0\|_\infty.$$

(Then setting $L^n = \{\mathbf{x}: B\mathbf{x} = \mathbf{0}\}$, we will have determined an optimal subspace for $d^n(\mathscr{A}_\infty; l_\infty^M)$.)

The optimal matrix B and subspace L^n are far less aesthetically pleasing than the optimal subspace for $d_n(\mathscr{A}_\infty; l_\infty^M)$.

We construct B as follows. Because $S^+(A\mathbf{x}^0) = S^-(A\mathbf{x}^0) = S^-(\mathbf{x}^0) = n$, we see that $(A\mathbf{x}^0)_1 (A\mathbf{x}^0)_M \neq 0$, and if $(A\mathbf{x}^0)_i = 0$ then $(A\mathbf{x}^0)_{i-1}(A\mathbf{x}^0)_{i+1} < 0$. Therefore at the ith sign change of $A\mathbf{x}^0$ one of two possibilities. Either

(a) $(A\mathbf{x}^0)_{k_i}(A\mathbf{x}^0)_{k_i+1} < 0$, or
(b) $(A\mathbf{x}^0)_{k_i} = 0$ and $(A\mathbf{x}^0)_{k_i-1}(A\mathbf{x}^0)_{k_i+1} < 0$,

where $1 \leqq k_1 < \ldots < k_n < M$.

For each $i\in\{1,\ldots,n\}$, we define $\mathbf{y}^i$ as follows. If (a) holds, set

$$(\mathbf{y}^i)_r = \begin{cases} |(A\mathbf{x}^0)_r|^{-1}, & r = k_i,\ k_i+1 \\ 0, & \textit{otherwise}. \end{cases}$$

If (b) holds, set $(\mathbf{y}^i)_r = \delta_{rk_i}$, $r = 1,\ldots, M$. Thus $(A\mathbf{x}^0, \mathbf{y}^i) = 0$, $i = 1,\ldots, n$, and, as is easily checked, the $M \times n$ matrix Y with ith column equal to $\mathbf{y}^i$ is TP and of rank n.

We now define a rank n, $M \times M$ matrix P by the condition that for every $\mathbf{x} \in \mathbb{R}^M$,

$$(\mathbf{x} - P\mathbf{x}, A^T\mathbf{y}^i) = 0, \quad i = 1,\ldots, n, \quad \text{and} \quad (P\mathbf{x})_r = 0, \ r \notin \mathbf{j}^0.$$

P exists since otherwise there would exist a $\mathbf{z} \in \mathbb{R}^M \backslash \{\mathbf{0}\}$ for which

$$(\mathbf{z}, A^T\mathbf{y}^i) = 0, \quad i = 1,\ldots, n, \quad \text{and} \quad (\mathbf{z})_r = 0, \ r \notin \mathbf{j}^0.$$

Since the matrix $A^T Y$ is an $M \times n$ STP matrix the first condition would imply, by Proposition 2.2 of Chapter III, that $S^-(\mathbf{z}) \geqq n$. However from the second condition $S^-(\mathbf{z}) \leqq n - 1$ since $\mathbf{z}$ has at most n nonzero coefficients. From this contradiction P exists. P is of rank at most n since $(P\mathbf{x})_r = 0, r \notin \mathbf{j}$. (Its range space is of dimension at most n.) Set $B = AP$.

Theorem 3.6. *For A, B and $\mathbf{x}^0$ as above, $0 < n < M$*

$$d^n(\mathscr{A}_\infty; l_\infty^M) = \delta_n(\mathscr{A}_\infty; l_\infty^M) = \| A\mathbf{x}^0 \|_\infty .$$

Furthermore,

(1) *$L^n = \{\mathbf{x}: B\mathbf{x} = \mathbf{0}\}$ is an optimal subspace for $d^n(\mathscr{A}_\infty; l_\infty^M)$.*
(2) *B is an optimal rank n matrix for $\delta_n(\mathscr{A}_\infty; l_\infty^M)$.*

Proof. It suffices to prove that

$$\sup_{\|\mathbf{x}\|_\infty \leqq 1} \| A\mathbf{x} - B\mathbf{x}\|_\infty \leqq \| A\mathbf{x}^0 \|_\infty .$$

Suppose there exists an $\mathbf{x}$ and r for which $\|\mathbf{x}\|_\infty \leqq 1$, and $|(A\mathbf{x} - B\mathbf{x})_r| > |(A\mathbf{x}^0)_r|$. Let $d = (A\mathbf{x}^0)_r/(A\mathbf{x} - B\mathbf{x})_r$, and set $\mathbf{z} = A\mathbf{x}^0 - d(A\mathbf{x} - B\mathbf{x})$. From the definition of P, $(A\mathbf{x} - B\mathbf{x}, \mathbf{y}^i) = 0$, $i = 1,\ldots, n$. If $(A\mathbf{x}^0)_r = 0$ then for some k, $(\mathbf{y}^k)_i = \delta_{ri}$ so that $(A\mathbf{x} - B\mathbf{x})_r = (A\mathbf{x} - B\mathbf{x}, \mathbf{y}^k) = 0$. Thus $d \neq 0$ and $\mathbf{z} \neq A\mathbf{x}^0$.

Now $(\mathbf{z}, \mathbf{y}^i) = 0$, $i = 1,\ldots, n$ and $(\mathbf{z})_r = 0$. From Proposition 2.2 of Chapter III, it therefore follows that $S^+(\mathbf{z}) \geqq n + 1$. However,

$$S^+(\mathbf{z}) = S^+(A\mathbf{x}^0 - d(A\mathbf{x} - B\mathbf{x})) \leqq S^-(\mathbf{x}^0 - d(\mathbf{x} - P\mathbf{x})).$$

Since, by assumption, $|d| < 1$, $\|\mathbf{x}\|_\infty \leqq 1$, and $(P\mathbf{x})_m = 0$, $m \notin \mathbf{j}^0$, we see that $\operatorname{sgn}(\mathbf{x}^0 - d(\mathbf{x} - P\mathbf{x}))_m = \operatorname{sgn}(\mathbf{x}^0)_m$, $m \notin \mathbf{j}^0$. Therefore $S^-(\mathbf{x}^0 - d(\mathbf{x} - P\mathbf{x})) \leqq n$. This contradiction proves that $\delta_n(\mathscr{A}_\infty; l_\infty^M) \leqq \| A\mathbf{x}^0 \|_\infty$. □

Corollary 3.7. *For A as above,*

$$d_n(\mathscr{A}_\infty; l_\infty^M)\, d_{M-n-1}(\mathscr{A}_\infty^{-1}; l_\infty^M) = 1$$

for $n = 0, 1, \ldots, M-1$, *where*

$$\mathscr{A}_\infty^{-1} = \{A^{-1}\mathbf{x}: \|\mathbf{x}\|_\infty \leqq 1\}.$$

Proof. The matrix A^{-1} is not STP. However $DA^{-1}D$ is STP where D is the $M \times M$ diagonal matrix with ith diagonal entry equal to $(-1)^i$, $i = 1, \ldots, M$. As such it follows that the results of this section apply to the calculation of $d_n(\mathscr{A}_\infty^{-1}; l_\infty^M)$, with the obvious changes. In particular, $d_{M-n-1}(\mathscr{A}_\infty^{-1}; l_\infty^M) = d^{M-n-1}(\mathscr{A}_\infty^{-1}; l_\infty^M)$. From Theorem 3.4 $d_n(\mathscr{A}_\infty; l_\infty^M) = b_n(\mathscr{A}_\infty; l_\infty^M)$. Thus

$$d_n(\mathscr{A}_\infty; l_\infty^M)\, d_{M-n-1}(\mathscr{A}_\infty^{-1}; l_\infty^M) = b_n(\mathscr{A}_\infty; l_\infty^M)\, d^{M-n-1}(\mathscr{A}_\infty^{-1}; l_\infty^M).$$

This latter quantity is equal to one by Proposition 1.2. □

Before concluding this section, we state the matrix analogue of the results of Section 5 of Chapter IV.

Assume that $A = (a_{ij})_{i,j=1}^M$ is an $M \times M$ STP matrix. (It actually suffices to assume that A is oscillatory (see Definition 2.4 of Chapter III).) From Theorem 2.8 of Chapter III the STP matrix $A^T A$ has eigenvalues $\lambda_1 > \ldots > \lambda_M > 0$ and associated eigenvectors $\mathbf{x}^1, \ldots, \mathbf{x}^M$. Set $\mathbf{y}^i = A\mathbf{x}^i$, $i = 1, \ldots, M$. Then

$$S^+(\mathbf{x}^k) = S^-(\mathbf{x}^k) = k-1, \quad \text{and} \quad S^+(\mathbf{y}^k) = S^-(\mathbf{y}^k) = k-1.$$

Thus $(\mathbf{x}^k)_1$, $(\mathbf{x}^k)_M \neq 0$, and if $(\mathbf{x}^k)_j = 0$ for $1 < j < M$, then $(\mathbf{x}^k)_{j-1}(\mathbf{x}^k)_{j+1} < 0$. Therefore at the ith sign change of $\mathbf{x}^{n+1}$ one of two possibilities occur. Either

(a) $(\mathbf{x}^{n+1})_{k_i}(\mathbf{x}^{n+1})_{k_i+1} < 0$, or
(b) $(\mathbf{x}^{n+1})_{k_i} = 0$, and $(\mathbf{x}^{n+1})_{k_i-1}(\mathbf{x}^{n+1})_{k_i+1} < 0$,

where $1 \leqq k_1 < \ldots < k_n < M$.

For each $i \in \{1, \ldots, n\}$, we define $\mathbf{z}^i \in \mathbb{R}^M$ as follows:
If (a) holds, set

$$(\mathbf{z}^i)_r = \begin{cases} |(\mathbf{x}^{n+1})_r|^{-1}, & r = k_i,\ k_i+1 \\ 0, & \textit{otherwise} \end{cases}$$

and if (b) holds, set $(\mathbf{z}^i)_r = \delta_{r,k_i}$, $r = 1, \ldots, M$.

Thus $(\mathbf{z}^i, \mathbf{x}^{n+1}) = 0$, $i = 1, \ldots, n$ and, as is easily checked, the $\{\mathbf{z}^i\}_{i=1}^n$ are linearly independent. From the vector $\mathbf{y}^{n+1}$, we similarly construct $\mathbf{w}^1, \ldots, \mathbf{w}^n \in \mathbb{R}^M$ such that $(\mathbf{w}^i, \mathbf{y}^{n+1}) = 0$, $i = 1, \ldots, n$. Then we have

Theorem 3.8. *For* $A \in \mathbb{R}^{M \times M}$ *STP*,

$$d_n(\mathscr{A}_2; l_2^M) = \lambda_{n+1}^{1/2}, \qquad n = 0, 1, \ldots, M-1.$$

Furthermore the following three subspaces are optimal.

(1) $X_n^1 = \text{span}\{\mathbf{y}^1, \ldots, \mathbf{y}^n\}$,
(2) $X_n^2 = \text{span}\{A\mathbf{z}^1, \ldots, A\mathbf{z}^n\}$,
(3) $X_n^3 = \text{span}\{AA^T\mathbf{w}^1, \ldots, AA^T\mathbf{w}^n\}$.

Also, if P *is the* $M \times M$ *matrix defined by requiring that* $P: \mathbb{R}^M \to X_n^2$ *and* $(\mathbf{x} - P\mathbf{x}, \mathbf{w}^i) = 0$, $i = 1, \ldots, n$ *for all* $\mathbf{x} \in \mathbb{R}^M$, *then*

$$\sup_{\|\mathbf{x}\|_2 \leq 1} \|A\mathbf{x} - PA\mathbf{x}\|_2 = \lambda_{n+1}^{1/2}.$$

Notes and References

Subsection 2.1. Theorem 2.1 for d_n and all p goes back at least to Mityagin, Tichomirov [1964]. The other portions of Theorem 2.1 are equally known, but the references are unclear. Theorem 2.2 was proved, independently, by Pietsch [1974] and Stesin [1975]. The proof presented here is due to Pietsch. Stesin's proof is rather laborious and brings in the Aleksandrov n-widths. Theorem 2.7 was originally announced, but never published by S. A. Smolyak. The first published proof is due to Sofman [1969]. Hutton, Morrell, Retherford [1976] also proved this result, unaware of previous work. For the case $D = I$, the identity matrix, the theorem may also be found in Stechkin [1954] and Solomjak, Tichomirov [1967]. See also the review paper of Pinkus [1979 b]. Many of the results of this subsection extend to infinite dimensional spaces, i.e., $M = \infty$. See, for example, Pietsch [1974], Ha [1974], Hutton [1974], Hutton, Morrell, Retherford [1976], and Johnson [1973].

Subsection 2.2. Theorem 2.13 is due to Pinkus [1979 b]. The proof given here is a somewhat simplified version of Melkman [1980]. Propositions 2.14 and 2.15 are due to Melkman [1981]. Theorem 2.16 is stated in Melkman [1981], where the connection between the lower bounds for n-widths and the topic of equiangular lines was noted. The proof of the theorem itself is due to Lemmens, Seidel [1973]. Proposition 2.17 is a result of Melkman [1981]. Proposition 2.18 was established for $k = n = 2^l$ by König [1977] (see also Pinkus [1979 b]). Theorem 2.19 is of course a result of Kashin [1974] and Theorem 2.20 (together with Proposition 2.21) is due to Höllig [1979 a]. Theorem 2.22 is also a result of Höllig [1979 a]. Proposition 2.25 may be found in Ismagilov [1974] (see also Kolli [1974]). Theorem 2.26 is due to Kashin [1977 b]. For $M = 2n$, Szarek [1978] has reproved, by a different method, Kashin's result.

Section 3. The material of this section is taken from Micchelli, Pinkus [1977 a]. The matrix analogue, for *STP* matrices, of the results of Section 5 of Chapter IV may be found in Melkman, Micchelli [1978].

Chapter VII. Asymptotic Estimates for n-Widths of Sobolev Spaces

1. Introduction

Let $W_p^{(r)}[0,1] = W_p^{(r)} = \{f : f \in C^{r-1}[0,1], f^{(r-1)} \text{ abs. cont.}, f^{(r)} \in L^p[0,1]\}$, and set $B_p^{(r)} = \{f : f \in W_p^{(r)}, \|f^{(r)}\|_p \leqq 1\}$. We are here concerned with the asymptotic behaviour of the values $d_n(B_p^{(r)}; L^q)$, $d^n(B_p^{(r)}; L^q)$ and $\delta_n(B_p^{(r)}; L^q)$ for $p, q \in [1, \infty]$ as n grows. It is only recently that this problem has been solved for all $p, q \in [1, \infty]$ (and almost all r). We first state the major result proven in this chapter.

We write $d_n(B_p^{(r)}; L^q) \asymp n^s$ to mean that there exist positive constants C and D, independent of n, although they may and generally do depend upon p, q and r, for which $Dn^s \leqq d_n(B_p^{(r)}; L^q) \leqq Cn^s$ for all n sufficiently large (with the analogous meaning for d^n and δ_n).

Theorem 1.1. *For $r \geqq 2$ and $1/p + 1/p' = 1$,*

$$(1)\quad d_n(B_p^{(r)}; L^q) \asymp \begin{cases} n^{-r}, & 1 \leqq q \leqq p \leqq \infty \\ n^{-r}, & 2 \leqq p \leqq q \leqq \infty \\ n^{-r+1/p-1/2}, & 1 \leqq p \leqq 2 \leqq q \leqq \infty \\ n^{-r+1/p-1/q}, & 1 \leqq p \leqq q \leqq 2. \end{cases}$$

$$(2)\quad d^n(B_p^{(r)}; L^q) \asymp \begin{cases} n^{-r}, & 1 \leqq q \leqq p \leqq \infty \\ n^{-r}, & 1 \leqq p \leqq q \leqq 2 \\ n^{-r+1/2-1/q}, & 1 \leqq p \leqq 2 \leqq q \leqq \infty \\ n^{-r+1/p-1/q}, & 2 \leqq p \leqq q \leqq \infty. \end{cases}$$

$$(3)\quad \delta_n(B_p^{(r)}; L^q) \asymp \begin{cases} n^{-r}, & 1 \leqq q \leqq p \leqq \infty \\ n^{-r+1/p-1/q}, & 1 \leqq p \leqq q \leqq 2 \\ n^{-r+1/p-1/q}, & 2 \leqq p \leqq q \leqq \infty \\ n^{-r+1/p-1/2}, & 1 \leqq p \leqq 2 \leqq q \leqq \infty,\ p' \geqq q \\ n^{-r+1/2-1/q}, & 1 \leqq p \leqq 2 \leqq q \leqq \infty,\ p' \leqq q. \end{cases}$$

(The condition on r may be further weakened, but we shall not concern ourselves with this problem.)

For convenience we present these results in graph form.

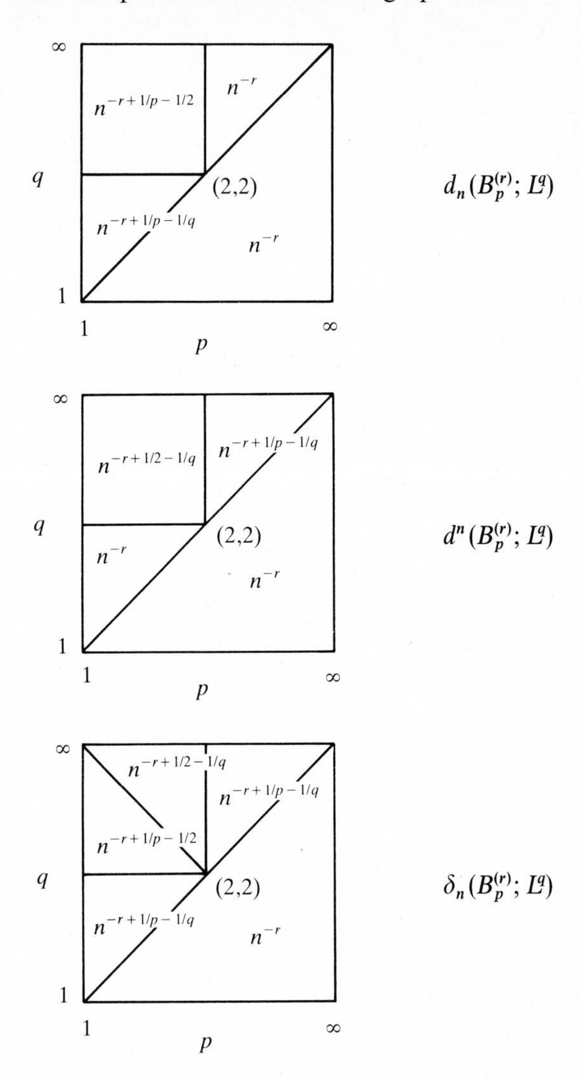

We mention some extensions of these results. Theorem 1.1 is stated for the Sobolev space of functions defined on $[0, 1]$. These results extend with minor modifications to the unit cube in $\mathbb{R}^N$, i.e., to $E^N = \{\mathbf{x}: \mathbf{x} = (x_1, \ldots, x_N), 0 \leqq x_i \leqq 1\}$. In this case we replace r by r/N in all statements in Theorem 1.1. It is then also necessary that the associated $B_p^{(r)}$ lies in L^q, i.e., that the Sobolev Embedding Theorem obtains, or in other words that $r/N - 1/p + 1/q > 0$. Theorem 1.1 can also be extended to more general compact domains in $\mathbb{R}^N$ while interpolation theory techniques also allow for an extension of these results to Besov spaces.

It transpires that the optimal lower bounds for all $p, q \in [1, \infty]$ are more easily obtained that the optimal upper bounds. The lower bounds are calculated in

Section 2 by a discretization technique. The upper bounds are proven in Section 3. The main tool used to analyse the more difficult cases is again a discretization technique (factorization and telescoping).

In Section 4 we present a different and rather interesting proof of the upper bound for $\delta_n(B_1^{(r)}; L^\infty)$.

For notational convenience the constant C will be used throughout this chapter as a generic constant independent of n (and of M, where appropriate).

2. Optimal Lower Bounds

The lower bound estimates follow from a discretization method and some simple known values of $d_n(\mathscr{I}_p; l_q^M)$ (as given in Chapter VI).

Theorem 2.1. *For $p, q \in [1, \infty]$,*

(1) $d_n(B_p^{(r)}; L^q) \geqq CM^{-r+1/p-1/q} d_n(\mathscr{I}_p; l_q^M)$,

(2) $d^n(B_p^{(r)}; L^q) \geqq CM^{-r+1/p-1/q} d^n(\mathscr{I}_p; l_q^M)$,

where M is any positive integer.

Proof. Let ϕ be any non-zero function in $C^\infty(\mathbb{R})$ with support in $[0, 1]$, i.e., $\operatorname{supp}(\phi) \subseteq [0, 1]$. Set $\phi_k(x) = \phi(xM - (k-1))$, $k = 1, \ldots, M$. Thus $\operatorname{supp}(\phi_k) \subseteq [k - 1/M, k/M]$, $k = 1, \ldots, M$.

For every $s \in [1, \infty]$, a simple change of variable argument shows that $\|\phi_k\|_s = M^{-1/s}\|\phi\|_s$, and $\|\phi_k^{(r)}\|_s = M^{r-1/s}\|\phi^{(r)}\|_s$. Furthermore, since the ϕ_k have essentially distinct support,

$$\left\| \sum_{k=1}^M a_k \phi_k \right\|_s = M^{-1/s} \|\mathbf{a}\|_s \|\phi\|_s,$$

where $\mathbf{a} = (a_1, \ldots, a_M)$ and $\|\mathbf{a}\|_s$ is the usual l_s^M norm on $\mathbf{a}$. (We are abusing notation by using $\|\cdot\|_s$ for both the $L^s[0, 1]$ and the l_s^M norms.) Similarly,

$$\left\| \sum_{k=1}^M a_k \phi_k^{(r)} \right\|_s = M^{r-1/s} \|\mathbf{a}\|_s \|\phi^{(r)}\|_s.$$

Now

$$\begin{aligned} d_n(B_p^{(r)}; L^q) &= \inf_{X_n} \sup_{\|f^{(r)}\|_p \leqq 1} \inf_{g \in X_n} \|f - g\|_q \\ &= \inf_{X_n} \sup_{\|f^{(r)}\|_p \leqq 1} \sup_{h \perp X_n,\ \|h\|_{q'} \leqq 1} (f, h) \end{aligned}$$

where $1/q + 1/q' = 1$. (See Proposition 6.1 of Chapter II.) Set $Y_M = \operatorname{span}\{\phi_1, \ldots, \phi_M\}$. Thus

$$d_n(B_p^{(r)}; L^q) \geqq \inf_{X_n} \sup_{\|f^{(r)}\|_p \leqq 1,\ f \in Y_M} \sup_{h \perp X_n,\ \|h\|_{q'} \leqq 1,\ h \in Y_M} (f, h).$$

Let $f = \sum_{k=1}^{M} a_k \phi_k$ and $h = \sum_{k=1}^{M} b_k \phi_k$. Thus $\|f^{(r)}\|_p = M^{r-1/p}\|\phi^{(r)}\|_p \|\mathbf{a}\|_p$, $\|h\|_{q'} = M^{-1/q'}\|\phi\|_{q'}\|\mathbf{b}\|_{q'}$, and $(f, h) = \sum_{k=1}^{M} a_k b_k \|\phi_k\|_2^2 = M^{-1}\|\phi\|_2^2 \sum_{k=1}^{M} a_k b_k$.

Assume $X_n = \operatorname{span}\{g_1, \ldots, g_n\}$ and define $d_{ij} = (g_i, \phi_j)$, $i = 1, \ldots, n$; $j = 1, \ldots, M$. Let $\mathbf{d}^i$, $i = 1, \ldots, n$, denote the ith row of the matrix $D = (d_{ij})$. Then

$$\begin{aligned} d_n(B_p^{(r)}; L^q) &\geqq C M^{-r+1/p+1/q'-1} \inf_{\mathbf{d}^1, \ldots, \mathbf{d}^n} \sup_{\|\mathbf{a}\|_p \leqq 1} \sup_{\substack{(\mathbf{d}^i, \mathbf{b}) = 0,\, i=1, \ldots, n \\ \|\mathbf{b}\|_{q'} \leqq 1}} (\mathbf{a}, \mathbf{b}) \\ &= C M^{-r+1/p-1/q} \inf_{\mathbf{d}^1, \ldots, \mathbf{d}^n} \sup_{\|\mathbf{a}\|_p \leqq 1} \inf_{\alpha_1, \ldots, \alpha_n} \left\| \mathbf{a} - \sum_{i=1}^{n} \alpha_i \mathbf{d}_i \right\|_q \\ &= C M^{-r+1/p-1/q} d_n(\mathscr{I}_q; l_q^M), \end{aligned}$$

where $C = \|\phi\|_2^2 \|\phi\|_{q'}^{-1} \|\phi^{(r)}\|_p^{-1}$.

An analogous and in fact somewhat simpler argument proves (2). $\square$

As a consequence of Theorem 2.1, we obtain the lower bounds of Theorem 1.1.

Theorem 2.2. *For given $p, q \in [1, \infty]$, $1/p + 1/p' = 1$, and r a positive integer,*

$$(1)\quad d_n(B_p^{(r)}; L^q) \geqq \begin{cases} C n^{-r}, & 1 \leqq q \leqq p \leqq \infty \\ C n^{-r}, & 2 \leqq p \leqq q \leqq \infty \\ C n^{-r+1/p-1/2}, & 1 \leqq p \leqq 2 \leqq q \leqq \infty \\ C n^{-r+1/p-1/q}, & 1 \leqq p \leqq q \leqq 2. \end{cases}$$

$$(2)\quad d^n(B_p^{(r)}; L^q) \geqq \begin{cases} C n^{-r}, & 1 \leqq q \leqq p \leqq \infty \\ C n^{-r}, & 1 \leqq p \leqq q \leqq 2 \\ C n^{-r+1/2-1/q}, & 1 \leqq p \leqq 2 \leqq q \leqq \infty \\ C n^{-r+1/p-1/q}, & 2 \leqq p \leqq q \leqq \infty. \end{cases}$$

$$(3)\quad \delta_n(B_p^{(r)}; L^q) \geqq \begin{cases} C n^{-r}, & 1 \leqq q \leqq p \leqq \infty \\ C n^{-r+1/p-1/q}, & 1 \leqq p \leqq q \leqq 2 \\ C n^{-r+1/p-1/q}, & 2 \leqq p \leqq q \leqq \infty \\ C n^{-r+1/p-1/2}, & 1 \leqq p \leqq 2 \leqq q \leqq \infty,\ p' \geqq q \\ C n^{-r+1/2-1/q}, & 1 \leqq p \leqq 2 \leqq q \leqq \infty,\ p' \leqq q. \end{cases}$$

Proof. First note that (3) is an immediate consequence of (1) and (2), and of the inequality $\delta_n(B_p^{(r)}; L^q) \geqq \max\{d_n(B_p^{(r)}; L^q), d^n(B_p^{(r)}; L^q)\}$. We prove (1) and (2) by applying Theorem 2.1 with $M = 2n$. We must therefore bound $d_n(\mathscr{I}_p; l_q^{2n})$ and $d^n(\mathscr{I}_p; l_q^{2n})$ from below by the appropriate power of n. From Proposition 1.1 of Chapter VI $d^n(\mathscr{I}_p; l_q^M) = d_n(\mathscr{I}_{q'}; l_{p'}^M)$ where $1/p + 1/p' = 1/q + 1/q' = 1$. It therefore suffices to bound $d_n(\mathscr{I}_p; l_q^{2n})$ from below.

For $1 \leqq s \leqq t \leqq \infty$, and $\mathbf{x} \in \mathbb{R}^{2n}$, $\|\mathbf{x}\|_t \leqq \|\mathbf{x}\|_s \leqq (2n)^{1/s-1/t} \|\mathbf{x}\|_t$.

We now consider each of the four cases as delineated in (1) and substitute the following estimates into Theorem 2.1.

1. $1 \leqq q \leqq p < \infty$. $d_n(\mathcal{I}_p; l_q^{2n}) = n^{1/q-1/p}$, by Theorem 2.2 of Chapter VI.

2. $2 \leqq p \leqq q \leqq \infty$. $d_n(\mathcal{I}_p; l_q^{2n}) \geqq (2n)^{1/q-1/p} d_n(\mathcal{I}_q; l_q^{2n}) = (2n)^{1/q-1/p}$, by Theorem 2.1 of Chapter VI.

3. $1 \leqq p \leqq 2 \leqq q \leqq \infty$. $d_n(\mathcal{I}_p; l_q^{2n}) \geqq d_n(\mathcal{I}_1; l_q^{2n}) \geqq (2n)^{1/q-1/2} d_n(\mathcal{I}_1; l_2^{2n}) = (2n)^{1/q-1/2} 2^{-1/2}$, by Corollary 2.11 of Chapter VI.

4. $1 \leqq p \leqq q \leqq 2$. $d_n(\mathcal{I}_p; l_q^{2n}) \geqq d_n(\mathcal{I}_1; l_2^{2n}) = 2^{-1/2}$, by Corollary 2.11 of Chapter VI.

This proves the theorem. □

3. Optimal Upper Bounds

Our first upper bound (Theorem 3.3) estimates $\delta_n(B_p^{(r)}; L^q)$ from above. As a consequence of Theorem 2.2 this upper bound is seen to be optimal for d_n, d^n and δ_n when $1 \leqq q \leqq p \leqq \infty$, for d_n and δ_n when $1 \leqq p \leqq q \leqq 2$, and for d^n and δ_n when $2 \leqq p \leqq q \leqq \infty$. The original proof of this upper bound is due to Birman and Solomjak [1967]. We prefer, however, to give a somewhat different proof more in line with the subsequent analysis.

Given k, subdivide the interval $[0, 1]$ into 2^k subintervals, each of length 2^{-k}. Let $\mathcal{S}_k^r$ denote the class of splines of degree r with simple knots at $i2^{-k}$, $i = 1, \ldots, 2^k - 1$. Thus $\mathcal{S}_k^r$ is a linear space of dimension $b_k = r + 2^k$. $\mathcal{S}_k^r$ is, in fact, spanned by b_k B-splines $\{M_i\}_{i=1}^{b_k}$. Normalize these B-splines so that $\sum_{i=1}^{b_k} M_i = 1$. Each $s \in \mathcal{S}_k^r$ may be uniquely written in the form $s = \sum_{i=1}^{b_k} a_i M_i$. For each such s, set $\mathbf{a} = (a_1, \ldots, a_{b_k}) \in \mathbb{R}^{b_k}$. The first fact necessary in the subsequent analysis is the following.

Theorem 3.1 (de Boor [1973]). *For* $s = \sum_{i=1}^{b_k} a_i M_i$, $\mathbf{a} = (a_1, \ldots, a_{b_k})$, $1 \leqq p \leqq \infty$, *there exists constants C and D, independent of k, for which*

$$D\, 2^{k/p} \|s\|_p \leqq \|\mathbf{a}\|_p \leqq C\, 2^{k/p} \|s\|_p.$$

Let P_k be any linear operator from $W_p^{(r)}$ to $\mathcal{S}_k^r$ such that

$$\|f - P_k f\|_p \leqq C\, 2^{-kr} \|f^{(r)}\|_p$$

for some C, independent of k. Many such operators exist, see e.g. de Boor and Fix [1973].

Set $T_0 = P_0$ and $T_k = P_k - P_{k-1}$, $k = 1, \ldots$. Then $f = \sum_{k=0}^{\infty} T_k f$ for $f \in W_p^{(r)}$. (This follows from the above estimate and the fact that $\sum_{k=\infty}^{N} T_k f = P_N f$.) We also have

Proposition 3.2. *For* $f \in W_p^{(r)}$, $1 \leqq p \leqq q \leqq \infty$, $r \geqq 2$,

$$\| T_k f \|_q \leqq C\, 2^{-k(r-1/p+1/q)} \| f^{(r)} \|_p .$$

Proof. For $1 \leqq p \leqq q \leqq \infty$ and $\mathbf{a} \in \mathbb{R}^{b_k}$, $\| \mathbf{a} \|_p \geqq \| \mathbf{a} \|_q$. Thus if $s \in \mathscr{S}_k^r$ then by Theorem 3.1,

$$\| s \|_q \leqq C\, 2^{-k/q} \| \mathbf{a} \|_q \leqq C\, 2^{-k/q} \| \mathbf{a} \|_p \leqq C\, 2^{-k(1/q-1/p)} \| s \|_p .$$

Furthermore

$$\| T_k f \|_p = \| P_k f - P_{k-1} f \|_p \leqq \| f - P_k f \|_p + \| f - P_{k-1} f \|_p \leqq C\, 2^{-kr} \| f^{(r)} \|_p .$$

Now, by definition, $T_k f \in \mathscr{S}_k^r$. Thus for $1 \leqq p \leqq q \leqq \infty$,

$$\| T_k f \|_q \leqq C\, 2^{-k(1/q-1/p)} \| T_k f \|_p \leqq C\, 2^{-k(r-1/p+1/q)} \| f^{(r)} \|_p . \quad \square$$

Theorem 3.3. *For* $p, q \in [1, \infty]$, *and* $r \geqq 2$,

$$\delta_n(B_p^{(r)}; L^q) \leqq \begin{cases} C n^{-r}, & 1 \leqq q \leqq p \leqq \infty \\ C n^{-r+1/p-1/q}, & 1 \leqq p \leqq q \leqq \infty . \end{cases}$$

Proof. For $1 \leqq q \leqq p \leqq \infty$, $\| f \|_q \leqq \| f \|_p$. It therefore suffices to prove that $\delta_n(B_p^{(r)}; L^q) \leqq C n^{-r+1/p-1/q}$ for $1 \leqq p \leqq q \leqq \infty$. Let k_0 be such that $b_{k_0} \leqq n < b_{k_0+1}$. Then

$$\delta_n(B_p^{(r)}; L^q) \leqq \delta_{b_{k_0}}(B_p^{(r)}; L^q) \leqq \sup_{f^{(r)} \neq 0} \frac{\| f - P_{k_0} f \|_q}{\| f^{(r)} \|_p},$$

where P_k is as above. Now,

$$f = \sum_{k=0}^{\infty} T_k f = \sum_{k=0}^{k_0} T_k f + \sum_{k=k_0+1}^{\infty} T_k f = P_{k_0} f + \sum_{k=k_0+1}^{\infty} T_k f .$$

Therefore,

$$\delta_n(B_p^{(r)}; L^q) \leqq \sup_{f^{(r)} \neq 0} \frac{\left\| \sum_{k=k_0+1}^{\infty} T_k f \right\|_q}{\| f^{(r)} \|_p} .$$

From Proposition 3.2, for $1 \leqq p \leqq q \leqq \infty$,

$$\left\| \sum_{k=k_0+1}^{\infty} T_k f \right\|_q \leqq \sum_{k=k_0+1}^{\infty} \| T_k f \|_q \leqq \sum_{k=k_0+1}^{\infty} C\, 2^{-k(r-1/p+1/q)} \| f^{(r)} \|_p .$$

Since $r - 1/p + 1/q > 0$,

$$\sum_{k=k_0+1}^{\infty} 2^{-k(r-1/p+1/q)} \leqq C 2^{-(k_0+1)(r-1/p+1/q)} \leqq C n^{-r+1/p-1/q}. \quad \square$$

It remains to obtain optimal upper bounds for $d_n(B_p^{(r)}; L^q)$ when $p < q$, $q > 2$; for $d^n(B_p^{(r)}; L^q)$ when $p < q$, $p < 2$; and for $\delta_n(B_p^{(r)}; L^q)$ when $p < 2 < q$. It was for a very long time generally thought that the upper bound of Theorem 3.3 was asymptotically optimal for all $p, q \in [1, \infty]$ and that the lower bound of Theorem 2.2 needed to be improved. In fact, of course, the opposite is true. We shall utilize a discretization technique in order to obtain the remaining upper bounds.

Theorem 3.4. *Let $r \geqq 2$ and let s_n denote any of the quantities d_n, d^n or δ_n. Then*

$$s_n(B_p^{(r)}; L^q) \leqq C \sum_{k=0}^{\infty} 2^{-k(r-1/p+1/q)} s_{n_k}(\mathscr{I}_p; l_q^{b_k}),$$

where $b_k = r + 2^k$, and the n_k are non-negative integers for which $\sum_{k=0}^{\infty} n_k \leqq n$.

Before proving this theorem we recall two facts (Proposition 7.6 of Chapter II) concerning the above s_n.

A) Let $T_i \in L(X, Y)$, $i = 1, 2$. Then

$$s_{n+m}(T_1 + T_2; X; Y) \leqq s_n(T_1; X; Y) + s_m(T_2; X; Y).$$

(Recall that $s_n(T; X; Y)$ is the appropriate n-width of the set $\{Tx \colon \|x\|_X \leqq 1\}$ as a subset of Y.)

B) Let $T_1 \in L(X, Z)$ and $T_2 \in L(Z, Y)$. Then $T = T_2 T_1 \in L(X, Y)$ and

$$s_{n+m}(T; X; Y) \leqq s_n(T_2; Z; Y) \cdot s_m(T_1; X; Z).$$

In particular, $s_n(T; X; Y) \leqq s_n(T_2; Z; Y) \| T_1 \|_{(X,Z)}$, where

$$s_0(T_1; X; Z) = \| T_1 \|_{(X,Z)} = \sup_{x \neq 0} \| T_1 x \|_Z / \| x \|_X.$$

In the above we considered the n-widths of the image of the unit ball of T in some normed linear space. The set $B_p^{(r)}$ is not precisely of this form. However the difference being only an r-dimensional subspace, this problem may be ignored. (Equivalently some authors define $B_p^{(r)} = \{f \colon f \in W_p^{(r)}, \|f\|_p + \|f^{(r)}\|_p \leqq 1\}$.)

Proof of Theorem 3.4. Since $f = \sum_{k=0}^{\infty} T_k f$ for $f \in W_p^{(r)}$ it follows from (A) that $s_n(B_p^{(r)}; L^q) \leqq C \sum_{k=0}^{\infty} s_{n_k}(T_k; W_p^{(r)}; L^q)$, where $\sum_{k=0}^{\infty} n_k \leqq n$. We wish to prove that $s_{n_k}(T_k; W_p^{(r)}; L^q) \leqq C 2^{-k(r-1/p+1/q)} s_{n_k}(\mathscr{I}_p; l_q^{b_k})$, for all $k = 0, 1, \ldots$.

We factor the operator $T_k: W_p^{(r)} \to L^q \cap \mathscr{S}_k^r$ as follows. $T_k = \tilde{T}_k L_k I L_k^{-1}$, where $\tilde{T}_k$ is simply the operator T_k, but taking $W_p^{(r)}$ to $L^p \cap \mathscr{S}_k^r$. Thus by Proposition 3.2, $\|\tilde{T}_k\|_{(W_p^{(r)}, L^p \cap \mathscr{S}_k^r)} \leqq C 2^{-kr}$. L_k denotes the map of $L^p \cap \mathscr{S}_k^r$ to $l_p^{b_k}$ such that $L_k s = \mathbf{a} \in \mathbb{R}^{b_k}$, where $s = \sum_{k=1}^{b_k} a_i M_i$. Thus by Theorem 3.1, $\|L_k\|_{(L^p \cap \mathscr{S}_k^r, l_p^{b_k})} \leqq C 2^{k/p}$. I denotes the identity map of $l_p^{b_k}$ into $l_q^{b_k}$, and L_k^{-1} is simply the inverse of L_k, but taking $l_q^{b_k}$ into $L^q \cap \mathscr{S}_k^r$. Thus $\|L_k^{-1}\|_{(l_q^{b_k}, L^q \cap \mathscr{S}_k^r)} \leqq C 2^{-k/q}$. Graphically,

$$W_p^{(r)} \xrightarrow{\tilde{T}_k} L^p \cap \mathscr{S}_k^r \xrightarrow{L_k} l_p^{b_k} \xrightarrow{I} l_q^{b_k} \xrightarrow{L_k^{-1}} L^q \cap \mathscr{S}_k^r .$$

Thus by (B),

$$s_{n_k}(T_k; W_p^{(r)}; L^q) \leqq \|\tilde{T}_k\|_{(W_p^{(r)}, L^p \cap \mathscr{S}_k^r)} \|L_k\|_{(L^p \cap \mathscr{S}_k^r, l_p^{b_k})} s_{n_k}(I; l_p^{b_k}; l_q^{b_k}) \|L_k^{-1}\|_{(l_q^{b_k}, L^q \cap \mathscr{S}_k^r)}.$$

Substituting the above estimates and since $s_{n_k}(I; l_p^{b_k}; l_q^{b_k})$ is, by definition, $s_{n_k}(\mathscr{I}_p; l_q^{b_k})$ we obtain

$$s_{n_k}(T_k; W_p^{(r)}; L^q) \leqq C 2^{-k(r-1/p+1/q)} s_{n_k}(\mathscr{I}_p; l_q^{b_k}).$$

This proves the theorem. □

With the result of Theorem 3.4, we can now complete the proof of Theorem 1.1. We divide the remaining cases into two parts.

Theorem 3.5. *For* $r \geqq 2$,

$$(1) \quad d_n(B_p^{(r)}; L^q) \leqq \begin{cases} C n^{-r}, & 2 \leqq p \leqq q \leqq \infty \\ C n^{-r+1/p-1/2}, & 1 \leqq p \leqq 2 \leqq q \leqq \infty . \end{cases}$$

$$(2) \quad d^n(B_p^{(r)}; L^q) \leqq \begin{cases} C n^{-r}, & 1 \leqq p \leqq q \leqq 2 \\ C n^{-r+1/2-1/q}, & 1 \leqq p \leqq 2 \leqq q \leqq \infty . \end{cases}$$

Proof. It suffices to prove (1). The proof of (2) is an immediate consequence of the proof of (1) and the equality $d_n(\mathscr{I}_p; l_q^M) = d^n(\mathscr{I}_{q'}; l_{p'}^M)$.

Since $\|f\|_t \leqq \|f\|_s$ for $1 \leqq t \leqq s \leqq \infty$, it follows that $d_n(B_p^{(r)}; L^q) \leqq d_n(B_2^{(r)}; L^\infty)$ for $2 \leqq p \leqq q \leqq \infty$, and $d_n(B_p^{(r)}; L^q) \leqq d_n(B_p^{(r)}; L^\infty)$ for $1 \leqq p \leqq 2 \leqq q \leqq \infty$. It is therefore sufficient to prove that $d_n(B_p^{(r)}; L^\infty) \leqq C n^{-r+1/p-1/2}$ for $1 \leqq p \leqq 2$.

The proof of the above inequality will depend on the following estimate from Theorem 2.26 of Chapter VI,

$$d_n(\mathscr{I}_2; l_\infty^M) \leqq C n^{-1/2}(1 + \ln(M/n))^{3/2}.$$

Thus for $1 \leqq p \leqq 2$ we also have

$$d_n(\mathscr{I}_p; l_\infty^M) \leqq C n^{-1/2}(1 + \ln(M/n))^{3/2}.$$

We apply the inequality of Theorem 3.4, namely

$$d_n(B_p^{(r)}; L^\infty) \leqq C \sum_{k=0}^{\infty} 2^{-k(r-1/p)} d_{n_k}(\mathscr{I}_p; l_\infty^{b_k}).$$

Since we are only interested in asymptotic upper bounds, we can and will assume that $b_k = 2^k$, all k, and $\sum_{k=0}^{\infty} n_k \leqq Cn$. Let $n \sim 2^m$, and set

$$n_k = \begin{cases} 2^k, & 0 \leqq k \leqq m \\ 2^{2m-k}, & m < k < 2m \\ 0, & 2m \leqq k. \end{cases}$$

Thus $\sum_{k=0}^{\infty} n_k = \sum_{k=0}^{m} 2^k + \sum_{k=m+1}^{2m-1} 2^{2m-k} \leqq 4 \cdot 2^m \leqq Cn$. For $0 \leqq k \leqq m$, $d_{2^k}(\mathscr{I}_p; l_\infty^{2^k}) = 0$ and for $2m \leqq k$, $d_0(\mathscr{I}_p; l_\infty^{2^k}) = 1$. For $m < k < 2m$,

$$\begin{aligned} d_{2^{2m-k}}(\mathscr{I}_p; l_\infty^{2^k}) &\leqq C 2^{-m+(k/2)} (1 + \ln(2^k/(2^{2m-k})))^{3/2} \\ &\leqq C 2^{-m+(k/2)} (k-m)^{3/2}. \end{aligned}$$

Thus,

$$d_n(B_p^{(r)}; L^\infty) \leqq C \sum_{k=m+1}^{2m-1} 2^{-k(r-1/p)} 2^{-m+(k/2)} (k-m)^{3/2} + C \sum_{k=2m}^{\infty} 2^{-k(r-1/p)}.$$

Now,

$$\begin{aligned} \sum_{k=m+1}^{2m-1} 2^{-k(r-1/p)-m+(k/2)} (k-m)^{3/2} &= \sum_{j=1}^{m-1} 2^{-(j+m)(r-1/p)-m+(j+m)/2} j^{3/2} \\ &= 2^{-m(r-1/p+1/2)} \sum_{j=1}^{m-1} 2^{-j(r-1/p-1/2)} j^{3/2} \\ &\leqq C n^{-r+1/p-1/2} \sum_{j=1}^{\infty} 2^{-j(r-1/p-1/2)} j^{3/2}. \end{aligned}$$

Since $r - 1/p - 1/2 > 0$, $\sum_{j=1}^{\infty} 2^{-j(r-1/p-1/2)} j^{3/2} < \infty$. Thus

$$\sum_{k=m+1}^{2m-1} 2^{-k(r-1/p)-m+(k/2)} (k-m)^{3/2} \leqq C n^{-r+1/p-1/2}.$$

Furthermore,

$$\sum_{k=2m}^{\infty} 2^{-k(r-1/p)} = 2^{-2m(r-1/p)} \sum_{j=0}^{\infty} 2^{-j(r-1/p)} \leqq C n^{-r+1/p-1/2}. \quad \square$$

Theorem 3.6. *For* $r \geqq 2$, $1/p + 1/p' = 1$,

$$\delta_n(B_p^{(r)}; L^q) \leqq \begin{cases} C n^{-r+1/p-1/2}, & 1 \leqq p < 2 < q,\ p' \geqq q, \\ C n^{-r+1/2-1/q}, & 1 \leqq p < 2 < q,\ p' \leqq q. \end{cases}$$

Proof. For $p' \geqq q$, $\delta_n(B_p^{(r)}; L^q) \leqq \delta_n(B_p^{(r)}; L^{p'})$, and for $p' \leqq q$, $\delta_n(B_p^{(r)}; L^q) \leqq \delta_n(B_{q'}^{(r)}; L^q)$. It therefore suffices to prove that $\delta_n(B_p^{(r)}; L^{p'}) \leqq C n^{-r+1/p-1/2}$ for $1 \leqq p < 2$. We apply Theorem 3.4 together with the estimate

$$\delta_n(\mathscr{I}_p; l_{p'}^M) \leqq C n^{1/2-1/p} (M/n)^{2/p'} (\ln M/\ln n)^{(2/p)-1}$$

for $1 \leqq p < 2$ (see Corollary 2.23 of Chapter VI).

From Theorem 3.4,

$$\delta_n(B_p^{(r)}; L^{p'}) \leqq C \sum_{k=0}^{\infty} 2^{-k(r-1/p+1/p')} \delta_{n_k}(\mathscr{I}_p; l_{p'}^{b_k}).$$

As in the proof of Theorem 3.5 it suffices to set $b_k = 2^k$, $n \sim 2^m$ and n_k non-negative integers such that $\sum_{k=\infty}^{\infty} n_k \leqq Cn$. As therein set

$$n_k = \begin{cases} 2^k, & 0 \leqq k \leqq m \\ 2^{2m-k}, & m < k < 2m \\ 0, & 2m \leqq k. \end{cases}$$

Thus $\sum_{k=0}^{\infty} n_k \leqq Cn$. For $0 \leqq k \leqq m$, $\delta_{2^k}(\mathscr{I}_p; l_{p'}^{2^k}) = 0$ and for $2m < k$, $\delta_0(\mathscr{I}_p; l_{p'}^{2^k}) = 1$ since $p < p'$. For $m < k < 2m$,

$$\delta_{2^{2m-k}}(\mathscr{I}_p; l_{p'}^{2^k}) \leqq C\, 2^{(2m-k)(1/2-1/p)}\, 2^{(2k-2m)\,2/p'} (k/(2m-k))^{2/p-1}.$$

As in the proof of Theorem 3.5 some simple computations lead to

$$\delta_n(B_p^{(r)}; L^{p'}) \leqq C n^{-r+1/p-1/2}. \quad \square$$

Theorems 2.2, 3.3, 3.5 and 3.6 together prove Theorem 1.1.

4. Another Look at $\delta_n(B_1^{(r)}; L^\infty)$

In the previous section we proved that $s_n(B_1^{(r)}; L^\infty) \asymp n^{-r+(1/2)}$, where s_n is any one of the three quantities d_n, d^n or δ_n. In this section we present a rather interesting and different proof of the upper bound (recall that this was the difficult part). That is, we construct a linear operator P_n of rank $\leqq n$ for which

$$\|f - P_n f\|_\infty \leqq C n^{-r+(1/2)} \|f^{(r)}\|_1$$

for all $f \in W_1^{(r)}$.

In the proof of this result we actually only construct a P_n for which

$$\|f - P_n f\|_\infty \leqq C n^{-(3/2)} \|f^{(2)}\|_1$$

i.e., $\delta_n(B_1^{(2)}; L^\infty) \leqq C n^{-(3/2)}$. This, however, together with the relaively simple estimate $\delta_n(B_\infty^{(r-2)}; L^\infty) \leqq C n^{-(r-2)}$ (see Theorem 3.3) and statement (B) of Section 3 suffices to prove the desired result.

The idea of the proof is to construct a suitable linear operaor S_N of large rank for which

$$\|f - S_N f\|_\infty \leqq C n^{-(3/2)} \|f^{(2)}\|_1$$

for all $f \in W_1^{(2)}$, and then to construct a linear operator P_n of rank $\leqq n$ for which

$$\| S_N f - P_n f \|_\infty \leqq C n^{-(3/2)} \| f^{(2)} \|_1 .$$

Thus

$$\| f - P_n f \|_\infty \leqq \| f - S_N f \|_\infty + \| S_N f - P_n f \|_\infty \leqq C n^{-(3/2)} \| f^{(2)} \|_1 .$$

The construction of the first operator S_N is fairly easy. The difficult problem will be in the construction of P_n.

Proposition 4.1. *Let $f \in W_1^{(2)}$ and let $S_N f$ denote the piecewise linear spline with knots i/N, $i = 1, \ldots, N-1$, which interpolates to f at i/N, $i = 0, 1, \ldots, N$. Then for $N = n^{3/2}$,*

$$\| f - S_N f \|_\infty \leqq n^{-(3/2)} \| f^{(2)} \|_1 ,$$

and $(S_N f)(x) = f(0) + \sum_{i=0}^{N-1} \alpha_i (x - (i/N))_+^1$, where

$$\sum_{i=0}^{N-1} |\alpha_i| \leqq 2 \| f^{(2)} \|_1 .$$

Proof. $(S_N f)(i/N) = f(i/N)$, $i = 0, 1, \ldots, N$, and $(S_N f)(x)$ is linear on each interval $[(i-1)/N, i/N]$, $i = 1, \ldots, N$. Thus $(S_N f)'(x) = N \int_{(i-1)/N}^{i/N} f'(s)\, ds$ for $x \in [(i-1)/N, i/N]$. For such x,

$$\begin{aligned}
|f(x) - (S_N f)(x)| &= \left| \int_{(i-1)/N}^{x} [f'(t) - (S_N f)'(t)]\, dt \right| \\
&= N \left| \int_{(i-1)/N}^{x} \int_{(i-1)/N}^{i/N} [f'(t) - f'(s)]\, ds\, dt \right| \\
&= N \left| \int_{(i-1)/N}^{x} \int_{(i-1)/N}^{i/N} \int_{s}^{t} f^{(2)}(r)\, dr\, ds\, dt \right| \\
&\leqq N \int_{(i-1)/N}^{i/N} \int_{(i-1)/N}^{i/N} \int_{(i-1)/N}^{i/N} |f^{(2)}(r)|\, dr\, ds\, dt \\
&\leqq N^{-1} \| f^{(2)} \|_1 .
\end{aligned}$$

This proves the first part of the proposition.

For $i = 1, \ldots, N-1$,

$$\begin{aligned}
\alpha_i &= N [f((i+1)/N) - 2 f(i/N) + f((i-1)/N)] \\
&= N \int_{i/N}^{(i+1)/N} [f'(t) - f'(t - 1/N)]\, dt .
\end{aligned}$$

Thus

$$|\alpha_i| \leqq N \int_{i/N}^{(i+1)/N} \int_{(i-1)/N}^{(i+1)/N} |f^{(2)}(s)|\, ds\, dt = \int_{(i-1)/N}^{(i+1)/N} |f^{(2)}(s)|\, ds ,$$

and therefore $\sum_{i=1}^{N-1} |\alpha_i| \leqq 2 \| f^{(2)} \|_1$. □

We wish to construct a linear operator P_n of rank $\leqq 2n+1$ (this will suffice since we are only concerned with asymptotic estimates) such that $P_n 1 = 1$, $P_n x = x$, and

$$\|(\cdot - (i/N))_+^1 - P_n((\cdot - (i/N))_+^1)\|_\infty \leqq C n^{-(3/2)},$$

for $i = 1, \ldots, N-1$, where $N = n^{3/2}$. This together with the above estimate $\sum_{i=1}^{N-1} |\alpha_i| \leqq 2\,\|f^{(2)}\|_1$ will prove the desired result. We construct the operator P_n only for $n = 2^{2r}$, every r. Again this is sufficient for our purposes.

Inherent in the construction of P_n is the following combinatorial problem.

Proposition 4.2. *Let $m = 2^r$. Then there exist subsets $B_1, \ldots, B_{m^2}$ of $\{1, \ldots, m^2\}$ satisfying*

(1) $|B_i| = m$, *all* i

(2) $|\{i: j \in B_i\}| = m$, *all* j

(3) $|B_i \cap B_j| \leqq 1$, *all* $i \neq j$.

(For any set B, $|B|$ denotes its cardinality.)

Proof. The existence of sets $B_1, \ldots, B_{m^2}$ satisfying (1) and (3) is a consequence of Proposition 2.21 of Chapter VI (where we here set the m and p of Proposition 2.21 equal to 2 and m, respectively). It remains to prove (2). This follows from a rereading of the proof of Proposition 2.21.

To be precise, let $GF(m)$ denote the field of order m with elements $\{0, 1, \ldots, m-1\}$ where 0 is the additive identity and 1 the multiplicative identity. Let us rewrite $1, \ldots, m^2$ as $\{(i, j): i, j = 0, 1, \ldots, m-1\}$. For each (i, j), we set

$$B_{(i,j)} = \{(i * k + j, k): k = 0, 1, \ldots, m-1\}$$

where $*$ is field multiplication and $+$ field addition. The m^2 sets $B_{(i,j)}$ are the $B_1, \ldots, B_{m^2}$ as above and satisfy (1) and (3). To prove that (2) holds we note that any $(\alpha, \beta) \in B_{(i,j)}$ if and only if $j = \alpha - i * \beta$. There are exactly $m\,(i, j)$'s as above which satisfy this equation. □

The above combinatorial problem is not used directly, but rather to solve another combinatorial problem.

Proposition 4.3. *Set $m = 2^r$, $n = m^2$, $N = m^3$. Let $A_{n+k} = \{(k-1)m + 1, \ldots, km\}$, $k = 1, \ldots, n$. Then there exists an additional partition $A_1, \ldots, A_n$ of $\{1, \ldots, N\}$ which satisfies*

(1) $|A_i| = m$, $i = 1, \ldots, n$

(2) $|A_i \cap A_{n+k}| \leqq 1$ *for all* $i, k \in \{1, \ldots, n\}$

(3) *if* $|A_i \cap A_{n+k}| = |A_j \cap A_{n+k}| = 1$, $i \neq j$, $i, j \in \{1, \ldots, n\}$, *then for all* $h \neq k$, $h \in \{1, \ldots, n\}$, *at least one of* $A_i \cap A_{n+h}$, $A_j \cap A_{n+h}$ *is empty.*

Proof. We construct the desired $A_1, \ldots, A_n$ from $B_1, \ldots, B_n$ as follows. First set $A_1 = \{(k-1)m + 1 : k \in B_1\}$. Since $|B_1| = m$, we have $|A_1| = m$. Now, construct A_i using $A_1, \ldots, A_{i-1}$ and B_i. For each $k \in B_i$, A_i is to contain the smallest integer in A_{n+k} *not* already contained in $A_1, \ldots, A_{i-1}$. By (2) and (3) of Proposition 4.2, such an integer exists.

By construction, $|A_i| = m$, $i = 1, \ldots, n$, and $|A_i \cap A_{n+k}| \leqq 1$, for $i, k \in \{1, \ldots, n\}$. It remains to prove (3). If $|A_i \cap A_{n+k}| = |A_j \cap A_{n+k}| = 1$ for $i \neq j$, then by construction $k \in B_i$ and $k \in B_j$. Since $|B_i \cap B_j| \leqq 1$, it follows that if $h \neq k$, $h \in \{1, \ldots, n\}$, then h cannot be in both B_i and B_j, i.e., at least one of $A_i \cap A_{n+h}$, $A_j \cap A_{n+h}$ is empty. □

As a result of Proposition 4.1 and the remarks thereafter, it suffices to construct a $(2n+1)$-dimensional subspace which contains the functions 1 and x, and functions $\{G_i\}_{i=1}^{N-1}$ such that

$$\|(\cdot - (i/N))_+^1 - G_i(\cdot)\|_\infty \leqq C n^{-(3/2)}, \qquad i = 1, \ldots, N-1.$$

Equivalently it suffices to construct functions $\{g_i\}_{i=0}^{N-1}$, where $g_0(x) = 1$, which lie in some $2n$-dimensional subspace X_{2n} and such that

$$\left| \int_0^x [(t - (i/N))_+^0 - g_i(t)]\, dt \right| \leqq C n^{-(3/2)},$$

$(g_i(t) = G_i'(t))$ for all $x \in [0, 1]$, $i = 1, \ldots, N-1$. This we now do.

Proposition 4.4. *Set $m = 2^r$, $n = m^2$ and $N = m^3$. Let $A_1, \ldots, A_{2n}$ be as in Proposition 4.3. For each μ, let $\delta_{A_\mu}(t)$ be the step function defined on $[0, 1]$ by*

$$\delta_{A_\mu}(t) = \sum_{\nu \in A_\mu} \delta_\nu(t),$$

where

$$\delta_\nu(t) = \begin{cases} 1, & (\nu - 1)/N < t \leqq \nu/N \\ 0, & \text{otherwise}. \end{cases}$$

Set $X_{2n} = \operatorname{span}\{\delta_{A_\mu}\}_{\mu=1}^{2n}$. Then the constant function $\left(\sum_{k=1}^{n} \delta_{A_{n+k}}(t) = 1\right)$ is in X_{2n}, and for each $i = 1, \ldots, N-1$, there exists a $g_i \in X_{2n}$ for which

$$\left| \int_0^x [(t - (i/N))_+^0 - g_i(t)]\, dt \right| \leqq n^{-(3/2)}$$

for all $x \in [0, 1]$.

Proof. Assume $(j-1)m < i < jm$. For $\nu = i+1, \ldots, jm$, let C_ν be the unique A_μ, $\mu \in \{1, \ldots, n\}$ which contains ν. Set

$$e_i(t) = \sum_{\nu = i+1}^{jm} \sum_{\mu \in C_\nu \setminus \{\nu\}} \delta_\mu(t),$$

and

$$m_{ik} = N \int_0^1 \delta_{A_{n+k}}(t)\, e_i(t)\, dt\,, \qquad i = 1, \dots, N-1;\ i \neq jm;\ k = 1, \dots, n\,.$$

Finally, let

$$g_i(t) = \sum_{h=j+1}^{n} \delta_{A_{n+h}}(t) + \sum_{v=i+1}^{jm} \delta_{C_v}(t) - \sum_{h=1}^{n} (m_{ih}/m)\, \delta_{A_{n+h}}(t)\,.$$

For $i = jm$, set $g_i(t) = \sum_{h=j+1}^{n} \delta_{A_{n+h}}(t)$. Thus $g_i \in X_{2n}$, $i = 1, \dots, N-1$.

Now, for $(j-1)m < i < jm$,

$$(t - (i/N))_+^0 = \sum_{v=i+1}^{N} \delta_v(t) = \sum_{v=i+1}^{jm} \delta_v(t) + \sum_{h=j+1}^{n} \delta_{A_{n+h}}(t)\,.$$

Thus

$$\begin{aligned}(t - (i/N))_+^0 - g_i(t) &= \sum_{h=1}^{n} (m_{ih}/m)\, \delta_{A_{n+h}}(t) - \sum_{v=i+1}^{jm} \delta_{C_v}(t) + \sum_{v=i+1}^{jm} \delta_v(t) \\ &= \sum_{h=1}^{n} (m_{ih}/m)\, \delta_{A_{n+h}}(t) - e_i(t)\,,\end{aligned}$$

while for $i = jm$, $(t - (i/N))_+^0 = g_i(t)$. We need only consider the former case. For each $k = 1, \dots, n$,

$$\begin{aligned}\int_{(k-1)m/N}^{km/N} [(t - (i/N))_+^0 - g_i(t)]\, dt &= \int_0^1 \delta_{A_{n+k}}(t) \left[\sum_{h=1}^{n} (m_{ih}/m)\, \delta_{A_{n+h}}(t) - e_i(t) \right] dt \\ &= (m_{ik}/m) \int_0^1 \delta_{A_{n+k}}(t)\, dt - \int_0^1 \delta_{A_{n+k}}(t)\, e_i(t)\, dt \\ &= (m_{ik}/m)(m/N) - (m_{ik}/N) = 0\,.\end{aligned}$$

Furthermore,

$$\begin{aligned}\int_{(k-1)m/N}^{km/N} |(t - (i/N))_+^0 - g_i(t)|\, dt &= \int_0^1 \delta_{A_{n+k}}(t) \left| \sum_{h=1}^{n} (m_{ih}/m)\, \delta_{A_{n+h}}(t) - e_i(t) \right| dt \\ &\leqq (m_{ik}/N) + (m_{ik}/N) = 2(m_{ik}/N)\,.\end{aligned}$$

Thus, if $(k-1)m/N < x \leqq km/N$,

$$\begin{aligned}\left| \int_0^x [(t - (i/N))_+^0 - g_i(t)]\, dt \right| &= \left| \int_{(k-1)m/N}^{x} [(t - (i/N))_+^0 - g_i(t)]\, dt \right| \\ &\leqq (m_{ik}/N)\,.\end{aligned}$$

$\left(\text{If } \int_a^b f(t)\, dt = 0,\ \int_a^b |f(t)|\, dt \leqq C, \text{ then } \left| \int_a^x f(t)\, dt \right| \leqq C/2 \text{ for all } x \in [a, b]. \right)$ Since $N = n^{3/2}$, it therefore suffices to prove that $m_{ik} \leqq 1$ for all i, k.

This is proven as a consequence of Proposition 4.3. Recall that

$$m_{ik} = N \int_0^1 \delta_{A_{n+k}}(t) \left(\sum_{v=i+1}^{jm} \sum_{\mu \in C_v \setminus \{v\}} \delta_\mu(t) \right) dt\,,$$

where $(j-1)m < i < jm$, and C_v is the unique A_μ, $\mu \in \{1, \ldots, n\}$ which contains v. If $k = j$, then since $\{i+1, \ldots, jm\} \in A_{n+j}$, it follows that $v \in C_v \cap A_{n+j}$. Hence by (2) of Proposition 4.3, $|(C_v \backslash \{v\}) \cap A_{n+j}| = 0$. Thus $m_{ij} = 0$. Assume now that $k \neq j$. $|C_v \cap A_{n+j}| = 1$ for all $v = i+1, \ldots, jm$. Thus by (3) of Proposition 4.3, all but at most one of $C_v \cap A_{n+k}$ is empty and the one possibly nonempty intersection has at most one element by (2). Therefore for $k \neq j$, $m_{ik} \leqq 1$. □

We have completed the proof of:

Theorem 4.5. *For* $r \geqq 2$, $\delta_n(B_1^{(r)}; L^\infty) \leqq C n^{-r+(1/2)}$.

Notes and References

The first result concerning asymptotic estimates for Sobolev spaces is contained in the paper of Kolmogorov [1936], where it is proven that $d_n(B_2^{(r)}; L^2) \asymp n^{-r}$. The upper bound on $\delta_n(B_p^{(r)}; L^q)$ for $1 \leqq q \leqq p \leqq \infty$, $1 \leqq p \leqq q \leqq 2$, and $2 \leqq p \leqq q \leqq \infty$, is to be found in the much referenced paper of Birman, Solomjak [1967]. The original proof of the lower bound for $1 \leqq q \leqq p \leqq \infty$ was given by Makovoz [1972] in a rather terse form and elaborated upon by Tichomirov [1976]. The first asymptotic result for $p < q$ seems to have been obtained by Rudin [1952], who essentially proved, in an elegent and interesting way, that $d_n(B_1^{(1)}; L^2) \asymp n^{-1/2}$. Stechkin [1954] generalized this result to obtain $d_n(B_1^{(r)}; L^2) \asymp n^{-r+1/2}$. Solomjak, Tichomirov [1967] proved that $d^n(B_p^{(r)}; L^\infty) \asymp n^{-r+1/p}$ for $p \geqq 2$, by essentially the method presented here. The proof of the asymptotic estimates of d_n and δ_n for $1 \leqq p \leqq q \leqq 2$ was given independently and by different methods by Scholz [1976] and Ismagilov [1974]. In Scholz [1974] and Ismagilov [1974] may also be found the asymptotics of δ_n for $2 \leqq p \leqq q \leqq \infty$. The special case $p = 1$, $q = \infty$, was investigated by Ismagilov [1974], who somewhat surprisingly obtained the estimate $d_n(B_1^{(2)}; L^\infty) \leqq C n^{-6/5}$. (Surprisingly because it had been assumed by many that $n^{-r+1/p-1/q}$ would be the asymptotic order of $d_n(B_p^{(r)}; L^q)$ for all $p < q$.) Glushkin [1974] improved this estimate and showed that $d_n(B_1^{(r)}; L^\infty) \asymp n^{-r+1/2}$. Finally, Kashin [1974], [1977b] was able to complete the picture for d_n, while Maiorov [1978a] and Höllig [1979a] did the same for δ_n.

Section 2. The discretization technique of Theorem 2.1 is due to Maiorov [1975]. A similar statement holds for δ_n (see Höllig [1979a]) but is not needed here.

Section 3. The approach taken in this section may be found in Höllig [1979a]. The discretization technique of Theorem 3.4 is due to Maiorov [1976b] (see also Glushkin [1974]). We follow, more closely, a proof due to Höllig [1979a].

Section 4. The material of this section is taken from de Boor, DeVore, Höllig [1980].

In this chapter we have discussed what we consider to be the main asymptotic estimates for n-widths of Sobolev spaces. However numerous generalizations and related results appear in the literature. The interested reader is referred to Babushka [1971], Desplanches [1977], Galeev [1981], Glushkin [1974], Grigorian [1975], Helfrich [1971], Ismagilov [1968], [1974], Ismagilov, Nasyrova [1977], Jerome [1967], [1968], [1970], [1972], Jerome, Schumaker [1968], [1974], Kashin [1981], Kolli [1971a], [1971b], [1974], König [1978], [1979], Kusainova [1980], Lorentz [1966a], Maiorov [1976a], [1978b], Nasyrova [1976], Oleinik [1976], Otelbaev [1976], Pukhov [1979b], Schock [1975], and Triebel [1978].

Chapter VIII. n-Widths of Analytic Functions

1. Introduction

In this chapter we consider n-widths of various classes of analytic functions. Our concern is with exact n-widths and the identification of optimal subspaces. Asymptotic estimates are relegated to the "Notes and References" at the end of the chapter.

Let $\Delta_R = \{z: |z| < R, z \in \mathbb{C}\}$, and let H_R^p denote the Hardy space of analytic functions on Δ_R which are L^p integrable in the Lebesgue sense on $\partial \Delta_R$. For fixed $m = 0, 1, 2, \ldots$, and $R \geqq 1$, set

$$A_R(m, p) = \{f: f^{(m)} \in H_R^p, \ \| f^{(m)} \|_{H_R^p} \leqq 1\}.$$

In Section 2 we determine the n-widths (d_n, d^n, δ_n and b_n) of $A_R(m, p)$ in H_1^p for all $p \in [1, \infty]$. The value of all these n-widths is $R^{m-n}(n-m)!/n!$, independent of p, and $\pi_{n-1} = \operatorname{span}\{1, z, \ldots, z^{n-1}\}$ is an optimal subspace for d_n. Some generalizations of this result are also presented.

In Section 3 we restrict ourselves to

$$A = A_1(0, 2) = \{f: f \in H_1^2, \ \| f \|_{H_1^2} \leqq 1\},$$

but determine the n-widths of A in $L^2(K, d\mu)$, where μ is a positive measure on a compact subset K of Δ_1. It follows from Theorem 2.2 of Chapter IV that an optimal subspace for d_n is spanned by the first n eigenfunctions of an appropriate eigenvalue problem and that the value of the n-width is the square root of the $(n+1)$st eigenvalue. However from a more considered analysis of this problem we are able to determine an additional optimal subspace (although in some cases these two optimal subspaces coincide).

Let

$$\mathscr{A} = A_1(0, \infty) = \{f: f \in H_1^\infty, \ \| f \|_{H_1^\infty} = \max_{z \in \Delta_1} |f(z)| \leqq 1\}.$$

Let K be a compact subset of Δ_1, and let $X = C(K)$ or $X = L^q(\nu)$ for some $q \in [1\ \infty)$ and some positive measure ν with support on K. In Section 4, we determine $d_n(\mathscr{A}; X)$ (and d^n, δ_n) and show that these quantities all equal

$$\min\{\| B \|_X: B \in \mathscr{B}_n\}$$

where $\mathscr{B}_n$ is the set of Blaschke products of degree $\leqq n$. Optimal subspaces are also constructed which depend on the zeros of the minimal norm Blaschke product.

Finally in Section 5 we consider the following class of functions: Let $\sigma, T > 0$, and

$$\begin{aligned} B(\sigma, T) = \{f : f \text{ entire, of exponential type } \sigma, \text{ and} \\ |f(x)| \leqq 1 \text{ for all } x \in (-\infty, -T) \cup (T, \infty)\}. \end{aligned}$$

We determine the n-widths ($d_n = d^n = \delta_n = b_n$) of $B(\sigma, T)$ in $C[-T, T]$, and exhibit optimal subspaces. This problem is, in a certain sense, the L^∞-analogue of Example 3.5 of Chapter IV.

2. n-Widths of Analytic Functions with Bounded mth Derivative

Set $\Delta_R = \{z : |z| < R, z \in \mathbb{C}\}$. Let f be analytic in Δ_R and for $r < R$, set

$$M_p(r, f) = \begin{cases} \left(1/2\pi \int_0^{2\pi} |f(re^{i\theta})|^p \, d\theta\right)^{1/p}, & 1 \leqq p < \infty \\ \max\{|f(re^{i\theta})| : 0 \leqq \theta < 2\pi\}, & p = \infty. \end{cases}$$

It is a well-known fact (Hardy's convexity theorem) that for f analytic in Δ_R, $p \in [1, \infty]$, $M_p(r, f)$ is a nondecreasing function of r. Set

$$H_R^p = H^p(\Delta_R) = \{f : f \text{ analytic in } \Delta_R, \lim_{r \uparrow R} M_p(r, f) < \infty\}.$$

Thus H_R^∞ is simply the set of bounded analytic functions on Δ_R, while H_R^2 is the class of power series $\sum_{n=0}^\infty a_n z^n$ for which $\sum_{n=0}^\infty R^{2n} |a_n|^2 < \infty$.

For $p \in [1, \infty]$, define

$$\|f\|_{H_R^p} = \lim_{r \uparrow R} M_p(r, f) = \begin{cases} \left(1/2\pi \int_0^{2\pi} |f(Re^{i\theta})|^p \, d\theta\right)^{1/p}, & 1 \leqq p < \infty \\ \operatorname{ess\,sup}\{|f(Re^{i\theta})| : 0 \leqq \theta < 2\pi\}, & p = \infty. \end{cases}$$

The extreme right-hand side is well-defined since if $f \in H_R^p$, then f has non-tangential limits on $\partial \Delta_R$ a.e., and $f(re^{i\theta}) \to f(\mathrm{R}e^{i\theta})$ as $r \uparrow R$ a.e. It is well-known that $\|\cdot\|_{H_R^p}$ is a norm on H_R^p, and H_R^p with this norm is a Banach space.

For fixed $m = 0, 1, \ldots,$ and $R \geqq 1$, set

$$A_R(m, p) = \{f : f^{(m)} \in H_R^p, \|f^{(m)}\|_{H_R^p} \leqq 1\}.$$

It is easily proven that $f^{(m)} \in H_R^p$ implies $f \in H_R^p$. We wish to determine the n-widths and optimal subspaces of $A_R(m, p)$ with respect to the space $H^p (= H_1^p)$.

In order to state the full result, we define $c_k = k!/(k+m)!$, $k = 0, 1, \ldots$, and

$$G_R(t) = R^{m-n} e^{int}\left[c_{n-m} + 2\sum_{k=1}^{\infty} R^{-k} c_{n-m+k} \cos kt\right].$$

Then for $f \in A_R(m, p)$, it is readily checked that

$$f(e^{it}) = p_f(e^{it}) + 1/2\pi \int_0^{2\pi} f^{(m)}(R e^{i\theta})\, e^{im\theta}\, G_R(t-\theta)\, d\theta,$$

where $p_f \in \pi_{n-1} = \operatorname{span}\{1, z, \ldots, z^{n-1}\}$ is dependent of f in a linear fashion (and $p_f = 0$ if $f^{(k)}(0) = 0$, $k = 0, 1, \ldots, n-1$). (The explicit form of p_f will be stated and used later in this section. It is not simply the first n terms of the power series of f.) For $f \in A_R(m, p)$, set $(P_n f)(z) = p_f(z)$. We can now state:

Theorem 2.1. *Let $R \geqq 1$, $p \in [1, \infty]$, $m = 0, 1, \ldots$. Then*

$$d_n(A_R(m,p); H^p) = d^n(A_R(m,p); H^p) = \delta_n(A_R(m,p); H^p) = b_n(A_R(m,p); H^p)$$
$$= \begin{cases} \infty, & n < m \\ R^{m-n}(n-m)!/n!, & n \geqq m. \end{cases}$$

Furthermore,

(1) $\pi_{n-1} = \operatorname{span}\{1, z, \ldots, z^{n-1}\}$ *is an optimal subspace for* d_n.
(2) $L^n = \{f\colon f \in H^p,\ f^{(k)}(0) = 0,\ k = 0, 1, \ldots, n-1\}$ *is optimal for* d^n.
(3) P_n, *as defined above, is optimal for* δ_n.
(4) π_n *is an optimal subspace for* b_n.

Proof. For $n < m$, there is nothing to prove. For $n \geqq m$, we prove that $\delta_n(A_R(m,p); H^p) \leqq R^{m-n}(n-m)!/n!$, and $b_n(A_R(m,p); H^p) \geqq R^{m-n}(n-m)!/n!$ The statements concerning the optimal subspaces are consequences of the proofs of these bounds.

Let P_n be as defined above. Then

$$\delta_n(A_R(m,p); H^p) \leqq \sup_{f \in A_R(m,p)} \| f - P_n f \|_{H^p}$$
$$= \sup_{f \in A_R(m,p)} \left\| 1/2\pi \int_0^{2\pi} f^{(m)}(R e^{i\theta})\, e^{im\theta}\, G_R(\cdot - \theta)\, d\theta \right\|_{L^p[0,2\pi]}.$$

Now,

$$\left(1/2\pi \int_0^{2\pi} \left| 1/2\pi \int_0^{2\pi} f^{(m)}(R e^{i\theta})\, e^{im\theta}\, G_R(t-\theta)\, d\theta \right|^p dt\right)^{1/p}$$
$$\leqq \left(1/2\pi \int_0^{2\pi} \left(1/2\pi \int_0^{2\pi} |f^{(m)}(R e^{i\theta})|\, |G_R(t-\theta)|\, d\theta\right)^p dt\right)^{1/p}$$
$$\leqq \left(1/2\pi \int_0^{2\pi} |G_R(t)|\, dt\right)\left(1/2\pi \int_0^{2\pi} |f^{(m)}(R e^{i\theta})|^p\, d\theta\right)^{1/p}.$$

This is a special case of Young's inequality which states that

$$\| f * g \|_{L^r[0,2\pi]} \leqq \| f \|_{L^p[0,2\pi]} \| g \|_{L^q[0,2\pi]}$$

where $1/r + 1 = 1/p + 1/q$. Thus,

$$\delta_n(A_R(m,p); H^p) \leqq 1/2\pi \int_0^{2\pi} |G_R(t)|\, dt.$$

Now,

$$|G_R(t)| = R^{m-n} \left| c_{n-m} + 2 \sum_{k=1}^{\infty} R^{-k} c_{n-m+k} \cos kt \right|.$$

Set

$$\tilde{G}_R(t) = c_{n-m} + 2 \sum_{k=1}^{\infty} R^{-k} c_{n-m+k} \cos kt.$$

We shall prove that $\tilde{G}_R(t) \geqq 0$ for all $t \in [0, 2\pi)$, and $R \geqq 1$, implying the upper bound

$$\delta_n(A_R(m,p); H^p) \leqq R^{m-n} c_{n-m} = R^{m-n} (n-m)!/n!$$

To prove that $\tilde{G}_R(t) \geqq 0$ for all $t \in [0, 2\pi)$ and $R \geqq 1$, we utilize the following lemmas.

Lemma 2.2. *Let* $H(t) = a_0 + 2 \sum_{k=1}^{\infty} a_k \cos kt$. *If* $\sum_{k=0}^{\infty} |a_k| < \infty$, *and* $\Delta^2 a_k = a_k - 2a_{k+1} + a_{k+2} \geqq 0$ *for all* $k = 0, 1, \ldots$, *then* $H(t) \geqq 0$ *for all* $t \in [0, 2\pi)$.

Proof. Let $D_k(t) = 1/2 + \cos t + \ldots + \cos kt$, and $\tilde{D}_k(t) = D_0(t) + \ldots + D_k(t)$. It is easily verified that

$$\tilde{D}_k(t) = [1 - \cos(k+1)t]/[4 \sin^2(t/2)] \geqq 0,$$

(see e.g. Lorentz [1966a]).

Set $\Delta a_k = a_k - a_{k+1}$. Then

$$\begin{aligned} a_0 + 2 \sum_{k=1}^{n} a_k \cos kt &= 2\left(\sum_{k=0}^{n-1} \Delta a_k D_k(t) + a_n D_n(t) \right) \\ &= 2\left(\sum_{k=0}^{n-2} \Delta^2 a_k \tilde{D}_k(t) + \Delta a_{n-1} \tilde{D}_{n-1}(t) + a_n D_n(t) \right). \end{aligned}$$

For $t \in (0, 2\pi)$ (i.e., $t \neq 0$) the last two terms on the right hand side tend to zero as $n \uparrow \infty$. Since $\Delta^2 a_k \geqq 0$ and $\tilde{D}_k(t) \geqq 0$, it follows that $H(t) = 2 \sum_{k=0}^{\infty} \Delta^2 a_k \tilde{D}_k(t) \geqq 0$, for all $t \in [0, 2\pi)$. □

Lemma 2.3. *Set* $H_r(t) = a_0 + 2 \sum_{k=1}^{\infty} a_k r^k \cos kt$, *where* $\sum_{k=0}^{\infty} |a_k| < \infty$. *If* $a_k \geqq 0$, $\Delta a_k \geqq 0$, *and* $\Delta^2 a_k \geqq 0$ *for all* k, *then* $H_r(t) \geqq 0$ *for all* $t \in [0, 2\pi)$ *and* $r \in [0, 1]$.

Proof. From Lemma 2.2, it suffices to show that $\Delta^2(r^k a_k) \geqq 0$ for all k. Now,

$$\begin{aligned}\Delta^2(r^k a_k) &= r^k a_k - 2r^{k+1} a_{k+1} + r^{k+2} a_{k+2} \\ &= r^k \Delta^2 a_k + 2r^k(1-r)\, \Delta a_{k+1} + r^k(1-r)^2 a_{k+2} \\ &\geqq 0,\end{aligned}$$

proving the result. □

Proof of Theorem 2.1 (*cont'd*). It follows from Lemma 2.3 that in order to prove that $\tilde{G}_R(t) \geqq 0$ it suffices to prove that $\Delta^2 c_k$, Δc_k, $c_k \geqq 0$ for all k, where $c_k = k!/(k+m)!$ Obviously $c_k > 0$ and

$$\begin{aligned}\Delta c_k = c_k - c_{k+1} &= k!/(k+m)! - (k+1)!/(k+m+1)! \\ &= m k!/(k+m+1)! \geqq 0.\end{aligned}$$

Similarly,

$$\begin{aligned}\Delta^2 c_k = \Delta c_k - \Delta c_{k+1} &= m k!/(k+m+1)! - m(k+1)!/(k+m+2)! \\ &= m(m+1)\, k!/(k+m+2)! \geqq 0.\end{aligned}$$

Thus $\tilde{G}_R(t) \geqq 0$ for all $t \in [0, 2\pi)$ and all $R \geqq 1$. This proves the upper bound.

To prove the lower bound, let $\pi_n = \operatorname{span}\{1, z, \ldots, z^n\}$ and set

$$M_{n+1} = \{q\colon q \in \pi_n,\ \|q\|_{H^p} \leqq R^{m-n}(n-m)!/n!\}.$$

To prove that $b_n(A_R(m,p); H^p) \geqq R^{m-n}(n-m)!/n!$, it suffices to prove that $M_{n+1} \subseteq A_R(m,p)$. In other words we must show that $\|q^{(m)}\|_{H^p_R} \leqq (R^{n-m} n!/(n-m)!)\, \|q\|_{H^p}$. We prove this inequality by an application of two well-known results, namely

(I) If $q \in \pi_n$, then for $R \geqq 1$,

$$\|q\|_{H^p_R} \leqq R^n\, \|q\|_{H^p}.$$

(II) If $q \in \pi_n$, then

$$\|q'\|_{H^p_R} \leqq (n/R)\, \|q\|_{H^p_R}.$$

If (I) and (II) hold, then

$$\|q^{(m)}\|_{H^p_R} \leqq (n!\, R^{-m}/(n-m)!)\, \|q\|_{H^p_R} \leqq (n!\, R^{n-m}/(n-m)!)\, \|q\|_{H^p},$$

and the lower bound is proved.

We first prove (I). Let $q \in \pi_n$, and set $f(z) = z^n q(1/z)$. f is analytic in $\mathbb{C}$, and $M_p(1/R, f) \leqq M_p(1, f)$ by the Hardy convexity theorem. Now,

$$M_p(1/R, f) = \begin{cases} \left(1/2\pi \int_0^{2\pi} |f(e^{i\theta}/R)|^p \, d\theta\right)^{1/p}, & 1 \leqq p < \infty \\ \max\{|f(e^{i\theta}/R)|: 0 \leqq \theta < 2\pi\}, & p = \infty, \end{cases}$$
$$= R^{-n} \|q\|_{H_R^p},$$

while $M_p(1, f) = \|q\|_{H^p}$. Thus (I) is proved.

The proof of (II) is more interesting. It is easy to see that (II) is implied by the inequality $\|t'\|_p \leqq n \|t\|_p$, for all $t \in T_n$ (trigonometric polynomials of degree $\leqq n$), where $\|\cdot\|_p$ is the usual $L^p[0, 2\pi]$ norm. This inequality is referred to as Bernstein's inequality (it was proved by Bernstein for $p = \infty$) and a proof may be found, for example, in Zygmund [1968, Chap. X]. We prefer, however, to present a direct proof of (II).

Set
$$K_n(t) = e^{int}\left[n + 2\sum_{k=1}^{n-1}(n-k)\cos kt\right].$$

Then it readily follows that for $q \in \pi_n$,

$$q'(Re^{it}) = \frac{e^{-it}}{2R\pi}\int_0^{2\pi} q(Re^{i\theta})\, K_n(t-\theta)\, d\theta.$$

Thus by Young's inequality,

$$\|q'\|_{H_R^p} \leqq (1/R)\, \|q\|_{H_R^p}\, \|K_n\|_{L^1[0,2\pi]}.$$

Now,

$$|K_n(t)| = \left|n + 2\sum_{k=1}^{n-1}(n-k)\cos kt\right|,$$

and

$$n + 2\sum_{k=1}^{n-1}(n-k)\cos kt = 2\tilde{D}_{n-1}(t) = (1 - \cos nt)/2\sin^2(t/2) \geqq 0$$

(see Lemma 2.2). Thus $\|K_n\|_{L^1[0,2\pi]} = n$, and $\|q'\|_{H_R^p} \leqq (n/R)\, \|q\|_{H_R^p}$, proving (II). This proves the theorem. □

Remark. We have actually proved a result stronger than that contained in the statement of Theorem 2.1. We have proved that $s_n(A_R(m,p); L^p) = s_n(A_R(m,p); H^p)$, where s_n is any of the quantities d_n, d^n, δ_n or b_n, and where L^p is the usual L^p-norm on $[0, 2\pi]$ (i.e., $\|f\|_{H^p} = \|f\|_{L^p}$ for all $f \in A_R(m,p)$). From general considerations (Proposition 3.2 of Chapter II) $d^n(A_R(m,p); H^p) = d^n(A_R(m,p); L^p)$. However the other equalities need not in general hold.

In addition to the space H^p, it is natural to consider the space $\mathscr{H}^p$ of analytic functions on Δ, whose norm is given by the $\mathscr{L}^p$ area norm, i.e., set

$$\|f\|_{\mathscr{L}^p} = \left(\iint_{|z|<1} |f(z)|^p \, dx\, dy\right)^{1/p} = \left(1/2\pi \int_0^{2\pi}\int_0^1 r\, |f(re^{i\theta})|^p \, d\theta\, dr\right)^{1/p},$$

for $1 \leqq p < \infty$, and

$$\mathscr{H}^p = \{f : f \text{ anal. on } \Delta,\ \| f \|_{\mathscr{L}^p} < \infty\},$$

$\| f \|_{\mathscr{H}^p} = \| f \|_{\mathscr{L}^p}$, $1 \leqq p < \infty$ (and $\| f \|_{\mathscr{H}^\infty} = \| f \|_{H^\infty}$ by the maximum modulus principle).

We wish to determine the n-widths of the set $A_R(m, p)$ with respect to the normed linear space $\mathscr{H}^p$. Unfortunately, the operator which we use to bound δ_n maps functions in $A_R(m, p)$ to functions in $\mathscr{L}^p$, but not in $\mathscr{H}^p$. This is highly unsatisfactory as we then only obtain the n-widths d_n and δ_n of $A_R(m, p)$ in $\mathscr{L}^p$. For d^n and b_n we do however obtain both the n-widths of $A_R(m, p)$ in $\mathscr{L}^p$ and of $A_R(m, p)$ in $\mathscr{H}^p$.

Set

$$G_{R,r}(t) = R^m (r/R)^n e^{int} \left[c_{n-m} + 2 \sum_{k=1}^{\infty} (r/R)^k c_{n-m+k} \cos kt \right]$$

where $c_k = k!/(k+m)!$ A calculation shows that

$$f(r e^{it}) - p_f(r e^{it}) = 1/2\pi \int_0^{2\pi} f^{(m)}(R e^{i\theta}) e^{im\theta} G_{R,r}(t - \theta)\, d\theta,$$

where for $f(z) = \sum_{j=0}^{\infty} a_j z^j$,

$$p_f(z) = \sum_{j=0}^{m-1} a_j z^j + \sum_{j=m}^{n-1} a_j z^j [1 - c_{j-m}^{-1} c_{2n-m-j} (|z|/R)^{2n-2j}].$$

Set $(P_n f)(z) = p_f(z)$.

Theorem 2.4. *Let $R \geqq 1$, $p \in [1, \infty]$, $m = 0, 1, \ldots$. Then,*

$$\begin{aligned} d_n(A_R(m,p); \mathscr{L}^p) &= \delta_n(A_R(m,p); \mathscr{L}^p) = d^n(A_R(m,p); \mathscr{L}^p) = d^n(A_R(m,p); \mathscr{H}^p) \\ &= b_n(A_R(m,p); \mathscr{L}^p) = b_n(A_R(m,p); \mathscr{H}^p) \\ &= \begin{cases} \infty, & n < m \\ R^{m-n}(n-m)!/n!(np+2)^{1/p}, & n \geqq m,\ 1 \leqq p < \infty, \\ R^{m-n}(n-m)!/n!, & n \geqq m,\ p = \infty. \end{cases} \end{aligned}$$

Furthermore,

(1) $X_n = \operatorname{span}\{\{z^j\}_{j=0}^{m-1}, \{[R^{2n-2j} - c_{j-m}^{-1} c_{2n-m-j} |z|^{2n-2j}] z^j\}_{j=m}^{n-1}\}$ *is optimal for* $d_n(A_R(m,p); \mathscr{L}^p)$.

(2) P_n, *as defined above, is optimal for* $\delta_n(A_R(m,p); \mathscr{L}^p)$.

(3) $L^n = \{f : f \in \mathscr{H}^p, f^{(k)}(0) = 0,\ k = 0, 1, \ldots, n-1\}$ *is optimal for* $d^n(A_R(m,p); \mathscr{H}^p)$.

(4) $\pi_n = \operatorname{span}\{1, z, \ldots, z^n\}$ *is optimal for* $b_n(A_R(m,p); \mathscr{H}^p)$.

Proof. For $n < m$, the result readily follows. Assume $n \geqq m$ and $1 \leqq p < \infty$. We shall prove the inequalities $\delta_n(A_R(m,p); \mathscr{L}^p) \leqq R^{m-n}(n-m)!/n!\,(np+2)^{1/p}$, and $b_n(A_R(m,p); \mathscr{H}^p) \geqq R^{m-n}(n-m)!/n!\,(np+2)^{1/p}$.

The equality $d^n(A_R(m,p); \mathscr{L}^p) = d^n(A_R(m,p); \mathscr{H}^p)$ follows from Proposition 3.2 of Chapter II. As a consequence of the definition of b_n, we see that $b_n(A_R(m,p); \mathscr{L}^p) \geqq b_n(A_R(m,p); \mathscr{H}^p)$. Thus it suffices to prove the two inequalities. The proofs of both the upper and lower bounds are variants of the proof of Theorem 2.1.

For P_n as above and $f \in A_R(m,p)$,

$$\| f - P_n f \|_{\mathscr{L}^p} = \left(1/2\pi \int_0^{2\pi}\int_0^1 \left|1/2\pi \int_0^{2\pi} f^{(m)}(R e^{i\theta})\, e^{im\theta}\, G_{R,r}(t-\theta)\, d\theta\right|^p r\,dr\,dt\right)^{1/p}.$$

Now,

$$1/2\pi \int_0^{2\pi} \left|1/2\pi \int_0^{2\pi} f^{(m)}(R e^{i\theta})\, e^{im\theta}\, G_{R,r}(t-\theta)\, d\theta\right|^p dt$$
$$\leqq \left(1/2\pi \int_0^{2\pi} |f^{(m)}(R e^{i\theta})|^p\, d\theta\right)\left(1/2\pi \int_0^{2\pi} |G_{R,r}(t)|\, dt\right)^p$$

by Young's inequality.

From the proof of Theorem 2.1 (see Lemmas 2.2 and 2.3),

$$1/2\pi \int_0^{2\pi} |G_{R,r}(t)|\, dt = R^m (r/R)^n\, c_{n-m}.$$

Thus

$$\delta_n(A_R(m,p); \mathscr{L}^p) \leqq R^{m-n}\, c_{n-m} \left(\int_0^1 r^{np}\, r\,dr\right)^{1/p}$$
$$= R^{m-n}(n-m)!/n!\,(np+2)^{1/p}.$$

Let $\pi_n = \operatorname{span}\{1, z, \ldots, z^n\}$. To prove the lower bound $b_n(A_R(m,p); \mathscr{H}^p) \geqq R^{m-n}(n-m)!/n!\,(np+2)^{1/p}$, it suffices to show that

$$\| q^{(m)} \|_{H_R^p} \leqq \| q \|_{\mathscr{H}^p}\, R^{n-m}\, n!\,(np+2)^{1/p}/(n-m)!$$

for all $q \in \pi_n$. From the proof of Theorem 2.1 we have

$$\| q^{(m)} \|_{H_R^p} \leqq \| q \|_{H^p}\, R^{n-m}\, n!/(n-m)!$$

for all $q \in \pi_n$. It therefore suffices to prove that

$$\| q \|_{H^p} \leqq (np+2)^{1/p}\, \| q \|_{\mathscr{H}^p}.$$

From (I) of the proof of Theorem 2.1, we see that for $r \in [0,1]$,

$$r^n\, \| q \|_{H^p} \leqq \| q \|_{H_r^p}.$$

Thus

$$\left(1/2\pi \int_0^{2\pi}\int_0^1 r^{np}\,|q(e^{it})|^p\, r\,dr\,dt\right)^{1/p} \leqq \left(1/2\pi \int_0^{2\pi}\int_0^1 |q(re^{it})|^p\, r\,dr\,dt\right)^{1/p},$$

and

$$\|q\|_{H^p}\left(\int_0^1 r^{np+1}\,dr\right)^{1/p} \leqq \|q\|_{\mathscr{H}^p}.$$

Therefore $\|q\|_{H^p} \leqq (np+2)^{1/p}\,\|q\|_{\mathscr{H}^p}$. □

Remark. The results of this theorem are unsatisfactory in that we are really interested in $\delta_n(A_R(m,p);\mathscr{H}^p)$ and $d_n(A_R(m,p);\mathscr{H}^p)$. It is expected that the values of these n-widths are the same as those given by the theorem. Moreover, this more general result is certainly true for $p=\infty$ (since $\|f\|_{\mathscr{H}^\infty} = \|f\|_{H^\infty}$) and is also true for $p=2$ for any number of reasons (e.g. $d_n(A_R(m,2);\mathscr{H}^2) = \delta_n(A_R(m,2);\mathscr{H}^2) = d^n(A_R(m,2);\mathscr{H}^2)$, see also Section 3).

There is at least one other result which easily follows from the above method of proof. Let $R \geqq 1$, $m = 0,1,\ldots$, and

$$\mathfrak{A}_R(m,p) = \{f\colon f\in\mathscr{H}^p_R,\ \|f^{(m)}\|_{\mathscr{H}^p_R} \leqq 1\}$$

where $\|f\|_{\mathscr{H}^p_R} = \left(1/2\pi \int_0^{2\pi}\int_0^R r\,|f(re^{it})|^p\,dr\,dt\right)^{1/p}$.

We would like to be able to calculate the n-widths of $\mathfrak{A}_R(m,p)$ in $\mathscr{H}^p$. Unfortunately we can only do this in the case $m=0$.

Set $F_R(t) = e^{int}\left[1 + 2\sum_{k=1}^{\infty} R^{-k}\cos kt\right]$. Then

$$f(re^{it}) - (Q_n f)(re^{it}) = 1/2\pi \int_0^{2\pi} f(Re^{i\theta})\,R^{-n}\,F_R(t-\theta)\,d\theta,$$

where for $f(z) = \sum_{j=0}^{\infty} a_j z^j$, $(Q_n f)(z) = \sum_{j=0}^{n-1} a_j(1 - R^{2j-2n})\,z^j$.

Proposition 2.5. *Let $R \geqq 1$. Then for $p\in[1,\infty]$,*

$$\begin{aligned} d_n(\mathfrak{A}_R(0,p);\mathscr{H}^p) &= d^n(\mathfrak{A}_R(0,p);\mathscr{H}^p) = \delta_n(\mathfrak{A}_R(0,p);\mathscr{H}^p)\\ &= b_n(\mathfrak{A}_R(0,p);\mathscr{H}^p) = R^{-n-2/p}. \end{aligned}$$

Furthermore,

(1) $\pi_{n-1} = \operatorname{span}\{1,z,\ldots,z^{n-1}\}$ *is optimal for* d_n.
(2) $L^n = \{f\colon f\in\mathscr{H}^p,\ f^{(k)}(0)=0,\ k=0,1,\ldots,n-1\}$ *is optimal for* d^n.
(3) Q_n, *as defined above, is optimal for* δ_n.
(4) $\pi_n = \operatorname{span}\{1,z,\ldots,z^n\}$ *is optimal for* b_n.

Proof. Let Q_n be as above. Then

$$1/2\pi \int_0^{2\pi} |(f - Q_n f)(re^{it})|^p \, dt \leqq 1/2\pi \int_0^{2\pi} \left(1/2\pi \int_0^{2\pi} |f(Rre^{i\theta})| \, R^{-n} \, |F_R(t-\theta)| \, d\theta\right)^p dt$$

$$\leqq R^{-np} \left(1/2\pi \int_0^{2\pi} |f(Rre^{i\theta})|^p \, d\theta\right) \left(1/2\pi \int_0^{2\pi} |F_R(t)| \, dt\right)^p$$

by Young's inequality. From the method of proof of Theorem 2.1,

$$1/2\pi \int_0^{2\pi} |F_R(t)| \, dt = 1.$$

Thus

$$1/2\pi \int_0^{2\pi} |(f - Q_n f)(re^{it})|^p \, dt \leqq R^{-np} \, 1/2\pi \int_0^{2\pi} |f(Rre^{i\theta})|^p \, d\theta$$

implying that

$$\|f - Q_n f\|_{\mathscr{H}^p} = \left(1/2\pi \int_0^{2\pi} \int_0^1 r \, |(f - Q_n f)(re^{it})|^p \, dr \, dt\right)^{1/p}$$

$$\leqq R^{-n} \left(1/2\pi \int_0^{2\pi} \int_0^1 r \, |f(Rre^{i\theta})|^p \, dr \, d\theta\right)^{1/p}$$

$$= R^{-n} \left(1/2\pi \int_0^{2\pi} \int_0^R (r/R^2) \, |f(re^{i\theta})|^p \, dr \, d\theta\right)^{1/p}$$

$$= R^{-n-2/p} \, \|f\|_{\mathscr{H}^p_R},$$

and therefore $\delta_n(\mathfrak{A}_R(0,p); \mathscr{H}^p) \leqq R^{-n-2/p}$.

To prove the lower bound, it suffices to show that $\|q\|_{\mathscr{H}^p_R} \leqq R^{n+2/p} \, \|q\|_{\mathscr{H}^p}$ for all $q \in \pi_n$. By (I) of Theorem 2.1,

$$\|q\|_{H^p_{Rr}} \leqq R^n \, \|q\|_{H^p_r}$$

for all $q \in \pi_n$. Thus

$$1/2\pi \int_0^{2\pi} |q(Rre^{it})|^p \, dt \leqq R^{np} \left(1/2\pi \int_0^{2\pi} |q(re^{it})|^p \, dt\right)$$

and so

$$1/2\pi \int_0^{2\pi} \int_0^1 r \, |q(Rre^{it})|^p \, dr \, dt \leqq R^{np} \left(1/2\pi \int_0^{2\pi} \int_0^1 r \, |q(re^{it})|^p \, dr \, dt\right).$$

Applying a change of variable argument to the left-hand side of the inequality, we obtain

$$\|q\|_{\mathscr{H}^p_R} \leqq R^{n+2/p} \, \|q\|_{\mathscr{H}^p}. \quad \square$$

3. n-Widths of Analytic Functions in H^2

In this section we consider n-widths of the set $A = A_1(0,2)$ (in the notation of the previous section) with respect to some Hilbert Space $L^2(K, d\mu)$. In other words we return to the setting of Section 2 of Chapter IV but develop, in more detail, the theory in this particular case.

Recall that $A = \{f\colon f \in H^2, \ \|f\|_{H^2} \leqq 1\}$, where H^2 is the class of functions analytic in Δ with norm given by

$$\|f\|_{H^2} = \left(\sum_{j=0}^{\infty} |a_j|^2\right)^{1/2},$$

for $f(z) = \sum_{j=0}^{\infty} a_j z^j$.

Let μ be a positive measure on a compact set K, $K \subset \Delta$. We wish to calculate $d_n(A; L^2(K, d\mu))$ and determine optimal subspaces for d_n (and d^n).

From Cauchy's integral formula,

$$f(z) = \frac{1}{2\pi} \int_0^{2\pi} \frac{f(e^{i\theta})}{1 - z e^{-i\theta}} \, d\theta.$$

For $z \in K$, set $Tf(z) = f(z)$. We regard T as a map from H^2 to $L^2(K, d\mu)$, and then

$$d_n(A; L^2(K, d\mu)) = d_n(T(H^2); L^2(K, d\mu))$$

in the notation of Section 2 of Chapter IV. From Theorem 2.2 of Chapter IV it follows that the n-widths d_n, d^n, δ_n and b_n are all equal; they equal the square root of the $(n+1)$st eigenvalue (arranged in decreasing order of magnitude to their multiplicity) of TT' $(T'T)$, where T' is the adjoint of T; and the n eigenfunctions of TT' corresponding to the n largest eigenvalues span an optimal subspace for d_n.

Let us be more precise in the determination of the eigenvalue-eigenfunction problem. To do this we must determine T'. T' maps $L^2(K, d\mu)$ to H^2 and satisfies

$$(f, Tg)_{L^2(K, d\mu)} = (T'f, g)_{H^2}$$

for all $f \in L^2(K, d\mu)$ and $g \in H^2$. It is an easy matter to check that

$$(T'f)(e^{i\theta}) = \int_K \frac{f(w)}{1 - \bar{w} e^{i\theta}} \, d\mu(w)$$

and

$$(TT'f)(z) = \int_K \frac{f(w)}{1 - \bar{w} z} \, d\mu(w).$$

The eigenvalue-eigenfunction problem thus takes the form

$$\lambda f(z) = \int_K \frac{f(w)}{1-\bar{w}z}\, d\mu(w), \qquad z \in \Delta.$$

(See the remark after Proposition 2.4 of Chapter IV). This problem has eigenvalues $\lambda_1 \geqq \lambda_2 \geqq \ldots \geqq 0$ and associated (orthogonal) eigenfunctions $f_1, f_2, \ldots$. For convenience we shall assume that μ has an infinite number of points of support so that $\lambda_k > 0$, $k = 1, 2, \ldots$. Furthermore, since K is a compact subset of Δ, it follows that each f_k is analytic in a region containing $\bar{\Delta}$. From Theorem 2.2 of Chapter IV, $d_n(A; L^2(K, d\mu)) = \lambda_{n+1}^{1/2}$ and span$\{f_1, \ldots, f_n\}$ is optimal for d_n.

In what follows we prove that the $\{\lambda_k\}_{k=1}^{\infty}$ are in fact distinct, and we obtain an additional optimal subspace for d_n and d^n (although in certain instances the "new" and "old" optimal subspaces may be identical). We are, in some sense, developing a theory similar to that presented in Section 5 of Chapter IV.

Proposition 3.1. *Let $\{\lambda_k\}_{k=1}^{\infty}$ and $\{f_k\}_{k=1}^{\infty}$ be as defined above. Then*

$$\lambda_1 > \lambda_2 > \ldots > 0.$$

Furthermore f_k has exactly $k-1$ zeros in Δ (counting multiplicities) and there is a v, dependent only on K, for which

$$|f_k(e^{it})| \geqq v \, \|f_k\|_{H^2}$$

for all $t \in [0, 2\pi)$, $k = 1, 2, \ldots$.

Proof. Let f_k be an eigenfunction with eigenvalue λ_k and let $g \in H^2$. From the eigenvalue-eigenfunction equation and Cauchy's integral formula we obtain

$$\frac{\lambda_k}{2\pi} \int_0^{2\pi} f_k(e^{i\theta})\, \bar{g}(e^{i\theta})\, d\theta = \int_K f_k(w)\, \bar{g}(w)\, d\mu(w).$$

Set $g(z) = f_k(z)\left(\dfrac{1+\bar{a}z}{1-\bar{a}z}\right)$ for any $a \in \Delta$. Thus

$$\frac{\lambda_k}{2\pi} \int_0^{2\pi} |f_k(e^{i\theta})|^2 \left(\frac{1+ae^{-i\theta}}{1-ae^{-i\theta}}\right) d\theta = \int_K |f(w)|^2 \left(\frac{1+a\bar{w}}{1-a\bar{w}}\right) d\mu(w).$$

Taking real parts on both sides, and since λ_k is real, and

$$\operatorname{Re}\left(\frac{1+ae^{-i\theta}}{1-ae^{-i\theta}}\right) = \frac{1-|a|^2}{|e^{i\theta}-a|^2},$$

we obtain

$$\frac{\lambda_k}{2\pi} \int_0^{2\pi} |f_k(e^{i\theta})|^2 \frac{1-|a|^2}{|e^{i\theta}-a|^2}\, d\theta = \int_K |f_k(w)|^2 \operatorname{Re}\left(\frac{1+a\bar{w}}{1-a\bar{w}}\right) d\mu(w).$$

Let a approach e^{it}. (Recall that f_k is analytic across $\partial \Delta$.) Then

$$\lambda_k |f_k(e^{it})|^2 = \int_K |f_k(w)|^2 \operatorname{Re}\left(\frac{1 + e^{it}\bar{w}}{1 - e^{it}\bar{w}}\right) d\mu(w).$$

For all $w \in K$ and $t \in [0, 2\pi)$, there exists a $v^2 > 0$, dependent only on the distance of K to $\partial \Delta$ for which

$$\operatorname{Re} \frac{1 + e^{it}\bar{w}}{1 - e^{it}\bar{w}} \geqq v^2.$$

Thus

$$\lambda_k |f_k(e^{it})|^2 \geqq v^2 \int_K |f_k(w)|^2 \, d\mu(w).$$

Since

$$\lambda_k f_k(z) = \int_K \frac{f_k(w)}{1 - \bar{w}z} \, d\mu(w)$$

we have

$$\lambda_k \| f_k \|_{H^2}^2 = \frac{\lambda_k}{2\pi} \int_0^{2\pi} |f_k(e^{i\theta})|^2 \, d\theta = \int_K |f_k(w)|^2 \, d\mu(w).$$

Therefore

$$\lambda_k |f_k(e^{it})|^2 \geqq v^2 \lambda_k \| f_k \|_{H^2}^2$$

implying

$$|f_k(e^{it})| \geqq v \| f_k \|_{H^2}$$

for all $t \in [0, 2\pi)$ and all $k = 1, 2, \ldots$.

If $\lambda_k = \lambda_{k+1}$ for some k, i.e., f_k, f_{k+1} are two linearly independent eigenfunctions with the same eigenvalue, then there exists a linear combination which vanishes at some point on $\partial \Delta$. This contradicts our previous result. Thus

$$\lambda_1 > \lambda_2 > \ldots > 0,$$

i.e., the eigenvalues are distinct.

It remains to prove that f_k has exactly $k - 1$ zeros, counting multiplicities, in Δ. We first prove this fact for a simple choice of K and $d\mu$ and then obtain the result for general K and $d\mu$ from continuity considerations.

Let $C = \{z : |z| = r\}$, $r \in (0, 1)$, and let $d\mu_0$ be the usual Lebesgue measure on C. It is a simple matter to prove that for this problem $\lambda_k = r^{2k-2}$, $k = 1, 2, \ldots$, and $f_k(z) = z^{k-1}$, $k = 1, 2, \ldots$. Thus f_k has $k - 1$ zeros in Δ, counting multiplicities.

For $t \in [0, 1]$, set $\mu_t = (1 - t)\mu_0 + t\mu$. Let $\lambda_{1,t} > \lambda_{2,t} > \ldots > 0$ denote the eigenvalues and $f_{1,t}, f_{2,t}, \ldots$ the associated eigenfunctions of the corresponding eigenvalue-eigenfunction problem with $d\mu_t$ replacing $d\mu$ and $K' = K \cup C$ replacing K. For convenience normalize $f_{k,t}$ so that $\| f_{k,t} \|_{H^2} = 1$ for all k. Note that there exists a v, dependent only on K', for which $|f_{k,t}(e^{i\theta})| \geqq v$ for all k, t, and θ.

We first claim that $\lambda_{k,t}$ is a continuous function of t. To see this, set

$$(S_t f)(z) = \int_{K'} \frac{f(w)}{1 - \bar{w} z} \, d\mu_t(w).$$

It is easily shown that

$$((S_t - S_s) f, f)_{H^2} = (t - s) \int_{K'} |f(w)|^2 \, d\sigma(w)$$

where $\sigma = \mu - \mu_0$. Thus

$$\begin{aligned} &\max_{\|f\|_{H^2} \leqq 1} ((S_t - S_s) f, f) \\ &\quad \leqq |t - s| \{ \max_{\|f\|_{H^2} \leqq 1} \int_{K'} |f(w)|^2 \, d\mu(w) + \max_{\|f\|_{H^2} \leqq 1} \int_{K'} |f(w)|^2 \, d\mu_0(w) \} \\ &\quad = |t - s| (\lambda_1 + 1). \end{aligned}$$

It now follows from the min-max characterization of $\lambda_{k,t}$ (and $\lambda_{k,s}$) that

$$|\lambda_{k,t} - \lambda_{k,s}| \leqq |t - s| (\lambda_1 + 1).$$

Thus $\lambda_{k,t}$ is a continuous function of t for each k.

A standard normal family argument proves that $f_{k,t} \to f_{k,s}$ as $t \to s$ uniformly on compact subsets of Δ. Since $|f_{k,t}(e^{i\theta})| \geqq v$ for all $\theta \in [0, 2\pi)$ and $t \in [0, 1]$, a simple application of Rouché's Theorem (see e.g., Ahlfors [1966, p. 152]) shows that the number of zeros of $f_{k,t}$ in Δ is independent of $t \in [0, 1]$, for each k. Thus $f_{k,1} = f_k$ has $k - 1$ zeros, counting multiplicities, in Δ. □

Before continuing we introduce the class of Blaschke products. This class will also prove to be important in the next section.

Definition 3.1. A *Blaschke product of degree m* is any function of the form

$$B(z) = \sigma \prod_{j=1}^{m} \frac{z - \alpha_j}{1 - \bar{\alpha}_j z}$$

where $|\alpha_j| < 1$, $j = 1, \ldots, m$, and $|\sigma| = 1$.

$\mathscr{B}_m$ shall denote the class of Blaschke products of degree m or less (i.e., $|\alpha_j| \leqq 1$, $j = 1, \ldots, m$). As is easily seen each $B \in \mathscr{B}_m$ satisfies $|B(e^{i\theta})| = 1$ for all $\theta \in [0, 2\pi)$.

For each positive measure μ on K, let

$$\lambda_1(d\mu) > \lambda_2(d\mu) > \ldots$$

denote the eigenvalues of the above-considered problem.

Let f_{n+1} be as above, and let $z_1^*, \ldots, z_n^*$ denote its n zeros in Δ. Set

$$B^*(z) = \prod_{j=1}^{n} \frac{z - z_j^*}{1 - \bar{z}_j^* z}.$$

Thus $B^* \in \mathscr{B}_n$ and $B^*(z_j^*) = 0$, $j = 1, \ldots, n$.

Proposition 3.2. *For B^* as above,*

$$\lambda_{n+1}(d\mu) = \lambda_1(|B^*|^2\, d\mu).$$

Proof. By definition

$$\lambda_{n+1}\, f_{n+1}(z) = \int_K \frac{f_{n+1}(w)}{1 - \bar{w} z}\, d\mu(w),$$

where $\lambda_{n+1} = \lambda_{n+1}(d\mu)$, and therefore $\lambda_{n+1}(f_{n+1}, f)_{H^2} = \int_K f_{n+1}(w)\, \bar{f}(w)\, d\mu(w)$ for all $f \in H^2$. Set $f_{n+1}(z) = g_{n+1}(z)\, B^*(z)$. Thus g_{n+1} is analytic in $\bar{\Delta}$ and zero free therein. Set $f(z) = B^*(z)\, g(z)$ for any $g \in H^2$ (and note that $\| f \|_{H^2} = \| g \|_{H^2}$). Since $|B^*(e^{i\theta})| = 1$,

$$\lambda_{n+1}(f_{n+1}, f)_{H^2} = \lambda_{n+1}(g_{n+1}, g)_{H^2},$$

and

$$\int_K f_{n+1}(w)\, \bar{f}(w)\, d\mu(w) = \int_K g_{n+1}(w)\, \bar{g}(w)\, |B^*(w)|^2\, d\mu(w).$$

Thus

$$\lambda_{n+1}(g_{n+1}, g)_{H^2} = \int_K g_{n+1}(w)\, \bar{g}(w)\, |B^*(w)|^2\, d\mu(w)$$

for all $g \in H^2$. Therefore

$$\lambda_{n+1}\, g_{n+1}(z) = \int_K \frac{g_{n+1}(w)}{1 - \bar{w} z}\, |B^*(w)|^2\, d\mu(w).$$

Since g_{n+1} is zero free, and $|B^*|^2\, d\mu$ is a positive measure, it follows from Proposition 3.1 that g_{n+1} is the eigenfunction associated with the largest eigenvalue of the measure $|B^*|^2\, d\mu$, i.e., $\lambda_{n+1}(d\mu) = \lambda_1(|B^*|^2\, d\mu)$. □

On the basis of Proposition 3.2, we can now state the main result.

Theorem 3.3. *If $z_1^*, \ldots, z_n^* \in \Delta$ are the n zeros of f_{n+1}, then*

$$L^n = \{g\colon g \in H^2,\ g(z_j^*) = 0,\ j = 1, \ldots, n\}$$

(with the understanding that the appropriate derivatives are taken in case of equal z_j^*'s) is also an optimal subspace for $d^n(A; L^2(K, d\mu))$. Let X_n denote the orthogonal complement of L^n in H^2. Then X_n is also an optimal subspace for $d_n(A; L^2(K, d\mu))$.

Proof. By definition

$$\begin{aligned}\lambda_1(|B^*|^2\, d\mu) &= \sup\Big\{\int_K |g|^2\, |B^*|^2\, d\mu\colon \|g\|_{H^2} \leqq 1\Big\}\\ &= \sup\Big\{\int_K |f|^2\, d\mu\colon \|f\|_{H^2} \leqq 1,\ f(z_j^*) = 0,\ j = 1, \ldots, n\Big\}.\end{aligned}$$

Thus from Proposition 3.2,

$$\lambda_{n+1}(=\lambda_{n+1}(d\mu)) = \sup\{\int_K |f|^2\, d\mu \colon \|f\|_{H^2} \leqq 1,\ f(z_j^*) = 0,\ j = 1,\ldots,n\}.$$

This implies (since we already have $d^n(A; L^2(K, d\mu)) = \lambda_{n+1}^{1/2}$) that L^n is optimal for $d^n(A; L^2(K, d\mu))$.

Let X_n be as above, and let P_n denote the orthogonal projection onto X_n. Then for every $g \in H^2$ we have $g - P_n g \in L^n$ and $\|g - P_n g\|_{L^2(K, d\mu)} \leqq \lambda_{n+1}^{1/2} \|g\|_{H^2}$. Thus X_n is optimal for $d_n(A; L^2(K, d\mu))$. □

Assume $z_1^*, \ldots, z_n^*$ are distinct points and set

$$M_j(z) = \frac{B^*(z)}{(z - z_j^*)\, B^{*\prime}(z_j)}, \qquad j = 1,\ldots,n.$$

Then $M_j(z_k^*) = \delta_{jk}, j, k = 1,\ldots,n$; the $\{M_j\}_{j=1}^n$ are linearly independent; and each M_j lies in the orthogonal complement to L^n, i.e., $(g, M_j)_{H^2} = 0$ if $g(z_j^*) = 0$, $j = 1,\ldots,n$, and $g \in H^2$. We may therefore take $X_n = \mathrm{span}\{M_1,\ldots,M_n\}$. A similar (but more complicated) formula exists when the $\{z_j^*\}_{j=1}^n$ are not all distinct.

Remark. If the eigenfunctions are $f_k(z) = z^{k-1}$, $k = 1, 2, \ldots$, as is the case when $K = C$ and $d\mu = d\mu_0$ (see the proof of Proposition 3.1), then $B^*(z) = z^n$, and $X_n = \mathrm{span}\{1, z, \ldots, z^{n-1}\}$, i.e., this method does not give us an additional optimal subspace for d_n (or, for that matter, for d^n).

4. n-Widths of Analytic Functions in H^∞

H^∞ is the Banach space of bounded analytic functions in Δ with norm

$$\|f\|_\infty = \|f\|_{H^\infty} = \sup\{|f(z)| \colon z \in \Delta\}.$$

Since the letter A has been overworked in the previous sections, we define

$$\mathscr{A} = A(0, \infty) = \{f \colon f \in H^\infty,\ \|f\|_\infty \leqq 1\}.$$

Let K be any compact subset of Δ. We shall concern ourselves with the n-widths of $\mathscr{A}$ in various normed linear spaces X. Specifically X will either be $C(K)$ (continuous functions on K with the usual L^∞-norm thereon) or $L^q(v)$ for some $q \in [1, \infty)$ and some positive measure v with support on K.

The main tool to be used in the determination of the n-widths is that of Blaschke products. These were defined in the last section. We redefine them here.

Definition 4.1. A *Blaschke product of degree* m is any function of the form

$$B(z) = \sigma \prod_{j=1}^{m} \frac{z - \alpha_j}{1 - \bar{\alpha}_j z},$$

where $|\alpha_j| < 1$, $j = 1, \ldots, m$, and $|\sigma| = 1$.

$\mathscr{B}_m$ shall denote the class of Blaschke products of degree m or less. We use two properties of $\mathscr{B}_m$. These are:

(1) If $B \in \mathscr{B}_m$, then $|B(e^{i\theta})| = 1$ for all $\theta \in [0, 2\pi)$.

(2) Given $m + 1$ distinct points $z_1, \ldots, z_{m+1}$ in Δ and a function $f \in H^\infty$, there exists a *unique* Blaschke product $B \in \mathscr{B}_m$, and a $\rho \geqq 0$, such that $\rho B(z_j) = f(z_j)$, $j = 1, \ldots, m+1$. Furthermore,

$$\rho = \inf\{\|g\|_\infty : g \in H^\infty,\ g(z_j) = f(z_j),\ j = 1, \ldots, m+1\}.$$

Remark. The proof of (1) is immediate. The deep result is, of course, (2). The minimality property of the interpolating Blaschke product follows, in a fairly straightforward manner, from Rouché's Theorem. Various proofs of the existence of the interpolating Blaschke product may be found in the literature. One proof is based on the classical Pick-Nevinlinna Theorem. Another proof, using a more functional analytic approach, is given in Fisher [1983].

Given n distinct points $\alpha_1, \ldots, \alpha_n$ in Δ, set

$$B(z) = \prod_{j=1}^{n} \frac{z - \alpha_j}{1 - \bar{\alpha}_j z}.$$

Let $\zeta \in K$ and define

$$R(z, \zeta) = \frac{B(\zeta)}{B(z)} \frac{1}{z - \zeta} \frac{1 - |\zeta|^2}{1 - \bar{\zeta} z}.$$

$R(z, \zeta)$, as a function of z, has poles of order 1 at $\alpha_1, \ldots, \alpha_n, \zeta$. Thus for $f \in H^\infty$,

$$\frac{1}{2\pi i} \int_{|z|=1} R(z, \zeta) f(z)\, dz = f(\zeta) - \sum_{j=1}^{n} f(\alpha_j)\, g_j(\zeta),$$

where

$$g_j(\zeta) = \frac{a_j B(\zeta)(1 - |\zeta|^2)}{(\alpha_j - \zeta)(1 - \bar{\zeta}\alpha_j)}, \qquad a_j \in \mathbb{C}.$$

If there is coalescence among the α_j's, then the above formulae are more complicated (but exist). (In particular if m α's are equal, then $f(\alpha)$, $f'(\alpha), \ldots, f^{(m-1)}(\alpha)$ will all appear on the right hand side of the above integral.)

Let $B^* \in \mathscr{B}_n$ satisfy

$$\|B^*\|_X = \inf\{\|B\|_X : B \in \mathscr{B}_n\}.$$

(The infimum is attained.) Since K is a compact subset on Δ, it follows that B^* has n (perhaps multiple) zeros in Δ. (Otherwise we could multiply B^* by z and reduce the norm.) Let $\alpha_1^*, \ldots, \alpha_n^*$ denote the zeros of B^* and $g_1^*, \ldots, g_n^*$ the corresponding functions, as above. We can now prove:

Theorem 4.1. *For $\mathscr{A}$ and X as above,*

$$d_n(\mathscr{A}; X) = d^n(\mathscr{A}; X) = \delta_n(\mathscr{A}; X) = \| B^* \|_X .$$

Furthermore,

(1) $X_n = \operatorname{span}\{g_1^*, \ldots, g_n^*\}$ *is optimal for* $d_n(\mathscr{A}; X)$.

(2) $L^n = \{f\colon f \in H^\infty,\ f(\alpha_j^*) = 0,\ j = 1, \ldots, n\}$ *is optimal for* $d^n(\mathscr{A}; X)$.

(3) $P_n f = \sum_{j=1}^{n} f(\alpha_j^*)\, g_j^*$ *is optimal for* $\delta_n(\mathscr{A}; X)$.

Remark. It is to be understood that in (2) and (3) successive derivatives of f (to the correct order) at α_j^* appear if α_j^* is a repeated root of B^*.

Remark. In the definition of $d^n(\mathscr{A}; X)$ we are taking the infimum over all subspaces of codimension n of H^∞. (See Section 7 of Chapter II.)

Proof. For given $\alpha_1, \ldots, \alpha_n$ in Δ we have

$$\frac{1}{2\pi i} \int_{|z|=1} R(z, \zeta)\, f(z)\, dz = f(\zeta) - \sum_{j=1}^{n} f(\alpha_j)\, g_j(\zeta)$$

where

$$R(z, \zeta) = \frac{B(\zeta)}{B(z)} \left(\frac{1}{z - \zeta}\right) \left(\frac{1 - |\zeta|^2}{1 - \bar{\zeta} z}\right)$$

and

$$B(z) = \prod_{j=1}^{n} \frac{z - \alpha_j}{1 - \bar{\alpha}_j z} .$$

Thus for $f \in \mathscr{A}$,

$$\left| f(\zeta) - \sum_{j=1}^{n} f(\alpha_j)\, g_j(\zeta) \right| \leqq \frac{1}{2\pi} \int_{|z|=1} |R(z, \zeta)|\, |dz| .$$

Now, for $|z| = 1$,

$$\begin{aligned} |R(z, \zeta)| &= \frac{|B(\zeta)|}{|B(z)|} \left(\frac{1}{|z - \zeta|}\right) \left(\frac{1 - |\zeta|^2}{|1 - \bar{\zeta} z|}\right) \\ &= |B(\zeta)| \left(\frac{1 - |\zeta|^2}{|1 - \bar{\zeta} z|^2}\right) \end{aligned}$$

and therefore

$$\begin{aligned} \frac{1}{2\pi} \int_{|z|=1} |R(z, \zeta)|\, |dz| &= |B(\zeta)| \frac{1}{2\pi} \int_{|z|=1} \frac{1 - |\zeta|^2}{|1 - \bar{\zeta} z|^2} |dz| \\ &= |B(\zeta)| . \end{aligned}$$

Hence for each $\zeta \in K$ and $f \in \mathscr{A}$,

$$\left| f(\zeta) - \sum_{j=1}^{n} f(\alpha_j)\, g_j(\zeta) \right| \leqq |B(\zeta)|,$$

and

$$\left\| f - \sum_{j=1}^{n} f(\alpha_j)\, g_j \right\|_X \leqq \| B \|_X.$$

Setting $B = B^*$, we obtain

$$\delta_n(\mathscr{A}; X) \leqq \| B^* \|_X.$$

The lower bounds are applications of the Borsuk Antipodality Theorem (Theorem 1.4 of Chapter II). We first prove the lower bound for d_n.

Suppose X_n is an n-dimensional subspace of X. Since

$$\| B^* \|_X = \inf\{ \| B \|_X : B \in \mathscr{B}_n \},$$

it suffices to prove, for each X_n, the existence of a $B \in \mathscr{B}_n$ for which

$$\inf_{g \in X_n} \| B - g \|_X = \| B \|_X,$$

i.e., with the zero function as a best approximant from X_n.

To this end, assume that $X = L^q(v)$ for some $q \in [1, \infty)$. (The cases $q = 1$ and $q = \infty$ ($X = C(K)$) are established by passing to the limit as $q \downarrow 1$ or $q \uparrow \infty$.) Let $X_n = \operatorname{span}\{g_1, \ldots, g_n\}$. For each $f \in \mathscr{A}$, let $\mathbf{c}(f) = (c_1(f), \ldots, c_n(f))$ denote the coefficient vector, with respect to the basis $\{g_1, \ldots, g_n\}$, of the unique best approximant to f from X_n. The mapping $Sf = \mathbf{c}(f)$ is an odd, continuous map of $\mathscr{A}$ into $\mathbb{C}^n$.

Now, fix $z_1, \ldots, z_{n+1}$, distinct points in Δ. Set

$$\Xi_{n+1} = \left\{ \mathbf{w} : \mathbf{w} = (w_1, \ldots, w_{n+1}),\ \sum_{j=1}^{n+1} |w_j| = 1 \right\}.$$

To each $\mathbf{w} \in \Xi_{n+1}$, there is a unique Blaschke product $B_{\mathbf{w}} \in \mathscr{B}_n$ and $\rho_{\mathbf{w}} > 0$ such that $\rho_{\mathbf{w}} B_{\mathbf{w}}(z_j) = w_j$, $j = 1, \ldots, n+1$. It is easily seen that $\rho_{\mathbf{w}}$ is a continuous function of $\mathbf{w} \in \Xi_{n+1}$, and therefore the mapping $T(\mathbf{w}) = B_{\mathbf{w}}$ is an odd, continuous map of Ξ_{n+1} into $\mathscr{B}_n$. Thus ST is an odd, continuous map of Ξ_{n+1} into $\mathbb{C}^n$. As such there exists a $\mathbf{w} \in \Xi_{n+1}$ for which $\mathbf{c}(B_{\mathbf{w}}) = \mathbf{0}$. This proves the lower bound $d_n(\mathscr{A}; X) \geqq \| B^* \|_X$.

To prove the lower bound for d^n, let L^n be any subspace of H^∞ of codimension n. Thus

$$L^n = \{ f : f \in \mathscr{A},\ l_j(f) = 0,\ j = 1, \ldots, n \}$$

for some linearly independent $l_j \in (H^\infty)'$. Let T be as above and set $Rf = (l_1(f), \ldots, l_n(f))$. RT is an odd, continuous map of Ξ_{n+1} into $\mathbb{C}^n$. By the

Borsuk Antipodality Theorem there exists a $B \in \mathscr{B}_n \cap L^n$. As such,

$$\| B^* \|_X \leqq \| B \|_X \leqq \sup_{f \in \mathscr{A} \cap L^n} \| f \|_X$$

which establishes the theorem. □

We previously indicated that B^* must necessarily use all n roots, i.e., $B^* \in \mathscr{B}_n \backslash \mathscr{B}_{n-1}$. We now consider various other properties of B^*. The first question to be discussed is the location of the zeros of B^*. Our problem is one of defining convexity in the correct way. This next result does this.

Proposition 4.2. *For K and X as defined above, let $B^* \in \mathscr{B}_n$ satisfy*

$$\| B^* \|_X = \inf \{ \| B \|_X : B \in \mathscr{B}_n \}.$$

Let $\alpha \in \Delta$. If there is some circle Γ through α and $1/\bar{\alpha}$ such that K is entirely inside or entirely outside Γ, then $B^*(\alpha) \neq 0$.

Proof. Assume to the contrary that $B^*(\alpha) = 0$ (i.e., $\alpha = \alpha_j^*$, some j). We prove the existence of a $\beta \in \Delta$ for which $\left| \frac{z - \beta}{1 - \bar{\beta} z} \right| < \left| \frac{z - \alpha}{1 - \bar{\alpha} z} \right|$ for all $z \in K$. Thus $B(z) = B^*(z) \left(\frac{1 - \bar{\alpha} z}{z - \alpha} \right) \left(\frac{z - \beta}{1 - \bar{\beta} z} \right) \in \mathscr{B}_n$, and $|B(z)| < |B^*(z)|$ for all $z \in K$, contradicting the minimality of B^*.

Set $K' = \{ w : w = (z - \alpha)/(1 - \bar{\alpha} z),\ z \in K \}$. The hypothesis of the theorem implies the existence of a γ' for which $|\gamma'| = 1$ and $\operatorname{Re} w/\gamma' > 0$ for all $w \in K'$. Since K' is closed (K is compact), $\operatorname{Re} w/\gamma' > \delta$ for some $\delta \in (0, 1)$. Because $|\alpha| < 1$, there exists a β, $|\beta| < 1$ such that $\gamma = (\beta - \alpha)/(1 - \bar{\alpha} \beta)$, $0 < |\gamma| < 1$, satisfies $\operatorname{Re} w/\gamma > 1$ for all $w \in K'$.

Let $w = (z - \alpha)/(1 - \bar{\alpha} z)$. Then $z = (w + \alpha)/(1 + \bar{\alpha} w)$, and

$$\left| \frac{z - \beta}{1 - \bar{\beta} z} \right| = \left| \frac{w - \gamma}{1 - \bar{\gamma} w} \right|.$$

It remains to prove that $|w| > \left| \frac{w - \gamma}{1 - \bar{\gamma} w} \right|$ for all w, γ satisfying $0 < |w|, |\gamma| < 1$, and $\operatorname{Re} w/\gamma > 1$. This is equivalent to establishing that $|w/\gamma - |w|^2| > |w/\gamma - 1|$. Since both terms have identical imaginary part and

$$\begin{aligned} \left| \operatorname{Re} \left(\left(\frac{w}{\gamma} \right) - |w|^2 \right) \right| &= \operatorname{Re} \left(\frac{w}{\gamma} \right) - |w|^2 \\ &> \operatorname{Re} \left(\frac{w}{\gamma} \right) - 1 \\ &= \left| \operatorname{Re} \left(\left(\frac{w}{\gamma} \right) - 1 \right) \right|, \end{aligned}$$

the result is proved. □

Remark. This proposition generalizes as follows. If Γ is some circle containing α and $1/\bar{\alpha}$ such that K is entirely outside Γ, then $B^*(\alpha) \neq 0$. Thus, for example, if K is a convex set containing the origin then the zeros of B^* all lie in K.

An immediate consequence of this proposition is:

Corollary 4.3. (1) *B^* has no zeros outside any disc containing K.*
(2) *If $K \subseteqq [a,b] \subset (-1,1)$ then all the zeros of B^* lie in $[a,b]$.*

Our next results deal only with the case $X = C(K)$ and are conerned with the question of uniqueness (up to multiplication by σ, $|\sigma| = 1$) of B^*. The two results presented here are rather particular. It is, as yet, an open question as to whether B^* is unique for any given compact set K of Δ.

Proposition 4.4. *Let $\tilde{B} \in \mathscr{B}_n \backslash \mathscr{B}_{n-1}$ and set $K = \{z: |\tilde{B}(z)| \leqq r\}$ for some $r \in (0,1)$. If $B \in \mathscr{B}_n$ and $\|B\|_{C(K)} \leqq r$, then $B = \sigma \tilde{B}$, $|\sigma| = 1$.*

Proof. Assume $B \in \mathscr{B}_n$ and $\|B\|_{C(K)} \leqq r$. By rotation we may assume that $B(1) = \tilde{B}(1)$ and $B \not\equiv \tilde{B}$. For each $\varepsilon > 0$, and all $z \in \partial K$,

$$|((1+\varepsilon)\tilde{B}(z) - B(z)) - ((1+\varepsilon)\tilde{B}(z))| = |B(z)| \leqq r < (1+\varepsilon) r = |(1+\varepsilon)\tilde{B}(z)|.$$

Thus by Rouché's Theorem the functions $(1+\varepsilon)\tilde{B} - B$ and $(1+\varepsilon)\tilde{B}$ have the same numbers of zeros in K. Since $(1+\varepsilon)\tilde{B}$ has n zeros in K, so does $(1+\varepsilon)\tilde{B} - B$. Passing to the limit as $\varepsilon \downarrow 0$ we see that $\tilde{B} - B$ has n zeros in K. Furthermore, $\tilde{B} - B$ vanishes at $z = 1$. $\tilde{B} - B$ cannot have n zeros in Δ and an additional zero at $z = 1$. (If $w \neq 0$ is a zero of $\tilde{B} - B$, then $1/\bar{w}$ is also a zero of $\tilde{B} - B$.) $\square$

Corollary 4.5. *Set $\bar{\Delta}_r = \{z: |z| \leqq r\}$, $r \in (0,1)$. Then*

$$d_n(\mathscr{A}; C(\bar{\Delta}_r)) = r^n$$

and $B^*(z) = z^n$ is the unique minimal Blaschke product on $\bar{\Delta}_r$ satisfying $B(1) = 1$. (See Theorem 2.1 with $q = \infty$, $m = 0$.)

We now consider $K = [a,b] \subset (-1,1)$ (or any rotation thereof).

Proposition 4.6. *Let $K = [a,b] \subset (-1,1)$. There is a unique $B^* \in \mathscr{B}_n$ of the form $B^*(z) = \prod_{j=1}^{n} \frac{z-\alpha_j}{1-\alpha_j z}$, $\alpha_j \in [a,b]$, for which $\|B^*\|_{C[a,b]} = \inf\{\|B\|_{C[a,b]}: B \in \mathscr{B}_n\}$. B^* is characterized by the property that there exist $n+1$ points, $a \leqq x_1 < \ldots < x_{n+1} \leqq b$ at which*

$$B^*(x_j) = (-1)^{j+n+1} \|B^*\|_{C[a,b]}, \quad j = 1, \ldots, n+1.$$

Proof. By Corollary 4.3, it follows that it suffices to consider $B \in \mathscr{B}_n \backslash \mathscr{B}_{n-1}$ of the form $B(z) = \prod_{j=1}^{n} \frac{z-\beta_j}{1-\beta_j z}$, with $a \leqq \beta_1 \leqq \ldots \leqq \beta_n \leqq b$. Assume that we have constructed B^* satisfying the conditions of the proposition. If B, as above, satisfies

$\|B\|_{C[a,b]} \leqq \|B^*\|_{C[a,b]}$, then since B^* and B are real on $[a,b]$, and B^* attains its norm, alternately at $n+1$ points in $[a,b]$, $B^* - B$ has at least n zeros in $[a,b]$. From the form of B^* and B, we see that $B^*(1) = B(1) = 1$. Thus $B^* - B$ has n zeros in $[a,b]$ and an additional zero at $z = 1$. It follows, as in the proof of Proposition 4.4, that $B^* = B$, proving the uniqueness.

It remains to establish the existence of B^*. Set

$$\Sigma_{n+1} = \left\{\mathbf{t}\colon \mathbf{t} = (t_1,\ldots,t_{n+1}),\ t_i \geqq 0,\ \sum_{i=1}^{n+1} t_i = b - a\right\}.$$

Define $\alpha_0(\mathbf{t}) = a$ and $\alpha_j(\mathbf{t}) = a + \sum_{i=1}^{j} t_i$, $j = 1,\ldots,n+1$. Thus $\alpha_0(\mathbf{t}) = a \leqq \alpha_1(\mathbf{t}) \leqq \ldots \leqq \alpha_{n+1}(\mathbf{t}) = b$. For each $\mathbf{t} \in \Sigma_{n+1}$ let

$$B_{\mathbf{t}}(x) = \prod_{j=1}^{n} \frac{x - \alpha_j(\mathbf{t})}{1 - \alpha_j(\mathbf{t})\, x}.$$

Thus $B_{\mathbf{t}}(\alpha_j(\mathbf{t})) = 0, j = 1,\ldots,n$, and $B_{\mathbf{t}}(x)\,(-1)^{j+n+1} > 0$ for $x \in (\alpha_{j-1}(\mathbf{t}), \alpha_j(\mathbf{t}))$, $j = 1,\ldots,n+1$. Set

$$g_j(\mathbf{t}) = \max\{(-1)^{j+n+1} B_{\mathbf{t}}(x)\colon x \in [\alpha_{j-1}(\mathbf{t}), \alpha_j(\mathbf{t})]\}$$
$$g(\mathbf{t}) = \max\{g_j(\mathbf{t})\colon j = 1,\ldots,n+1\} = \|B_{\mathbf{t}}\|_{C[a,b]}$$

and

$$h_j(\mathbf{t}) = g(\mathbf{t}) - g_j(\mathbf{t}), \qquad j = 1,\ldots,n+1.$$

Note that $h_j(\mathbf{t}),\ g_j(\mathbf{t}) \geqq 0,\ j = 1,\ldots,n+1$; $g_j(\mathbf{t}) = 0$ if and only if $t_j = 0$; $g(\mathbf{t}) = \|B_{\mathbf{t}}\|_{C[a,b]} > 0$ for all $\mathbf{t} \in \Sigma_{n+1}$; and for each $\mathbf{t} \in \Sigma_{n+1}$ there exists at least one $j \in \{1,\ldots,n+1\}$ for which $h_j(\mathbf{t}) = 0$.

Assume that $\sum_{j=1}^{n+1} h_j(\mathbf{t}) > 0$ for all $\mathbf{t} \in \Sigma_{n+1}$. Consider the mapping

$$t_j \to \frac{h_j(\mathbf{t})\,(b-a)}{\sum_{i=1}^{n+1} h_i(\mathbf{t})}, \qquad j = 1,\ldots,n+1.$$

Since $h_j(\mathbf{t}) \geqq 0, j = 1,\ldots,n+1$, this mapping is a continuous map of the simplex Σ_{n+1} into itself. The fixed point theorem of Brouwer affirms the existence of a $\tilde{\mathbf{t}} \in \Sigma_{n+1}$ for which

$$\tilde{t}_j = \frac{h_j(\tilde{\mathbf{t}})}{\sum_{i=1}^{n+1} h_i(\tilde{\mathbf{t}})}, \qquad j = 1,\ldots,n+1.$$

Let $k \in \{1,\ldots,n+1\}$ be such that $h_k(\tilde{\mathbf{t}}) = 0$. Then $\tilde{t}_k = 0$ and $g_k(\tilde{\mathbf{t}}) = g(\tilde{\mathbf{t}}) = \|B_{\tilde{\mathbf{t}}}\|_{C[a,b]} > 0$. However, $\tilde{t}_k = 0$ implies that $g_k(\tilde{\mathbf{t}}) = 0$. This contradiction implies the existence of a $\mathbf{t}^* \in \Sigma_{n+1}$ for which $h_j(\mathbf{t}^*) = 0$, $j = 1,\ldots,n+1$, i.e., $g_j(\mathbf{t}^*) = g(\mathbf{t}^*) = \|B_{\mathbf{t}^*}\|_{C[a,b]} > 0, j = 1,\ldots,n+1$. Set $B^* = B_{\mathbf{t}^*}$. Thus there exist $x_j \in (\alpha_{j-1}(\mathbf{t}^*), \alpha_j(\mathbf{t}^*))$ satisfying $B^*(x_j) = (-1)^{j+n+1}\,\|B^*\|_{C[a,b]}$. □

Remark. The above proposition is easily generalized to any compact subset K of $(-1,1)$ (and not necessarily an interval).

Remark. In the above analysis we started with Δ. Since H^∞ is preserved under conformal mappings, it suffices to start with any simply-connected domain Ω.

5. n-Widths of a Class of Entire Functions

Let $B(\sigma,t)$ denote the class of entire functions f of exponential type σ for which $|f(x)| \leqq 1$ for all $x \in (-\infty, -T) \cup (T, \infty)$. Let $B_R(\sigma, T)$ denote the set of functions of $B(\sigma, T)$ which are real on the real axis. In this section we determine the n-widths of $B(\sigma, T)$ as a subset of $C[-T, T]$. This problem is, in a certain sense, the L^∞-analogue of Example 3.5 of Chapter IV.

The following theorem is basic to our understanding of $B(\sigma, T)$. Let $\{x_i\}_{i=1}^m$ be given distinct points in $[-T, T]$, and $\{a_{ij}\}_{i=1}^m{}_{j=0}^{\mu_i}$ given *real* numbers. Set $l = \sum_{i=1}^m (\mu_i + 1)$, and

$$Lf = \sum_{i=1}^m \sum_{j=0}^{\mu_i} a_{ij}\, f^{(j)}(x_i).$$

We will consider $\max\{|Lg| : g \in B(\sigma, T)\}$.

Theorem 5.1. *There is a unique $f \in B_R(\sigma, T)$ for which*

$$Lf = \max\{|Lg| : g \in B(\sigma, T)\}.$$

This f is either a constant or of exact type σ. If f is not a constant, then it satisfies the differential equation

$$\frac{(f'(z))^2}{1-(f(z))^2} = \sigma^2 \frac{(p(z))^2}{q(z)}$$

where p and q are monic polynomials with real coefficients. f is therefore of the form

$$f(z) = \sin\psi(z), \quad \psi(z) = \sigma\int_0^z (q(w))^{-1/2}\, p(w)\, dw + \sin^{-1}(f(0)).$$

Furthermore, let $\ldots < \lambda_{-2} < \lambda_{-1} \leqq -T$, $T \leqq \lambda_1 < \lambda_2 \ldots$ denote the points in $(-\infty, -T] \cup [T, \infty)$ at which $|f(\lambda_i)| = 1$. If $\lambda_{-1} = -T$ and $f'(-T) \neq 0$, set $s_-(z) = z + T$. Otherwise $s_-(z) = 1$. Similarly, if $\lambda_1 = T$ and $f'(T) \neq 0$, set $s_+(z) = z - T$. Otherwise set $s_+(z) = 1$. Define $s(z) = s_+(z)\, s_-(z)$, and $v = \text{degree}\, s$. Then

(1) $\deg p \leqq l - 2 + v$ *and the zeros of p are precisely those zeros of f' different from λ_i. Thus f nearly equioscillates outside $(-T, T)$ in the sense that $f(\lambda_i) f(\lambda_{i+1}) = -1$ for all $i \leqq -2$, $i \geqq 1$, with the exception of at most $l - 2 - v$ values*

(2) *f vanishes simply between those λ_i, $i \leqq -2$, $i \geqq 1$, for which $f(\lambda_i) f(\lambda_{i+1}) = -1$, and has at most $l-1$ additional zeros.*

A proof of Theorem 5.1 may be found in Melkman [1982]. It is very much based on a result of Boas, Schaeffer [1957]. Melkman [1982] is interested in determining the n-widths of $B(\sigma, T)$ in $C[-T, T]$ and also in the optimal recovery of functions in $B(\sigma, T)$ from their values at n given points in $[-T, T]$. He uses the latter problem to attack the former.

Assume that we are given n points $-T \leqq t_1 \leqq \ldots \leqq t_n \leqq T$. Then

Theorem 5.2. *There is a unique (up to multiplication by -1) function $f \in B_R(\sigma, T)$ such that for every $t \in [-T, T]$,*

$$|f(t)| = \max\{|g(t)| : g \in B(\sigma, T),\ g(t_i) = 0,\ i = 1, \ldots, n\}.$$

(In case of coincident points, appropriate derivative evaluation should be taken). f satisfies the following:

(1) *f vanishes at the t_i, while all its other zeros are real, simple and outside $[-T, T]$.*

(2) *f equioscillates outside $(-T, T)$. If $\ldots < \lambda_{-2} < \lambda_{-1} \leqq T$, $T \leqq \lambda_1 < \lambda_2 < \ldots$ denote the points at while $|f(\lambda_i)| = 1$, then $f(\lambda_i) = \varepsilon(-1)^i$ for $i > 0$, and $f(\lambda_i) = \varepsilon(-1)^{n+1}$ for $i < 0$, with $\varepsilon \in \{-1, 1\}$, fixed.*

(3) *(i) If $-T < t_2$, $t_{n-1} < T$, then f' vanishes in $(-T, T)$ at precisely $n-1+v$ points $\{\mu_i\}_{i=1}^{n-1+v}$ (these may be multiple if we have coincident t_i's) separating the t_i, with $-T < \mu_1 < t_1$ if $s_-(z) \not\equiv 1$, and $t_n < \mu_{n-1+v} < T$ if $s_+(z) \not\equiv 1$.*
(ii) If $-T = t_2$, $t_{n-1} < T$, then the above statement holds on $[-T, T)$, i.e., $\mu_1 = -T$.
(iii) If $-T < t_2$, $t_{n-1} = T$, then the above statement holds on $(-T, T]$.
(iv) If $-T = t_2$, $t_{n-1} = T$, then the above statement holds on $[-T, T]$.

(4) *Set*

$$h(z) = \frac{f'(z)\, s(z)}{\prod_{i=1}^{n-1+v} (z - \mu_i)}.$$

$h(z)$ vanishes only at the λ_i, $i = \pm 1, \pm 2, \ldots$, and the functions $\{h(t)\, t^j\}_{j=0}^{n-1}$ are linearly independent on the $\{t_i\}_{i=1}^n$. For given $g \in B(\sigma, T)$, let Sg denote the interpolant to g at the $\{t_i\}_{i=1}^n$ from span $\{h(t)\, t^j\}_{j=0}^{n-1}$. *Then*

$$|f(t)| = \max\{|g(t) - (Sg)(t)| : g \in B(\sigma, T)\}$$

for all $t \in [-T, T]$. That is, if $L_i(t_j) = \delta_{ij}$, $L_i \in \pi_{n-1}$, $i, j = 1, \ldots, n$, then for every $g \in B(\sigma, T)$, and every $t \in [-T, T]$

$$|f(t)| \geqq \left| g(t) - \sum_{i=1}^{n} g(t_i) \frac{h(t)}{h(t_i)} L_i(t) \right|.$$

(5) *The function f is a continuous function of* $\mathbf{t} = (t_1, \ldots, t_n)$, $-T \leqq t_1 \leqq \ldots \leqq t_n \leqq T$.

The proof of the theorem is essentially an application of Theorem 5.1. It depends on the fact, proven in Micchelli and Rivlin [1977], that in the search for optimal coefficients a_i^* and an extremal function f achieving

$$\left| f(t) - \sum_{i=1}^{n} a_i^* f(t_i) \right| = \min_{a_i} \max_{g \in B(\sigma, T)} \left| g(t) - \sum_{i=1}^{n} a_i g(t_i) \right|,$$

the optimal f is at the same time extremal for

$$\max \{ |g(t)| : g \in B(\sigma, T),\ g(t_i) = 0,\ i = 1, \ldots, n \}.$$

The inequality of (4) in the statement of the theorem follows from the fact that every $g \in B(\sigma, T)$ has the representation

$$g(z) = \sum_{i=1}^{n} g(t_i) \frac{h(z) L_i(z)}{h(t_i)} + \sum_{\substack{k=-\infty \\ k \neq 0}}^{\infty} \frac{g(\lambda_k) h(z) w(z)}{(z - \lambda_k) h'(\lambda_k) w(\lambda_k)}$$

where $w(z) = \prod_{i=1}^{n} (z - t_i)$, and from the formula

$$f(t) = h(t) w(t) \sum_{\substack{k=-\infty \\ k \neq 0}}^{\infty} \frac{1}{|(t - \lambda_k) h'(\lambda_k) w(\lambda_k)|}.$$

We use Theorem 5.2 to construct the function which will be extremal in our n-width problem.

Theorem 5.3. *There exists a function $F_n \in B_R(\sigma, T)$ with the following properties:*

(1) *F_n equioscillates in $[-T, T]$ between the values $\pm \|F_n\|_{C[-T,T]}$ exactly $n+1$ times at the points $\rho_1 < \ldots < \rho_{n+1}$, i.e., $F_n(\rho_i) = (-1)^i \|F_n\|_{C[-T,T]}$, $i = 1, \ldots, n+1$.*

(2) *F_n equioscillates outside $(-T, T)$ between ± 1.*

(3) *If $\|F_n\|_{C[-T,T]} < 1$, then $|F_n(\pm T)| = \|F_n\|_{C[-T,T]}$ and otherwise $|F_n(\pm T)| = 1$.*

(4) *F_n has only the real, simple zeros implied by (1) and (2).*

(5) $$\min_{-T \leqq t_1 \leqq \ldots \leqq t_n \leqq T} \max_{\substack{g \in B(\sigma, T) \\ g(t_i) = 0,\ i = 1, \ldots, n}} \|g\|_{C[-T,T]} = \|F_n\|_{C[-T,T]}.$$

Remark. The above F_n is actually unique (up to multiplication by -1). However we do not need this fact in the subsequent analysis.

Proof. We first prove the existence of F_n satisfying (1), (2), (3), and (4) using the method of proof of Proposition 4.6. Let $\Sigma_{n+1} = \left\{ \mathbf{s} : \mathbf{s} = (s_1, \ldots, s_{n+1}),\ s_i \geqq 0,\ \sum_{i=1}^{n+1} s_i = 2T \right\}$. For each $\mathbf{s} \in \Sigma_{n+1}$, set $t_0(\mathbf{s}) = -T$,

$t_i(\mathbf{s}) = -T + \sum_{j=1}^{i} s_j$, $i = 1, \ldots, n+1$. Thus $t_0(\mathbf{s}) = -T \leqq t_1(\mathbf{s}) \leqq \ldots \leqq t_n(\mathbf{s}) \leqq t_{n+1}(\mathbf{s}) = T$.

From Theorem 5.2 there exists a unique $f_\mathbf{s} \in B_R(\sigma, T)$ satisfying (2) and (4) and such that $f_\mathbf{s}(t_i(\mathbf{s})) = 0$, $i = 1, \ldots, n$, and $f_\mathbf{s}(t)(-1)^i > 0$ for $t_{i-1}(\mathbf{s}) < t < t_i(\mathbf{s})$, $i = 1, \ldots, n+1$. For $\mathbf{s} \in \Sigma_{n+1}$, set

$$g_i(\mathbf{s}) = \max\{|f_\mathbf{s}(t)|: t \in [t_{i-1}(\mathbf{s}), t_i(\mathbf{s})]\}$$

and

$$g(\mathbf{s}) = \max\{g_i(\mathbf{s}): i = 1, \ldots, n+1\} = \|f_\mathbf{s}\|_{C[-T,T]}.$$

Note that for each $\mathbf{s} \in \Sigma_{n+1}$, $\|f_\mathbf{s}\|_{C[-T,T]} > 0$, and $g_i(\mathbf{s}) = 0$ if and only if $t_{i-1}(\mathbf{s}) = t_i(\mathbf{s})$, i.e., $s_i = 0$. Define

$$h_i(\mathbf{s}) = g(\mathbf{s}) - g_i(\mathbf{s}).$$

For each $\mathbf{s} \in \Sigma_{n+1}$, $h_i(\mathbf{s}) = 0$ for some i. Since $f_\mathbf{s}(t)$ changes sign at $t_i(\mathbf{s})$, it follows that the existence of F_n satisfying (1) is equivalent to the existence of an $\mathbf{s} \in \Sigma_{n+1}$ for which $\sum_{i=1}^{n+1} h_i(\mathbf{s}) = 0$.

Assume that $\sum_{i=1}^{n+1} h_i(\mathbf{s}) > 0$ for all $\mathbf{s} \in \Sigma_{n+1}$. Consider the mapping

$$s_j \to \frac{2Th_j(\mathbf{s})}{\sum_{i=1}^{n+1} h_i(\mathbf{s})}, \qquad j = 1, \ldots, n+1.$$

By property (5) of Theorem 5.2, this is a continuous map of the simplex Σ_{n+1} into itself. The Brouwer fixed point theorem affirms the existence of an $\mathbf{s}^* \in \Sigma_{n+1}$ for which

$$s_j^* = \frac{2Th_j(\mathbf{s}^*)}{\sum_{i=1}^{n+1} h_i(\mathbf{s}^*)}, \qquad j = 1, \ldots, n+1.$$

Let k be such that $h_k(\mathbf{s}^*) = 0$. Thus, on the one hand, $s_k^* = 0$ and $g_k(\mathbf{s}^*) = 0$. On the other hand $h_k(\mathbf{s}^*) = 0$ implies that $\|f_{\mathbf{s}^*}\|_{C[-T,T]} = g(\mathbf{s}^*) = g_k(\mathbf{s}^*)$. Since $\|f_{\mathbf{s}^*}\|_{C[-T,T]} > 0$, this is a contradiction. Thus there exists a function F_n satisfying (1), (2), and (4). It now follows from Theorem 5.2 that F_n must also satisfy (3).

We now prove (5). Let $\{t_i^*\}_{i=1}^n$ denote the n unique zeros of F_n in $(-T, T)$. Thus

$$-T \leqq \rho_1 < t_1^* < \rho_2 < \ldots < \rho_n < t_n^* < \rho_{n+1} \leqq T.$$

For given $-T < t_1 < \ldots < t_n < T$, let $f(t)$ be as in Theorem 5.2, i.e., $|f(t)| = \max\{|g(t)|: g \in B(\sigma, T),\ g(t_i) = 0,\ i = 1, \ldots, n\}$ for each $t \in [-T, T]$. Assume that $\|F_n\|_{C[-T,T]} = |F_n(\rho_i)| > \|f\|_{C[-T,T]}$, $i = 1, \ldots, n+1$. From statement (4) of Theorem 5.2.

$$|f(t)| \geqq |F_n(t) - h(t)\,p(t)|$$

where $h(t)$ is as defined in Theorem 5.2, and p is some polynomial of degree $\leqq n-1$, i.e., $p \in \pi_{n-1}$. In particular

$$|F_n(\rho_i)| > |f(\rho_i)| \geqq |F_n(\rho_i) - h(\rho_i)\, p(\rho_i)|$$

for $i = 1, \ldots, n+1$. Thus

$$\operatorname{sgn} h(\rho_i)\, p(\rho_i) = -\operatorname{sgn} F_n(\rho_i) = (-1)^{i+1}, \; i = 1, \ldots, n+1.$$

Since h is of one sign on the ρ_i, it follows that $p(\rho_i)\, p(\rho_{i+1}) < 0, i = 1, \ldots, n$. Thus p has at least n zeros in (ρ_1, ρ_{n+1}). But $p \in \pi_{n-1}$, and therefore $p \equiv 0$. A contradiction now ensues, proving (5). □

Theorem 5.4. *Let F_n, $\{\rho_i\}_{i=1}^{n+1}$ and $\{t_i^*\}_{i=1}^{n}$ be as defined above. Then*

$$\begin{aligned} d_n(B(\sigma, T); C[-T, T]) &= d^n(B(\sigma, T); C[-T, T]) = \delta_n(B(\sigma, T); C[-T, T]) \\ &= b_n(B(\sigma, T); C[-T, T]) = \|F_n\|_{C[-T, T]}. \end{aligned}$$

Furthermore,

(1) $X_n = \operatorname{span}\{h(t)\, t^j\}_{j=0}^{n-1}$ *is an optimal subspace for* d_n, *where*

$$h(t) = \frac{F_n'(t)\,(t^2 - T^2)}{\prod_{i=1}^{n+1} (t - \rho_i)}.$$

(2) $L^n = \{g \colon g \in B(\sigma, T),\; g(t_i^*) = 0,\; i = 1, \ldots, n\}$ *is optimal for* d^n.

(3) *Interpolation at the* $\{t_i^*\}_{i=1}^{n}$ *from* X_n *is an optimal operator for* δ_n.

(4) $Y_{n+1} = \left\{\frac{F_n(z)\, p(z)}{w(z)} : p \in \pi_n,\; w(z) = \prod_{i=1}^{n} (z - t_i^*)\right\}$ *is an optimal subspace for* b_n.

Proof. It suffices to prove that $\delta_n(B(\sigma, T); C[-T, T]) \leqq \|F_n\|_{C[-T, T]}$ and $b_n(B(\sigma, T); C[-T, T]) \geqq \|F_n\|_{C[-T, T]}$. The upper bound is a consequence of statement (4) of Theorem 5.2. It remains to prove the lower bound. In other words it suffices to show that if $g \in Y_{n+1}$, and $\|g\|_{C[-T, T]} \leqq \|F_n\|_{C[-T, T]}$, then $g \in B(\sigma, T)$.

Let $g = F_n p / w \in Y_{n+1}$. Then g is of exponential type σ. We must therefore show that $\left\|\frac{F_n p}{w}\right\|_{C[-T, T]} \leqq \|F_n\|_{C[-T, T]}$ implies the inequality $\left|\frac{F_n(t)\, p(t)}{w(t)}\right| \leqq 1$ for all $|t| > T$. Since $\left|\frac{F_n(\rho_i)\, p(\rho_i)}{w(\rho_i)}\right| \leqq \|F_n\|_{C[-T, T]} = |F_n(\rho_i)|$, we have $|p(\rho_i)| \leqq |w(\rho_i)|$ for all $i = 1, \ldots, n+1$. Now for all t

$$|p(t)| = \left|\sum_{i=1}^{n+1} p(\rho_i)\, L_i(t)\right| \leqq \sum_{i=1}^{n+1} |p(\rho_i)\, L_i(t)| \leqq \sum_{i=1}^{n+1} |w(\rho_i)\, L_i(t)|,$$

where $L_i \in \pi_n$ and $L_i(\rho_j) = \delta_{ij}$, $i, j = 1, \ldots, n+1$. For $|t| \geqq T$,

$$\sum_{i=1}^{n+1} |w(\rho_i)\, L_i(t)| = \left|\sum_{i=1}^{n+1} w(\rho_i)\, L_i(t)\right| = |w(t)|.$$

Thus $|p(t)| \leqq |w(t)|$ for all $|t| \geqq T$, which implies that $|F_n(t)\, p(t)/w(t)| \leqq |F_n(t)| \leqq 1$ for all $|t| \geqq T$, i.e., $F_n p/w \in B(\sigma, T)$. $\square$

Remark. Let $N(\sigma, T)$ denote the smallest n for which $d_n(B(\sigma, T); C[-T, T]) \leqq 1$. It is not difficult to ascertain that if $2m < 2\sigma T/\pi \leqq 2m+1$, then $N(\sigma, T) = 2m+1$; while if $2m-1 < 2\sigma T/\pi \leqq 2m$, then $N(\sigma, T) = 2m$, i.e., $N(\sigma, T)$ is essentially $2\sigma T/\pi$. (If $2\sigma T/\pi = 2m$, then $F_{2m}(t) = \cos \sigma t$, while if $2\sigma T/\pi = 2m-1$, then $F_{2m-1}(t) = \sin \sigma t$.) Assume that we are given the class of entire functions of exponential type σ bounded by ε off $(-T, T)$, i.e., off $(-T, T)$ this class is indistinguishable from the zero function to within ε. Then there exists a subspace of dimension $N(\sigma, T)$ (independent of ε) such that on $[-T, T]$ all functions of the class are within ε of this finite dimensional subspace.

Notes and References

Section 2. The upper bound in Theorem 2.1 for $p = \infty$ was proved by Babenko [1958]. The lower bound for this same p was obtained by Tichomirov [1960a] as one of the first applications of Proposition 1.6 of Chapter II. The generalization to all $p \in [1, \infty]$, as presented here, is due to Taikov [1967b]. The important property in the upper bound argument is the positivity of the kernel $\tilde{G}_R(t)$. This was noted by Scheick [1966] and by Taikov [1967b]. Subsequently Taikov [1977b] used these methods to obtain upper bounds for other classes of functions and also presented criteria under which the upper bound is also the n-width.

Section 3. The material of this section is taken from Fisher, Micchelli [1984], see also Fisher [1983].

Section 4. The material of this section is taken from Fisher, Micchelli [1980], see also Fisher [1983], except for Proposition 4.6 which is new. The proof of the existence of the B^* of Proposition 4.6 follows from a method of proof of Karlin, Studden [1966, p. 68]. A different method of proof of Proposition 4.6 is via an application of the Borsuk Antipodality Theorem (see the proof of Theorem 2.5 of Chapter V). In the paper Fisher, Micchelli [1980] there are to be found generalizations to finitely connected domains.

Section 5. The material of this section is from Melkman [1982].

The n-widths of another class of analytic functions was determined in Chapter IV (see Example 6.1) and Chapter V (see the Notes and References to Section 4). This is the class $\tilde{H}^p_\beta$ of functions f which are analytic in $S_\beta = \{z\colon |\operatorname{Im} z| < \beta\}$, real and 2π-periodic on the x-axis, and which satisfy $\|\operatorname{Re} f(\cdot + i\beta)\|_{L^p[0, 2\pi]} \leqq 1$. The n-widths of $\tilde{H}^2_\beta$ in $L^2[0, 2\pi]$ were determined in Section 6 of Chapter IV, while those of $\tilde{H}_\beta^\infty$ in L^q, $q \in [1, \infty]$, and $\tilde{H}_\beta^p$ in L^1, $p \in [1, \infty]$, are given by the results of Section 4 of Chapter V. All this follows from the fact that $f \in \tilde{H}_\beta^p$ has the representation

$$f(z) = 1/2\pi \int_0^{2\pi} K_\beta(z-t) \operatorname{Re}[f(t+i\beta)]\,dt$$

where $K_\beta(z) = 1 + 2 \sum_{k=1}^{\infty} (\cos kz)/(\cosh k\beta)$, and K_β is CVD on $[0, 2\pi)$ (see Section 4 of Chapter III).

Let Ω be a domain in C with finitely many complementary components (and with additional minor assumptions to be found in the references given below). Let $\mathscr{A} = \{f : f \text{ analytic in } \Omega, \| f \|_{C(\Omega)} \leqq 1\}$, and let K be a compact subset of Ω. The major asymptotic result for n-widths of analytic functions is that

$$\lim_{n \to \infty} [d_n(\mathscr{A}; C(K))]^{1/n} = \exp(-1/C(K, \Omega)),$$

where $C(K, \Omega)$ denotes the capacity of K relative to Ω. This result was proven for Ω simply connected and K a continuum by Erokhin [1968]. The full result was obtained by Widom [1972], and a simpler proof is due to Fisher, Micchelli [1980]. Other asymptotic estimates for classes of n-widths of analytic functions may be found in Fisher, Micchelli [1984], Oleinik [1975], and Konovalov [1978].

Bibliography

Abramov, A. M.
[1972] On sets with equal diameters after Aleksandrov, Vestnik Mosk. Univ. Mat. *27*, 15–17; see also Moscow Univ. Math. Bull. *27*, 80–81

Ahlfors, L. V.
[1966] *Complex Analysis*, 2nd ed., McGraw-Hill, New York

Allahverdiev, D. E.
[1957] On the rate of approximation of completely continuous operators by finite-dimensional operators, Azerb. Gos. Univ. Ucen. Zap. *2*, 27–35

Anselone, P. M., Lee, J. W.
[1974] Spectral properties of integral operators with nonnegative kernels, Linear Algebra Appl. *9*, 67–87

Aubin, J. P.
[1968] Approximation of non-homogeneous Neumann problems: regularity of the convergence and estimates in terms of n-width, Math. Res. Center TSR 924, Madison, Wisc.
[1972] *Approximation of elliptic boundary-value problems*, Wiley-Interscience, New York

Babenko, K. I.
[1958] On the best approximation of a class of analytic functions, Izv. Akad. Nauk SSSR *22*, 631–640

Babushka, I.
[1971] The rate of convergence for the finite element method, SIAM J. Numer. Anal. *8*, 304–315

Bakhalov, N. S.
[1962] On optimal methods of specifying information during the solution of differential equations, Z. Vycisl. Mat. i. Mat. Fiz. *2*, 569–592; see also USSR Comp. Math. and Math. Phys. *2*, 608–640 (1963)

Bergh, J., Löfstrom, J.
[1976] *Interpolation Spaces; An Introduction*, Springer, Berlin

Bessaga, C., Pelczyński, A., Rolewicz, S.
[1961] On diametral approximative dimension and linear homogeneity of F-spaces, Bull. Acad. Polon. Sci. *9*, 677–683

Birman, M. S., Solomjak, M. Z.
[1966] Approximation of the functions of the class W_p^α by piecewise polynomial functions, Dokl. Akad. Nauk SSSR *171*, 1015–1018; see also Soviet. Math. Dokl. *7*, 1573–1577
[1967] Piecewise polynomial approximations of functions of the class W_p^α, Mat. Sb. (N.S.) *73*, 331–355; see also Math. USSR Sb. *2*, 295–317

Boas, R. P., Jr., Schaeffer, A. C.
[1957] Variational methods in entire functions, Amer. J. Math. *79*, 857–884

de Boor, C.
[1963] Best approximation properties of spline functions of odd degree, J. Math. Mech. *12*, 747–750

[1973] The quasi-interpolant as a tool in elementary polynomial spline theory, in *Approximation Theory*, ed. G. G. Lorentz, 269–276, Academic Press, New York

de Boor, C., DeVore, R., Höllig, K.
[1980] Mixed norm n-widths, Proc. Amer. Math. Soc. *80*, 577–583

de Boor, C., Fix, G.
[1973] Spline approximation by quasi-interpolants, J. Approx. Theory *7*, 19–45

Borsuk, K.
[1933] Drei Sätze über die n-dimensionale euklidische Sphäre, Fund. Math. *20*, 177–191

Boyanov, T. P.
[1973] On the widths of the space of continuous functions on a metric compactum, Annuaire Univ. Sofia *65*, 25–32

Boyanov, T. P., Popov, V. A.
[1970] On the widths of the space of continuous functions in the metric of Hausdorff, Annuaire Univ. Sofia *63*, 167–185

Brown, A. L.
[1964] Best n-dimensional approximation to sets of functions, Proc. London Math. Soc. *14*, 577–594
[1982] Finite rank approximations to integral operators which satisfy certain total positivity conditions, J. Approx. Theory *34*, 42–90

Chui, C. K., Smith, P. W.
[1975] Some nonlinear spline approximation problems related to N-widths, J. Approx. Theory *13*, 421–430

Chzhan, G. T.
[1962] On the minimum number of interpolation points in the numerical integration of the heat-conduction equation, Z. Vycisl. Mat. i. Mat. Fiz. *2*, 80–88; see also USSR Comp. Math. and Math. Phys. *2*, 78–87 (1963)

Courant, R., Hilbert, D.
[1953] *Methods of Mathematical Physics*, Wiley-Interscience, New York

Davie, A. M.
[1973] The approximation problem for Banach spaces, Bull. London Math. Soc. *5*, 261–266

Desplanches, R.
[1977] n-ieme diametre d'une classe d'espace des Sobolev a poids sur une demi-droite, C.R. Acad. Sci. Paris *284*, 1377–1380

Dubinsky, E.
[1979] *The structure of nuclear Frechet spaces*, Springer-Verlag, Lect. Notes in Math., No. 720

Dunford, N., Schwartz, J. T.
[1957] *Linear Operators, Part I*, Wiley-Interscience, New York

Duren, P. L.
[1970] *Theory of H^p spaces*, Academic Press, New York

Dyn, N.
[1983] Perfect splines of minimum norm for monotone norms and norms induced by inner-products, with applications to tensor product approximation and n-widths of integral operators, J. Approx. Theory *38*, 105–138

Erokhin, V. D.
[1968] Best linear approximation of functions analytically continuable from a given continuum into a given region, Uspehi Mat. Nauk *23*, 91–132; see also Russian Math. Surveys *23*, 93–135

Fisher, S. D.
[1978] Quantitative approximation theory, Amer. Math. Monthly *85*, 318–332
[1983] *Function Theory on Planar Domains; A Second Course in Complex Analysis*, Wiley-Interscience, New York

Fisher, S. D, Micchelli, C. A.
[1980] The n-widths of sets of analytic functions, Duke Math. J. *47*, 789–801
[1984] Optimal sampling of holomorphic functions, Amer. J. Math. *106*, 593–609

Forst, W.
[1977] Über die Breite von Klassen holomorpher periodischer Funktionen, J. Approx. Theory *19*, 325–331

Frum Ketkov, R. L.
[1965] On the metric diameter of function spaces, Uspehi Mat. Nauk *20*, 176–180

Gabdulhaev, B. G.
[1977] Diameters and optimal quadrature formulas for singular integrals, Dokl. Akad. Nauk SSSR *234*, 513–516; see also Soviet. Math. Dokl. *18*, 700–705

Galeev, E. M.
[1981] The Kolmogorov diameters of the intersection of classes of periodic functions and of finite-dimensional sets, Mat. Zametki *29*, 749–760; see also Math. Notes *29*, 382–388

Gantmacher, F. R., Krein, M. G.
[1960] *Oszillationsmatrizen, Oscillationskerne und kleine Schwingungen mechanischer Systeme*, Akademie-Verlag, Berlin

Garkavi, A. L.
[1960] On existence of a best net and a best diameter of a set in a Banch space, Uspehi Mat. Nauk *15*, 210–211
[1962] On best net and best section of a set in a normed space, Izv. Akad. Nauk SSSR, Ser. Math. *26*, 87–106

Garling, D. J. H., Gordon, Y.
[1971] Relations between some constants associated with finite dimensional Banach spaces, Israel J. Math. *9*, 346–361

Glushkin, E. D.
[1974] On a problem concerning diameters, Dokl. Akad. Nauk SSSR *219*, 527–530; see also Soviet. Math. Dokl. *15*, 1592–1596
[1981] On some finite-dimensional problems of width theory, Vest. Leningrad Univ. Mat. *3*, 5–10

Gohberg, I. C., Krein, M. G.
[1969] *Introduction to the theory of linear nonselfadjoint operators*, Transl. Math. Monographs, *18*, Amer. Math. Soc., Providence, R.I.

von Golitschek, M.
[1979] On n-widths and interpolation by polynomial splines, J. Approx. Theory *26*, 132–141

Golomb, M.
[1965] Optimal approximating manifolds in L_2-spaces, J. Math. Anal. Appl. *12*, 505–512
[1967] Splines, n-widths and optimal approximations, Math. Res. Center TSR 784, Madison, Wisc.
[1976] Interpolation operators as optimal recovery schemes for classes of analytic functions, in *Optimal Estimation in Approximation Theory*, eds. C. A. Micchelli and T. J. Rivlin, 93–138, Plenum, New York

Grigorian, Yu. I.
[1973] Diameters of certain sets in function spaces, Mat. Zametki *13*, 637–646; see also Math. Notes *13*, 383–388
[1975] Widths of certain sets in functional spaces, Uspehi Mat. Nauk *30*, 161–162

Ha, C. W.
[1974] Approximation numbers of linear operators and nuclear spaces, J. Math. Anal. Appl. *46*, 292–311

Hardy, G. W., Littlewood, J. E., Polya, G.
[1952] *Inequalities*, Cambridge Univ. Press, 2nd ed., Cambridge

Helfrich, H.-P.
[1971] Optimale lineare Approximation beschränkter Mengen in normierten Räumen, J. Approx. Theory *4*, 165–182

Hobby, C. R., Rice, J. R.
[1965] A moment problem in L_1 approximation, Proc. Amer. Math. Soc. *16*, 665–670

Höllig, K.
[1979a] Approximationszahlen von Sobolev-Einbettungen, Math. Ann. *242*, 273–281
[1979b] *Approximationszahlen von Sobolev-Einbettungen*, Dissertation, Bonn
[1980] Diameters of classes of smooth functions, in *Quantitative Approximation*, eds. R. A. DeVore and K. Scherer, 163–175, Academic Press, New York

Hutton, C. V., Morrell, J. S., Retherford, J. R.
[1974] Approximation numbers and Kolmogoroff diameters of bounded linear operators, Bull. Amer. Math. Soc. *80*, 462–466
[1976] Diagonal operators, approximation numbers, and Kolmogoroff diameters, J. Approx. Theory *16*, 48–80

Hutton, C. V.
[1974] On the approximation numbers of an operator and its adjoint, Math. Ann. *210*, 277–280

Ioffe, A. D., Tichomirov, V. M.
[1968a] Duality in problems of the calculus of variations, Dokl. Akad. Nauk SSSR *180*, 789–792; see also Soviet. Math. Dokl. *9*, 685–688
[1968b] Duality of convex functions and extremal problems, Uspehi Mat. Nauk *23*, 51–116; see also Russian Math. Surveys *23*, 53–124

Ismagilov, R. S.
[1968] On n-dimensional diameters of compacts in a Hilbert space, Funktsional. Anal. i Prilozhen. *2*, 32–39; see also Functional Anal. Appl. *2*, 125–132
[1974] Diameters of sets in normed linear spaces and the approximation of functions by trigonometric polynomials, Uspehi Mat. Nauk, *29*, 161–178; see also Russian Math. Surveys *29*, 169–186

Ismagilov, R. S., Nasyrova, Kh.
[1977] Diameters of a class of smooth functions in the space L_2, Mat. Zametki *22*, 671–678; see also Math. Notes *22*, 865–870

Jagerman, D.
[1969] ε-Entropy and approximation of bandlimited functions, SIAM J. Appl. Math. *17*, 362–377
[1970] Information theory and approximation of bandlimited functions, Bell System Tech. J. *49*, 1911–1941

Jentzsch, R.
[1912] Über Integralgleichungen mit positivem Kern, J. Math. Crelle *141*, 235–244

Jerome, J. W.
[1967] On the L_2 n-width of certain classes of functions of several variables, J. Math. Anal. Appl. *20*, 110–123
[1968] Asymptotic estimates of the L_2 n-width, J. Math. Anal. Appl. *22*, 449–464
[1970] On n-widths in Sobolev spaces and applications to elliptic boundary value problems, J. Math. Anal. Appl. *29*, 201–215
[1972] Asymptotic estimates of the n-widths in Hilbert space, Proc. Amer. Math. Soc. *33*, 367–372
[1973] Topics in multivariate approximation theory, in *Approximation Theory*, ed., G. G. Lorentz, 151–198, Academic Press, New York

Jerome, J. W., Schumaker, L. L.
[1969] Applications of ε-entropy to the computation of n-widths, Proc. Amer. Math. Soc. *22*, 719–722
[1974] On the distance to a class of generalized splines, in ISNM 25, eds. P. L. Butzer and B. Sz.-Nagy, 503–517, Birkhäuser Verlag, Basel

Johnson, P. D., Jr.
[1973] *Approximation numbers of diagonal maps between normed sequence spaces*, Doctoral dissertation, Univ. of Michigan

Kadec, M., Snobar, S.
[1971] Certain functionals on the Minkowski compactum, Mat. Zametki *10*, 453–457; see also Math. Notes *10*, 694–696

Karlin, S.
[1968] *Total Positivity, Vol. I*, Stanford Univ. Press, Stanford, Ca.

Karlin, S., Studden, W. J.
[1966] *Tchebycheff Systems: with Applications in Analysis and Statistics*, Interscience, New York

Karlovitz, L. A.
[1973] On a class of Kolmogorov n-width problems, Atti. Accad. Naz. Lincei, Roma *53*, 241–245
[1976] Remarks on variational characterizations of eigenvalues and n-width problems, J. Math. Anal. Appl. *53*, 99–110

Kashin, B. S.
[1974] On Kolmogorov diameters of octahedra, Dokl. Akad. Nauk SSSR *214*, 1024–1026; see also Soviet. Math. Dokl. *15*, 304–307
[1975] On diameters of octahedra, Uspehi Mat. Nauk *30*, 251–252
[1977a] Orders of the widths of certain classes of smooth functions, Uspehi Mat. Nauk *32*, 191–192
[1977b] Diameters of some finite-dimensional sets and classes of smooth functions, Izv. Akad. Nauk SSSR *41*, 334–351; see also Math. USSR Izv. *11*, 317–333
[1979] General orthonormal systems and certain problems of approximation theory, Mat. Zametki *26*, 299–315; see also Math. Notes *26*, 641–650
[1981] Diameters of Sobolev classes of small order smoothness, Vestnik Moskov. Univ. *5*, 50–54

Kolli, A. El.
[1971a] n-ieme epaisseur dans les espaces de Sobolev, C.R. Acad. Sci. Paris *272*, 537–539
[1971b] n-ieme epaisseur dans les espaces de Sobolev avec poids, C.R. Acad. Sci. Paris *273*, 450–453
[1974] n-ieme epaisseur dans les espaces de Sobolev, J. Approx. Theory *10*, 268–294

Kolmogoroff, A.
[1963] Über die beste Annäherung von Funktionen einer gegebenen Funktionenklasse, Annals of Math. *37*, 107–110

König, H.
[1977] *s-Zahlen und Eigenwertverteilung von Operatoren in Banachräumen*, Dissertation, Bonn
[1978] Approximation numbers of Sobolev imbeddings over unbounded domains, J. Funct. Anal. *29*, 74–87
[1979] s-Numbers of Besov-Lorentz imbeddings, Math. Nachr. *91*, 389–400

Konovalov, V. N.
[1978] Problem of the diameters of classes of analytic functions, Ukrain. Mat. Z. *30*, 668–670; see also Ukrainian Math. J. *30*, 511–513

Korneichuk, N. P.
[1961] The best uniform approximation on certain classes of continuous functions, Dokl. Akad. Nauk SSSR *140*, 748–751; see also Soviet. Math. Dokl. *2*, 1254–1257
[1963] Exact value of the best approximations and of the diameters of certain classes of functions, Dokl. Akad. Nauk SSSR *150*, 1218–1220; see also Soviet Math. Dokl. *4*, 856–859

[1971 a] Extreme values of functionals and best approximation on classes of periodic functions, Izv. Akad. Nauk SSSR *35*, 93–124; see also Math. USSR Izv. *5*, 97–129
[1971 b] On the diameters of classes of continuous functions in the space L_p, Mat. Zametki *10*, 493–500; see also Math. Notes *10*, 719–723
[1974] On methods of investigating extremal problems in the theory of best approximation, Uspehi Mat. Nauk *29*, 9–42; see also Russian Math. Surveys *29*, 7–43
[1976 a] *Extremal Problems of the Theory of Approximation*, Nauka, Moscow
[1976 b] Approximation of periodic functions by spline functions, in *Fourier Analysis and Approximation Theory*, eds. G. Alexitis and P. Turan, 465–471, Budapest
[1977 a] Some extremal problems in approximation theory, A.M.S. Transl. (2), *109*, 51–57
[1977 b] Exact error bounds of approximation by interpolating splines in L-metric on the class W_p^r, $1 \leqq p < \infty$, of periodic functions, Anal. Math. *3*, 109–117
[1979] Widths in L_p of classes of continuous and differentiable functions and optimal reconstruction of functions and their derivatives, Dokl. Akad. Nauk SSSR *244*, 1317–1321; see also Soviet Math Dokl. *20*, 229–233
[1981] Widths in L_p of classes of continuous and of differentiable functions, and optimal methods of coding and recovering functions and their derivatives, Izv. Akad. Nauk SSSR *45*, 266–290; see also Math. USSR Izv. *18*, 227–247 (1982)

Krein, M. G., Krasnosel'ski, M. A., Milman, D. P.
[1948] On deficiency numbers of linear operators in Banach spaces and on some geometric problems, Sb. Trudov Inst. Mat. Akad. Nauk SSSR *11*, 97–112

Krein, M. G., Nudel'man, A. A.
[1977] *The Markov Moment Problem and Extremal Problems*, Transl. Math. Monographs, *50*, Amer. Math. Soc., Providence, R.I.

Kusainova, L. K.
[1980] Estimates of widths of the unit ball of the function space $L_p^l(\Omega, v)$ in $L_q(\Omega)$, Dokl. Akad. Nauk SSSR *251*, 791–794; see also Soviet Math. Dokl. *21*, 519–522

Landau, H. J., Pollack, H. O.
[1961] Prolate spheroidal wave functions, Fourier analysis and uncertainty – II, Bell System Tech. J. *40*, 65–84
[1962] Prolate spheroidal wave functions, Fourier analysis and uncertainty – III: The dimension of the space of essentially time- and band-limited signals, Bell System Tech. J. *41*, 1295–1336

Lemmens, P. W. H., Seidel, J. J.
[1973] Equiangular lines, J. Algebra *24*, 494–512

Levin, A. L., Tichomirov, V. M.
[1968] On a theorem of Erokhin, appendix to article of Erokhin [1968]

Ligun, A. A.
[1976] Inequalities for upper bounds of functionals, Anal. Math. *2*, 11–40
[1980] Diameters of certain classes of differentiable periodic functions, Mat. Zametki *27*, 61–75; see also Math. Notes *27*, 34–41

Lindenstrauss, J., Rosenthal, H. P.
[1969] The L_p-spaces, Israel J. Math. *7*, 325–349

Lorentz, G. G.
[1960] Lower bounds for the degree of approximation, Trans. Amer. Math. Soc. *97*, 25–34
[1962] Metric entropy, widths, and superposition of functions, Amer. Math. Monthly *69*, 469–485
[1965] Russian literature on Approximation in 1958-1964, in *Approximation of Functions*, ed. H. L. Garabedian, 191–215, Elsevier, New York
[1966 a] *Approximation of Functions*, Holt, Rinehart and Winston, New York
[1966 b] Metric entropy and approximation, Bull. Amer. Math. Soc. *72*, 903–937

Lubitz, C.
[1978] *s-Zahlen von Sobolev Einbettungen*, Diplom-Arbeit, Bonn

Maiorov, V. E.
[1975] Discretization of the problem of diameters, Uspehi Mat. Nauk *30*, 179–180
[1976a] Theorems of representation and best approximation in the classes W_p^r and H_p^r, Dokl. Akad. Nauk SSSR *228*, 293–296; see also Soviet Math. Dokl. *17*, 708–712
[1976b] The best approximation of classes $W_1^r(I^s)$ in the space $L_\infty(I^s)$, Mat. Zametki *19*, 699–706; see also Math. Notes *19*, 420–424
[1978a] On linear diameters of Sobolev classes, Dokl. Akad. Nauk SSSR *243*, 1127–1130; see also Soviet Math. Dokl. *19*, 1491–1494
[1978b] Various widths of the class H_p^r in the space L_q, Izv. Akad. Nauk SSSR *42*, 773–788; see also Math. USSR Izv. *13*, 73–87
[1979] Extremal chains of subspaces for Kolmogorov and linear diameters, Funktsional. Anal. i Prilozhen. *13*, 91–92; see also Functional Anal. Appl. *13*, 231–232

Makovoz, Yu. I.
[1969] Diameters of certain function classes in the space L, Vesci Akad. Navuk BSSR Ser. Fiz.-Mat. Navuk *4*, 19–28
[1972] On a method for estimation from below of diameters of sets in Banach spaces, Mat. Sb. (N.S.) *87*, 136–142; see also Math. USSR Sb. *16*, 139–146
[1979] Diameters of Sobolev classes and splines deviating least from zero, Mat. Zametki *26*, 805–812; see also Math. Notes *26*, 897–901

Melkman, A. A.
[1976] n-Width under restricted approximations, in *Approximation Theory II*, eds. G. G. Lorentz, C. K. Chui, and L. L. Schumaker, 463–468, Academic Press, New York
[1977] n-Widths and optimal interpolation of time- and band-limited functions, in *Optimal Estimation in Approximation Theory*, eds. C. A. Micchelli and T. J. Rivlin, 55–68, Plenum Press, New York
[1980] n-Widths of octahedra, in *Quantitative Approximation*, eds. R. A. DeVore and K. Scherer, 209–216, Academic Press, New York
[1981] The distance of a subspace of R^m from its axes and n-widths of octahedra, preprint
[1982] n-Widths and optimal interpolation of time- and band-limited functions, II, preprint

Melkman, A. A., Micchelli, C. A.
[1978] Spline spaces are optimal for L^2 n-width, Illinois J. Math. *22*, 541–564
[1979] Optimal estimation of linear operators in Hilbert spaces from inaccurate data, SIAM J. Numer. Anal. *16*, 87–105

Micchelli, C. A., Pinkus, A.
[1976a] The exact asymptotic value for the n-width of smooth functions in L^∞, in *Approximation Theory II*, eds. G. G. Lorentz, C. K. Chui and L. L. Schumaker, 469–474, Academic Press, New York
[1976b] On n-widths and optimal recovery in M^r, in *Approximation Theory II*, eds. G. G. Lorentz, C. K. Chui, and L. L. Schumaker, 475–478, Academic Press, New York
[1977a] On n-widths in L^∞, Trans. Amer. Math. Soc. *234*, 139–174
[1977b] Total positivity and the exact n-width of certain sets in L^1, Pacific J. Math. *71*, 499–515
[1978] Some problems in the approximation of functions of two variables and n-widths of integral operators, J. Approx. Theory *24*, 51–77
[1979] The n-width of rank $n+1$ kernels, J. Integral Equations *1*, 111–130

Micchelli, C. A., Rivlin, T. J.
[1977] A survey of optimal recovery, in *Optimal Estimation in Approximation Theory*, eds. C. A. Micchelli and T. J. Rivlin, 1–54, Plenum Press, New York

Milota, J.
[1976] Interpolation in a Banach space, Czechoslovak Math. J. *26*, 84–92

Mityagin, B. S.
[1961] Approximate dimension and bases in nuclear spaces, Uspehi Mat. Nauk *16*, 63–132; see also Russian Math. Surveys *16*, 59–127
[1962] Approximation of functions in L^p and C space on the torus, Mat. Sb. (N.S.) *58*, 397–414

Mityagin, B. S., Henkin, G. M.
[1963] Inequalities between n-diameters, in *Proc. of the Seminar on Functional Analysis 7*, Voronezh, 97–103

Mityagin, B. S., Pelczyński, A.
[1968] Nuclear operators and approximative dimension, *Proc. Inter. Congr. Math.* (Moscow, 1966), 366–372, "Mir" Moscow; see also A.M.S. Transl. (2), *70*, 137–145 (1968)

Mityagin, B. S., Tichomirov, V. M.
[1964] Asymptotic characteristic of compact sets in linear spaces, *Proc. Fourth All-Union Soviet Math. Congress, vol. II*; sectional lecture, "Nauka" Leningrad, 299–308

Mityagin, B., Torok, J.
[1982] Optimal subspaces for n-widths of p-ellipsoids, J. Approx. Theory *34*, 91–96

Motornyi, V. P., Ruban, V. I.
[1975] Diameters of some classes of differentiable functions in the space L, Mat. Zametki *17*, 531–543; see also Math. Notes *17*, 313–320

Nasyrova, Kh.
[1976] Asymptotic formulas for n-diameters of certain compacta in L_2 [0,1], Mat. Zametki *20*, 331–340; see also Math. Notes *20*, 745–750

Netravali, A.
[1973] A note on optimal approximating manifolds of a function class, Bell System Tech. J. *52*, 1237–1242

Newman, D. J.
[1976] n-Widths of function spaces, J. Approx. Theory *16*, 81–84

Oleinik, V. L.
[1975] Estimates of the widths of compact sets of analytic functions in L^p with a weight, Vestnik Leningrad Univ. Mat. *7*, 47–51; see also Vestnik Leningrad Univ. Math. *8*, 219–224 (1980)
[1976] Estimates for the n-widths of compact sets of differentiable functions in spaces with weight functions, Zap. Naučn. Seminar *Leningrad Otdel. Mat. Inst. V.A. Steklov AN SSSR 59*, 117–132; see also J. of Soviet Math. *10*, 286–298, (1978)

Otelbaev, M.
[1976] Two-sided estimates for diameters and applications, Dokl. Akad. Nauk SSSR *231*, 810–813; see also Soviet Math. Dokl. *17*, 1655–1659

Parks, T. W.
[1974] The use of signal properties for signal representation, J. Franklin Inst. *297*, 229–242

Parks, T. W., Meier, R. G.
[1971] Reconstruction of signals of a known class from a given set of linear measurements, IEEE Trans. Inform. Theory *17*, 37–44

Pietsch, A.
[1963] Einige neue Klassen von kompakten linearen Abbildungen, Revue Roumaine Math. Pures Appl. *8*, 427–447
[1972] *Nuclear Locally Convex Spaces*, Springer-Verlag, Band 66, Berlin
[1974] s-Numbers of operators in Banach spaces, Studia Math. *51*, 201–223
[1980] *Operator Ideals*, North-Holland Publ. Co., Amsterdam

Pinkus, A.
[1976] A simple proof of the Hobby-Rice theorem, Proc. Amer. Math. Soc. *60*, 82–84
[1979a] On n-widths of periodic functions, J. Analyse Math. *35*, 209–235
[1979b] Matrices and n-widths, Linear Algebra Appl. *27*, 245–278

Prosser, R. T.
[1966] The ε-entropy and ε-capacity of certain time-varying channels, J. Math. Anal. Appl. *16*, 553–573
[1971] Determinable classes of channels, II, Indiana Univ. Math. J. *20*, 789–806

Pukhov, S. V.
[1979a] Inequalities between the Kolmogorov and the Bernstein diameters in a Hilbert space, Mat. Zametki *25*, 619–628; see also Math. Notes *25*, 320–326
[1979b] Approximation of weight classes of Sobolev type, Uspehi Mat. Nauk *34*, 211–212; see also Russian Math. Surveys *34*, 215–216

Riesz, F., Sz.-Nagy, B.
[1955] *Functional Analysis*, F. Ungar, New York

Rivlin, T. J.
[1969] *An Introduction to the Approximation of Functions*, Blaisdell, Waltham, Mass.

Rolewicz, S.
[1972] *Metric Linear Spaces*, PWN-Polish Scientific Publishers, Warszawa

Ruban, V. I.
[1974] Even diameters of the class $W^r H_\omega$ in the space $C_{2\pi}$, Mat. Zametki *15*, 387–392; see also Math. Notes *15*, 222–225
[1975] Extremal subspaces in the problem about widths of classes H_ω [a,b] in the space C[a,b], Anal. Math. *1*, 131–139
[1980] Widths of sets in spaces of periodic functions, Dokl. Akad. Nauk SSSR *255*, 34–35; see also Soviet Math. Dokl. *22*, 658–659

Rudin, W.
[1952] L^2-approximation by partial sums of orthogonal developments, Duke Math. J *19*, 1–4

Sattes, U.
[1980] *Beste Approximation durch glatte Funktionen und Anwendungen in der intermediären Approximation*, Dissertation, Universität Erlangen-Nürnberg

Scepin, E. V.
[1974] On a problem of L.A. Tumarkin, Dokl. Akad. Nauk SSSR *217*, 42–43; see also Soviet Math. Dokl. *15*, 1024–1026

Scheick, J. T.
[1966] Polynomial approximation of functions analytic in a disk, Proc. Amer. Math. Soc. *17*, 1238–1243

Schmidt, E.
[1907] Zur Theorie der linearen und nichtlinearen Integralgleichungen. I, Math. Ann. *63*, 433–476

Schock, E.
[1975] Approximation numbers of bounded operators, J. Math. Anal. Appl. *51*, 440–448

Scholz, R.
[1974] Abschätzungen linearer Durchmesser in Sobolev- und Besov-Räumen, Manuscripta Math. *11*, 1–14
[1976] Durchmesserabschätzungen für die Einheitskugel des Sobolev-Raumes $W_q^r(\Omega)$ in $L_p(\Omega)$, Applicable Anal. *5*, 257–264

Schultz, M. H.
[1974] The complexity of linear approximation algorithms, in *SIAM-AMS Proceedings 7*, 135–148, ed. R. E. Karp, Amer. Math. Soc., Providence, R.I.

Sendov, B.
[1969] Some questions of the theory of approximations of functions and sets in the Hausdorff metric, Uspehi Mat. Nauk *24*, 141–178; see also Russian Math. Surveys *24*, 143–184

Sendov, Bl., Penkov, B.
[1964] On widths of the space of continuous functions, C.R. Acad. Bulgare Sci. *17*, 689–691

Shapiro, H. S.
[1971] *Topics in Approximation Theory*, Springer-Verlag, Lect. Notes in Math., No. 187
[1979] Stefan Bergman's theory of doubly-orthogonal functions: An operator-theoretic approach, Proc. Royal Irish Acad., Sect. A *79*, 49–58

Sharygin, I. F.
[1972] A lower bound for n-diameters, Mat. Zametki *12*, 413–419; see also Math. Notes *12*, 680–684

Singer, I.
[1970] *Best Approximation in Normed Linear Spaces by Elements of Linear Subspaces*, Springer-Verlag, Band 171

Slepian, D., Pollack, H. O.
[1961] Prolate spheroidal wave functions. Fourier analysis and uncertainty, I., Bell System Tech. J *40*, 43–64

Sofman, L. B.
[1969] Diameters of octahedra, Mat. Zametki *5*, 429–436; see also Math. Notes *5*, 258–262
[1973] Diameters of an infinite-dimensional octahedron, Vestnik Mosk. Univ. Mat. *28*, 54–56; see also Moscow Univ. Math. Bull *28*, 45–47

Solomjak, M. E., Tichomirov, V. M.
[1967] Some geometric characteristics of the embedding map from W_p^a into C, Izv. Vysš. Učebn. Zaved. Mat. *10*, 76–82

Stechkin, S. R.
[1954] The best approximation of given classes of functions, Uspehi Mat. Nauk *9*, 133–134

Stehling, W.
[1979] Approximationszahlen für Diagonaloperatoren zwischen Orlicz-Räumen, Math. Nachr. *93*, 165–176

Stesin, M. I.
[1973a] On Aleksandrov diameters of the ball in Banach space, Uspehi Mat. Nauk *27*, 219–220
[1973b] On the Aleksandrov diameters of finite dimensional octahedra, Vestnik Mosk. Univ. Mat. *28*, 30–35; see also Moscow Univ. Math. Bull. *28*, 24–28
[1974] On Aleksandrov diameters of balls, Dokl. Akad. Nauk SSSR *217*, 31–33; see also Soviet Math. Dokl. *15*, 1011–1014
[1975] Aleksandrov diameters of finite-dimensional sets and classes of smooth functions, Dokl. Akad. Nauk SSSR *220*, 1278–1281; see also Soviet Math. Dokl. *16*, 252–256

Subbotin, Yu. N.
[1970] Diameters of class $W^r L$ in $L(0, 2\pi)$ and spline function approximation, Mat. Zametki *7*, 43–52; see also Math. Notes *7*, 27–32
[1971] Approximation by spline functions and estimates of diameters, Trudy Mat. Inst. Steklov *109*, 35–60; see also Proc. Steklov Inst. Math. *109*, 39–67

Szarek, S. J.
[1978] On Kashin's almost Euclidean orthogonal decomposition of l_n^1, Bull. Acad. Polon., Sci. *26*, 691–694

Taikov, L. V.
[1967a] On approximating some classes of periodic functions in mean, Trudy Mat. Inst. Steklov *88*, 61–70; see also Proc. Steklov Inst. Math. *88*, 65–74
[1967b] On the best approximation in the mean of certain classes of analytic functions, Mat. Zametki *1*, 155–162; see also Math. Notes *1*, 104–109
[1976] Inequalities concerning best approximation and the modulus of continuity of functions in L_2, Mat. Zametki *20*, 433–438; see also Math. Notes *20*, 797–800
[1977a] Best approximation of differentiable functions in the metric of the space L_2, Mat. Zametki *22*, 535–542; see also Math. Notes *22*, 789–794
[1977b] Diameters of certain classes of analytic functions, Mat. Zametki *22*, 285–295; see also Math. Notes *22*, 650–656
[1979] Structural and constructive characteristics of functions in L_2, Mat. Zametki *25*, 217–223; see also Math. Notes *25*, 113–116
[1980] The best approximation in $L_2(0, 2\pi)$ of classes of periodic functions with derivatives of bounded variation, Mat. Zametki *28*, 239–242; see also Math. Notes *28*, 582–584

Tichomirov, V. M.
[1960 a] Diameters of sets in function spaces and the theory of best approximations, Uspehi Mat. Nauk *15*, 81–120; see also Russian Math. Surveys *15*, 75–111
[1960 b] On n-dimensional diameters of certain functional classes, Dokl. Akad. Nauk SSSR *130*, 734–737; see also Soviet Math. Dokl. *1*, 94–97
[1965 a] A remark on n-dimensional diameters of sets in Banach spaces, Uspehi Mat. Nauk *20*, 227–230
[1965 b] Some problems of approximation theory, Dokl. Akad. Nauk SSSR *160*, 774–777; see also Soviet Math. Dokl. *6*, 202–205
[1966] Remarks on a paper of L. A. Tumarkin "On widths of infinite dimensional compacts", Vestnik Mosk. Univ. Mat. *3*, 73–74
[1969] Best methods of approximation and interpolation of differentiable functions in the space $C[-1,1]$, Mat. Sbornik *80*, 290–304; see also Math. USSR Sbornik *9*, 275–289
[1971] Some problems in approximation theory, Mat. Zametki *9*, 593–607; see also Math. Notes *9*, 343–350
[1976] *Some Problems in the Theory of Approximation*, Nauka, Moscow
[1979] Theory of extremal problems and approximation theory, in *Approximation Theory-Banach Center Publications*, *4*, 273–286, PWN, Polish Scientific Publishers, Warszawa

Tichomirov, V. M., Babadjanov, S. B.
[1967] On the width of a functional class in the space $L_p (p \geqq 1)$, Izv. Akad. Nauk UzSSR Ser. Fiz.-Mat. Nauk *2*, 24–30

Timan, A. F.
[1960] A geometric problem in the theory of approximations, Dokl. Akad. Nauk SSSR *140*, 307–310; see also Soviet Math. Dokl. *2*, 1208–1211

Traub, J. F., Wozniakowski, H.
[1980] *A General Theory of Optimal Algorithms*, Academic Press, New York

Triebel, H.
[1970] Interpolationseigenschaften von Entropie- und Durchmesseridealen Kompakter Operatoren, Studia Math. *34*, 89–107
[1975] Interpolation properties of ε-entropy and diameters. Geometric characteristics of imbedding for function spaces of Sobolev-Besov type, Mat. Sb. (N.S.) *98*, 27–41; see also Math. USSR Sb. *27*, 23–37
[1978] *Interpolation Theory, Function Spaces, Differential Operators*, North-Holland Publ. Co., Amsterdam

Tumarkin, L. A.
[1966] On widths of infinite dimensional compacts, Vestnik Mosk. Univ. Mat. *3*, 67–72

Weinstein, A., Stenger, W.
[1972] *Methods of Intermediate Problems for Eigenvalues*, Academic Press, New York

Whitley, R.
[1982] Markov and Bernstein's inequalities, and compact and singular operators, J. Approx. Theory *34*, 277–285

Widom, H.
[1972] Rational approximation and n-dimensional diameters, J. Approx. Theory *5*, 343–361

Zensykbaev, A. A.
[1976] On the best quadrature formulas on the class $W^r L_p$, Dokl. Akad. Nauk SSSR *227*, 277–279; see also Soviet Math. Dokl. *17*, 377–380

Zielke, R.
[1979] *Discontinuous Cebysev System*, Lect. Notes in Math., No. 707

Zygmund, A.
[1968] *Trigonometric Series*, Cambridge Univ. Press, Cambridge

Glossary of Selected Symbols

Symbol	Meaning	Page
$E(x; X_n)$	distance of X_n to element x	1
$E(A; X_n)$	distance of X_n to set A	1
$d_n(A; X)$	Kolmogorov n-width	2
T_{n-1}	trigonometric polynomials of degree $\leq n-1$	3
$\delta_n(A; X)$	linear n-width	5
P_n	continuous linear operator of rank $\leq n$	5
$d^n(A; X)$	Gel'fand n-width	7
L^n	subspace of codimension $\leq n$	7
$S(X)$	unit ball of normed linear space X	11
$b_n(A; X)$	Bernstein n-width	13
$L(X, Y)$	set of continuous linear operators from X to Y	29
$K(X, Y)$	set of compact operators in $L(X, Y)$	30
$F(X, Y)$	closure of set of finite rank opertors in $L(X, Y)$	30
$S(f)$	number of sign changes of f	40
$Z(f)$	number of zeros of f	41
$\tilde{Z}(f)$	number of zeros of f, where nodal zeros are counted once, and nonnodal zeros are counted twice	42
$Z^*(f)$	number of zeros of f counting multiplicities	42
$S^-(\mathbf{x})$	number of strict sign changes of $\mathbf{x}$	45
$S^+(\mathbf{x})$	number of weak sign changes of $\mathbf{x}$	45
$S_c^-(\mathbf{x})$	number of cyclic strict sign changes of $\mathbf{x}$	59
$S_c(f)$	number of sign changes of periodic f	60
$s_n(T)$	nth singular value (s-number) of T	65
Δ_R	open unit ball of radius R in $\mathbb{C}$	67
$\tilde{\mathscr{K}}_p$	$\{k * \mathrm{h}: \|h\|_p \leq 1\}$	96
$\tilde{\mathscr{B}}_p$	$\{a + G * h: \|h\|_p \leq 1, a \in \mathbb{R}\}$	96
$\mathscr{K}_p$	$\{\int_0^1 K(x, y) h(y) dy: \|h\|_p \leq 1\}$	138
$\mathscr{K}_p^r$	$\{\sum_{i=1}^r a_i k_i(x) + \int_0^1 K(x, y) h(y) dy: \|h\|_p \leq 1, a_i \in \mathbb{R}\}$	138
Λ_m	open simplex in $([0, 1])^m$	140
$\mathscr{A}_p$	$\{A\mathbf{x}: \|\mathbf{x}\|_p \leq 1\}$ where A is an $M \times M$ matrix	198
$\mathscr{D}_p$	$\{D\mathbf{x}: \|\mathbf{x}\|_p \leq 1\}$ where D is an $M \times M$ diagonal matrix	201
$\mathscr{I}_p$	unit ball in l_p^M	210
$\mathscr{B}_m$	class of Blaschke products of degree $\leq m$	261

Author Index

Subject Index

Ergebnisse der Mathematik und ihrer Grenzgebiete, 3. Folge

A Series of Modern Surveys in Mathematics

Volume 1
A. Fröhlich

Galois Module Structure of Algebraic Integers

1983. X, 262 pages. ISBN 3-540-11920-5

Contents: Introduction. - Notation and Conventions. - Survey of Results. - Classgroups and Determinants. - Resolvents, Galois Gauss Sums, Root Numbers, Conductors. - Congruences and Logarithmic Values. - Root Number Values. - Relative Structure. - Appendix. - Literature List. - List of Theorems. - Some Further Notation. - Index.

Volume 2
W. Fulton

Intersection Theory

1984. XI, 470 pages. ISBN 3-540-12176-5

Contents: Introduction. - Rational Equivalence. - Divisors. - Vector Bundles and Chern Classes. - Cones and Segre Classes. - Deformation to the Normal Cone. - Intersection Products. - Intersection Multiplicities. - Intersections on Non-singular Varieties. - Excess and Residual Intersections. - Families of Algebraic Cycles. - Dynamic Intersections. - Positivity. - Rationality. - Degeneracy Loci and Grassmannians. - Riemann-Roch for Non-singular Varieties. - Correspondences. - Bivariant Intersection Theory. - Riemann-Roch for Singular Varieties. - Algebraic, Homological, and Numerical Equivalence. - Generalizations. - Appendix A: Algebra. - Appendix B: Algebraic Geometry (Glossary). - Bibliography. - Notation. - Index.

Volume 3
J. C. Jantzen

Einhüllende Algebren halbeinfacher Lie-Algebren

1983. V, 298 Seiten. ISBN 3-540-12178-1

Inhaltsübersicht: Einleitung. - Einhüllende Algebren. - Halbeinfache Lie-Algebren. - Zentralisatoren in Einhüllenden halbeinfacher Lie-Algebren. - Moduln mit einem höchsten Gewicht. - Annullatoren einfacher Moduln mit einem höchsten Gewicht. - Harish-Chandra-Moduln. - Primitive Ideale und Harish-Chandra-Moduln. - Gel'fand-Kirillov-Dimension und Multiplizität. - Die Multiplizität von Moduln in der Kategorie $\mathcal{O}$. - Gel'fand-Kirillov-Dimension von Harish-Chandra-Moduln. - Lokalisierungen von Harish-Chandra-Moduln. - Goldie-Rang und Konstants Problem. - Schiefpolynomringe und der Übergang zu den $\mathfrak{m}$-Invarianten. - Goldie-Rang-Polynome und Darstellungen der Weylgruppe. - Induzierte Ideale und eine Vermutung von Gel'fand und Kirillov. Kazhdan-Lusztig-Polynome und spezielle Darstellungen der Weylgruppe. - Assoziierte Varietäten. - Literatur. - Verzeichnis der Notationen. - Sachregister.

Springer-Verlag Berlin Heidelberg New York Tokyo

Ergebnisse der Mathematik und ihrer Grenzgebiete, 3. Folge

A Series of Modern Surveys in Mathematics

Volume 4
W. Barth, C. Peters, A. Van de Ven

Compact Complex Surfaces

1984. X, 304 pages. ISBN 3-540-12172-2

Contents: Introduction. – Standard Notations. – Preliminaries. – Curves on Surfaces. – Mappings of Surfaces. – Some General Properties of Surfaces. – Examples. – The Enriques-Kodaira Classification. – Surfaces of General Type. – $K3$-Surfaces and Enriques Surfaces. – Bibliography. – Notations. – Subject Index.

Volume 5
K. Strebel

Quadratic Differentials

1984. 74 figures. XII, 184 pages.
ISBN 3-540-13035-7

Contents: Background Material on Riemann Surfaces. – Quadratic Differentials. – Local Behaviour of the Trajectories and the φ- Metric. – Trajectory Structure in the Large. – The Metric Associated with a Quadratic Differential. – Quadratic Differentials with Closed Trajectories. – Quadratic Differentials of General Type. – References. – Subject Index.

Volume 6
M. Beeson

Foundations of Constructive Mathematics

Metamathematical Studies
1984. Approx. 450 pages. ISBN 3-540-12173-0

Contents: Practice and Philosophy of Constructive Mathematics: Examples of Constructive Mathematics. Informal Foundations of Constructive Mathematics. Some Different Philosophies of Constructive Mathematics. Recursive Mathematics: Living with Church's Thesis. The Role of Formal Systems in Foundational Studies. – Formal Systems of the Seventies: Theories of Rules. Realizability. Constructive Set Theories. The Existence Property in Constructive Set Theory. Theories of Rules, Sets, and Classifications. Constructive Type Theories. – Metamathematical Studies: Constructive Models of Set Theory. Proof-Theoretic Strength. Some Formalized Metamathematics and Church's Rule. Forcing. Continuity. – Metaphilosophical Studies: Theories of Rules and Proof. Historical Appendix. – References. – Index of System and Axioms. – Index of Names. – Index.

Forthcoming titles:

K. Diederich, J. E. Fornaess, R. P. Pflug

Convexity in Complex Analysis
ISBN 3-540-12174-9

E. Freitag, R. Kiehl

Etale Cohomology and the Weil Conjecture
ISBN 3-540-12175-7

M. Gromov

Partial Differential Relations
ISBN 3-540-12177-3

G. A. Margulis

Discrete Subgroups of Lie Groups
ISBN 3-540-12179-X

Springer-Verlag
Berlin
Heidelberg
New York
Tokyo